D0849110

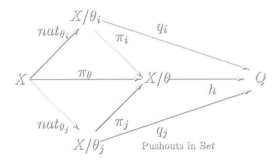

Pushouts in Set

# Universal
# Algebra
# and Coalgebra

# Universal
# Algebra
## and Coalgebra

**Klaus Denecke**
Universität Potsdam, Germany

**Shelly L Wismath**
University of Lethbridge, Canada

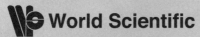

**World Scientific**

NEW JERSEY · LONDON · SINGAPORE · BEIJING · SHANGHAI · HONG KONG · TAIPEI · CHENNAI

*Published by*

World Scientific Publishing Co. Pte. Ltd.

5 Toh Tuck Link, Singapore 596224

*USA office:* 27 Warren Street, Suite 401-402, Hackensack, NJ 07601

*UK office:* 57 Shelton Street, Covent Garden, London WC2H 9HE

**Library of Congress Cataloging-in-Publication Data**
Denecke, Klaus.
   Universal algebra and coalgebra / Klaus Denecke, Shelly L Wismath.
   p. cm.
   Includes bibliographical references and index.
   ISBN-13 978-981-283-745-5 (hbk.)
   ISBN-10 981-283-745-0 (hbk.)
   1. Algebra, Universal.   I. Wismath, Shelly L.
                                                    2009281398

**British Library Cataloguing-in-Publication Data**
A catalogue record for this book is available from the British Library.

First published 2009
Reprinted 2010

Printed in Singapore by World Scientific Printers

*Les chiens aboient, les chats miaulent,*
*c'est leur nature;*
*nous nous sommes mathématicienne,*
*c'est la notre.*

# Preface

A mathematical theory is considered to be elegant if it allows us to model various notions with only a few basic concepts. From this point of view, category theory is elegant. The concepts of a category, a functor, a natural transformation and limits provide a strong expressive power. Universal coalgebra is another powerful and elegant theory, based on category theory. Categorical coalgebras are the dual objects of algebras; they are used to model several kinds of automata and more generally to model transition and dynamical systems. The basic notions of algebras, homomorphisms and congruences from universal algebra correspond to coalgebras, homomorphisms of coalgebras and bisimulations, respectively in the theory of coalgebras.

Universal algebra and coalgebra theory are studied here by a category theory approach, using $F$-coalgebras and the dual concept of $F$-algebras for a set-valued functor $F$. This approach allows similar ideas and proofs for the basic results for both $F$-algebras and $F$-coalgebras, but it is not as general as category theory itself. But also there are some structures which do not admit any representation as coalgebras. The concept of a clone, for instance, while one of the most important concepts of universal algebra, has no counterpart in the general theory of coalgebras. Clones appear only in a particular case, for coalgebras of type $\tau$, as clones of co-operations. This was the motivation for our comparison of clones of operations and clones of co-operations.

Coalgebras based on a functor were first used to dualize the Eilenberg-Moore construction ([60], [3]). In [5] a coalgebraic approach was used to describe the dynamics of deterministic automata, and later coalgebraic methods were applied to infinite data structures in [6]. Aczel and Mendler suggested in [1] a way to specify semantics of processes by means of final coalgebras. This idea is expressed more explicitly in [2], and culminates in the argument that syntax is an algebraic phenomenon while semantics are coalgebraic.

Not all aspects of the theory of $F$-coalgebras are covered in this book: for instance, the duality between the equational theory of universal algebra and modal logic is not considered. We believe however that we have presented a clear overview of the area, from which further study may proceed. The reader is also encouraged to tackle the problems listed as exercises in each chapter, to further consolidate the theory and applications presented.

The material of this book is based on lectures and research seminars given by the first author and his students at the University of Potsdam (Germany), and the authors are grateful for the critical input of a number of students. Some of the material was also presented as a part of the DAAD-funded project "Centre of Excellence for Applications of Mathematics" at the South-West-University Blagoevgrad, Bulgaria. The work of the second author was funded by the NSERC of Canada.

*K. Denecke and S. L. Wismath*

# Contents

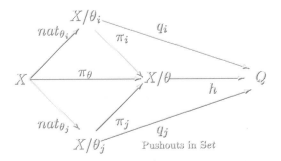

Pushouts in Set

# Universal
# Algebra
# and Coalgebra

# Introduction

The purpose of this book is to study the structures needed to model objects in the two areas of algebra and theoretical computer science. Universal or general algebra is used to describe algebraic structures, while coalgebras are used to model state-based or finite-state machines in computer science.

Most mathematicians first encounter algebraic structures in the classical examples of groups, rings, fields and vector spaces. In each of these areas, common themes arise: we have sets of objects which are closed under one or more operations performed on the objects, and we are interested in subsets which inherit the structure (subgroups, subspaces, etc.), in mappings which preserve the structure (group homomorphisms, linear transformations, etc.) and construction of new structures from old, for instance by Cartesian products or quotients. We can also classify our structures according to the laws or identities they satisfy, as for instance with commutative groups or groups of order four.

In universal algebra we abstract and generalize from these examples to a core structure of *an algebra*: a set $A$ of objects, with one or more operations defined on the set. We study substructures, homomorphisms and product algebras, and we classify algebras according to the identities they satisfy. To study such algebras we also need to know how many operation symbols our algebra has, and the arity of each one. This information is called the *type* or *signature* of the algebra. In general we assume a type indexed by some set $I$: for each $i \in I$ we have an operation symbol $f_i$, of arity $n_i \geq 0$, and we write the type as $\tau = (n_i)_{i \in I}$.

While universal algebras can be used to model most algebraic structures, they are not as useful in modelling state-based systems. The main reason for this is the following. An $n_i$-ary operation on set $A$ is a mapping $f^A : A^{n_i} \mapsto A$, which combines $n_i$ "input" elements of $A$ into one output element. In

1

a state-based system however, we often have the opposite situation: we need to map a single state to an output which carries several pieces of information, for instance to a state-output-symbol pair. That is, we need mappings from set $A$ to some more complex set involving $A$. Originally a co-operation on set $A$ was defined to be a mapping from $A$ to a copower $A^{\sqcup n}$ of $A$, that is, a disjoint union of $n$ copies of $A$ for some number $n$ which is the arity of the co-operation. A co-algebra was then a structure consisting of a set $A$ with one or more co-operations defined on it. Again we have a type for the coalgebra, indexed by some set $I$, and we look at coalgebras of type $\tau$.

Coalgebras of type $\tau$ are thus dual to algebras of type $\tau$. The first papers on coalgebras (see [33], [31]) were motivated by this theoretical interest in dualizing concepts and theorems from universal algebra, and were largely forgotten. However, in the last decade coalgebras have become important as the fundamental structures needed to model state-based systems and state-based dynamics, and universal coalgebra has emerged as a general theory of such systems. The connection between algebras and coalgebras also provides a way to connect static data-oriented systems with dynamical behaviour-oriented systems. This duality has been informally known for a long time, with algebras used to describe data types and coalgebras used to describe abstract systems or machines.

In order to model dynamic state-based systems, we need to generalize the original definition of a co-algebra to allow other codomains than just copowers of $A$. For instance, the codomain might be a product of the form $A \times \Sigma$ where $\Sigma$ is an input or output language of the machine. In general, we use some functor $F$ to describe this structure, and consider mappings $f : A \mapsto F(A)$. This leads to the definition of an $F$-coalgebra, for a functor $F$, as a structure with a base set $A$ together with one or more mappings from $A$ to $F(A)$. There is an algebraic dual of this concept too: an $F$-algebra is a set $A$ with one or more mappings from $F(A)$ to $A$.

It is evident that duality plays a key role in the study of algebras and coalgebras, and a main goal of this book is to explain this duality in a clear and accessible way. We have already mentioned the use of a functor in defining the general version of $F$-algebras and $F$-coalgebras, and indeed category theory plays a key role in formalizing the duality we need. For this reason we introduce categorical notions early in our development of the theory. Let $\mathbf{C}$ be a category, and let $F : \mathbf{C} \to \mathbf{C}$ be a functor on this category. An $F$-coalgebra over the category $\mathbf{C}$ is defined as an object $A$ of $\mathbf{C}$ equipped with a morphism $\alpha_A : A \to F(A)$. When $\mathbf{C}$ is a category of sets, the morphism $\alpha_A$ is a mapping from the set $A$ to the set $F(A)$. For a special set-valued functor $F$ the mapping $\alpha_A$ can encode the indexed or non-indexed set of fundamental co-operations of a coalgebra $\mathcal{A}$. Dually, $F$-algebras are pairs consisting of a set $A$ and a mapping $\beta_A : F(A) \to A$, where $\beta_A$ encodes the indexed or non-indexed set of fundamental operations of the algebra $\mathcal{A}$. In this most general sense, the concept of an $F$-coalgebra generalizes that of a coalgebra of type $\tau$, and similarly for $F$-algebras and algebras of type $\tau$. It turns out that the concept of an $F$-coalgebra models many kinds of structures occurring in mathematics and computer science. $F$-coalgebras over categories of algebras are also of interest in the area of concurrency theory (see [16]).

In Chapter 1 of this book we introduce the topic of universal algebra, beginning with algebras and operations. We describe the subalgebra, homomorphic image, product and quotient constructions, and define a variety as a class of algebras closed under the formation of subalgebras, products and homomorphic images. We also define terms and identities, and an equational class as a class of algebras which all satisfy a common set of identities. The chapter culminates in Section 1.4 where we study the connection between the algebraic-structural approach of varieties and the more model-theoretical, logical approach of equational classes. The main theorem here is Birkhoff's Theorem, which tells us that the two approaches are equivalent: this theorem says that any variety is an equational class, and vice versa.

As a lead-in to the modelling of state-based systems, we show that any algebra on set $A$ with only unary fundamental operations $f_1^A, \ldots, f_r^A$ can be regarded as an automaton without output. The set of unary fundamental operations forms the input alphabet of the automaton, while the set of states is the universe set $A$ of the algebra. A non-empty subset $A'$ of the set $A$ must be chosen as the set of final states, and one designated element $a_0$ from the universe is used as an initial state. Then a word $f_{i_1}^A \cdots f_{i_r}^A$ is recognized if and only if the output $f_{i_1}^A(f_{i_2}^A(\ldots(f_{i_r}^A(a_0))\ldots))$ is in $A'$.

To model automata with output we also need structures with more than one set of objects. These are called multi-based or multi-sorted algebras. In the last section of Chapter 1 we describe such multi-based algebras. An important example of this structure is a clone, which has a disjoint set of objects for each $n \geq 1$. For example, the collection of all finitary operations on a base set $A$ forms a clone, with the set $O_n(A)$ for each $n \geq 1$ consisting of all the $n$-ary operations on $A$.

The main purpose of Chapter 2 is to illustrate through a number of examples how the notion of coalgebra models a large class of state-based systems. In general, state-based systems satisfy the following criteria:

- the behaviour of the system depends on internal states, which are not visible to the user of the system;
- the system is reactive, interacts with its environment, and is not necessarily terminating;
- the interaction is performed by a set of operations.

For each example presented of a state-based system, we show how it may be regarded as a coalgebra, how to express structure-preserving maps or homomorphisms between such systems as coalgebra homomorphisms, and how to model behavioural equivalence of states.

As noted above, our general approach to the duality of algebras and coalgebras is based on categorical ideas, and in Chapter 3 we give a brief introduction to and overview of category theory. But our approach is limited mainly to the category of sets: we consider the interpretation of each abstractly defined categorical notion in the category of sets, and we consider mainly what are called concrete categories, where the objects are sets equipped with some structure such as operations, co-operations, relations, or topologies.

In Chapter 4 we define $F$-coalgebras and their homomorphisms, and study the properties of the category of all $F$-coalgebras. The main question is the categorical one of whether there exist coproducts, coequalizers and pushouts, or colimits generally, in this category. It turns out that some properties, dual to well-known and easy to prove results in universal algebra, work for $F$-coalgebras only if we restrict to special kinds of functors $F$. These are what are called standard functors, functors which preserve weak pullbacks. In this chapter we also define the dual to a variety of algebras, the concept of a covariety of coalgebras, as a class of $F$-coalgebras which is closed under the formation of homomorphic images, subcoalgebras and co-products. We also introduce the concept of a bisimulation, a relation which models behavioural equivalence of state-based systems; this is a concept not studied in universal algebra.

Noting that coalgebras are dual to algebras, and having generalized coalgebras to $F$-coalgebras in Chapter 4, we turn in Chapter 5 to the question of the dual of $F$-coalgebras. Here we construct $F$-algebras as generalizations of algebras of type $\tau$.

Chapter 6 takes a further step in abstraction to a single structure which encompasses both $F$-algebras and $F$-coalgebras. This can be done by using two functors $F_1$ and $F_2$, instead of a single functor $F$. A functorial system or $(F_1, F_2)$-structure $((F_1, F_2)$-coalgebra), for functors $F_1$ and $F_2$, consists of a set $A$ and mappings $f : F_1(A) \mapsto F_2(A)$. This expresses both $F$-algebras, by taking $F_1 = F$ and $F_2$ to be the identity functor, and $F$-coalgebras, when $F_2 = F$ and $F_1$ is the identity functor. But this concept also models other interesting algebraic structures as well, such as power algebras (also called hyperstructures) and power coalgebras, and tree automata. This structure thus allows us to unify results from the two different areas of algebra and theoretical computer science. We also investigate in this chapter conditions on $F_1$ and $F_2$ under which limits and colimits in the category of all $(F_1, F_2)$-coalgebras exist.

A key concept in the study of algebras of type $\tau$ is the free algebra of the type over a countably infinite set of generators. This algebra has a universal mapping property, expressed in categorical terms by the fact that it is an initial object in the category $Alg(\tau)$ of all algebras of type $\tau$. In Chapter 7 we consider the dual concept in the category of $F$-coalgebras, a terminal $F$-coalgebra. Elements of a terminal coalgebra can be viewed as interpreting the behaviour of state-based systems.

As mentioned in Chapter 1, a key theorem in universal algebra is Birkhoff's result that varieties are the same as equational classes. The varietal approach to coalgebras is discussed here in Chapter 4, and in Chapter 8 we study the equational approach. We introduce coequations (as patterns of behaviour) and prove a dual of Birkhoff's Theorem.

To consider the dual of terms and identities, we turn in Chapter 9 to the study of coalgebras of type $\tau$, as $F$-coalgebras for a particular functor $F$. In this setting we can define coterms, coidentities and covarieties, analogous to the terms, identities and varieties of universal algebra. The set of all coterms of type $\tau$ induces a clone of co-operations.

The study of clones is another important feature of this work. Clones occur in universal algebra as clones of term operations or polynomial operations of an algebra or variety. Also well known is the clone of all operations on a base set $A$, and its lattice of subclones. When $A$ has cardinality three or more, this lattice is uncountably infinite. The remaining case, when $A$ is a two-element set, is usually called the Boolean case, and the countably infinite lattice of all Boolean clones has been well-studied, particularly by Post. In Chapter 10 we compare Post's lattice of all Boolean clones of operations with the lattice of all clones of co-operations defined on a two-element set, which turns out to be a finite lattice.

The superposition or composition of operations used in clones of operations or co-operations can be modified slightly to give an associative binary operation on sets of operations or co-operations. This leads to the study in Chapter 11 of semigroups of operations and co-operations. Finally in Chapter 12 we show how the theory of hypersubstitutions and hyperidentities, developed in universal algebra, can be dualized for coalgebras of type $\tau$.

# Chapter 1

# Algebras and Identities

A common theme underlying many different areas of mathematics is the study of sets of objects with some operations for combining the objects into new ones. These objects can be numbers, under the usual combining operations of addition, subtraction, multiplication and division; matrices with the operations of matrix addition or multiplication; vectors with the operations of vector addition and scalar multiplication; functions with the operations of composition and differentiation; Boolean functions with operations of disjunction, conjunction and negation, and so on. In each case we are interested not only in the properties of the objects but in the properties or identities satisfied by the combining operations.

In this chapter we introduce the basics of universal algebra, which is the study of algebras and identities. We define an *algebra* as a set of objects together with a list of one or more operations on the objects. We classify algebras according to the identities (the laws or axioms) that they satisfy. For any collection of algebras, we can form the collection of all identities that are satisfied by all those algebras. Conversely, for any collection of identities we can form the collection of all algebras that satisfy those identities; such a collection is called a *variety*. This back-and-forth interplay between algebras and identities is called a *Galois connection*.

This Galois connection means that there are two main approaches to studying algebras. First, one can focus on algebraic constructions of new algebras from old, particularly the construction of subalgebras, homomorphic images and product algebras. These constructions are introduced in Sections 1.1 and 1.2. We can also study algebras by means of the identities they satisfy, as discussed in Section 1.3. Our Galois connection between algebras and identities, described in Section 1.4, essentially means that these two approaches are equivalent. That is, they both focus on the same classes

7

of algebras: the classes of algebras which are closed under the formation of homomorphic images, subalgebras and products are precisely the classes determined by some set of identities.

In the final section in this Chapter, Section 5, we extend the basic definition of an algebra to multi-based algebras. These are structures with two or more sets of objects as universes, along with some operations on these sets. We briefly discuss clones of operations on a base set as an example of multi-based algebras; further examples will be given in Chapter 2 when we discuss state-based systems.

## 1.1   Subalgebras and Homomorphic Images

Let $A$ be a set, and let $n \geq 1$ be a natural number. An *n-ary operation* on $A$ is a function $f^A : A^n \to A$ from the $n$-th Cartesian power of $A$ into $A$. The number $n$ is called the arity of the function $f^A$. A *finitary operation* on $A$ is an operation on $A$ which is $n$-ary for some $n \geq 1$. We let $O^n(A)$ be the set of all $n$-ary operations defined on set $A$ and let $O(A) := \bigcup_{n=1}^{\infty} O^n(A)$ be the set of all finitary operations defined on $A$. An *algebra* is a non-empty set $A$ together with a set of finitary operations defined on this set.

We remark that an $n$-ary operation $f^A$ on $A$ can also be regarded as an $(n+1)$-ary relation on $A$ called the *graph* of $f^A$. This relation is defined by $\{(a_1, \ldots, a_{n+1}) \in A^{n+1} \mid f^A(a_1, \ldots, a_n) = a_{n+1}\}$.

Although we have defined $n$-ary operations for $n \geq 1$, we can extend the definition of an operation to include nullary operations. It is customary to set $A^0 := \{\emptyset\}$, and we define a nullary operation to be a function $f^A : \{\emptyset\} \to A$. In this case there is a unique element $f^A(\emptyset) \in A$ determined or selected by $f^A$. Conversely, when $A$ is non-empty, every element $a \in A$ determines a unique mapping $f_a^A : \{\emptyset\} \to A$, with $f_a^A(\emptyset) = a$. This means that a nullary operation may be thought of as selecting an element from the set $A$. Of course if $A = \emptyset$ then there can be no nullary operations on $A$.

There are two possible ways to define an algebra. In the *non-indexed* form, an algebra is a pair $\mathcal{A} := (A; F^{\mathcal{A}})$, consisting of a set $A$ and a set $F^{\mathcal{A}}$ of operations defined on $A$. We define an *indexed algebra* using some index set $I$ to label our operations on $A$. The main advantage of this approach is that the fundamental operations form a sequence in the indexed case, allowing the possibility of repetition of operations.

**Definition 1.1.1.** Let $A$ be a non-empty set. Let $I$ be some non-empty index set, and let $(f_i^{\mathcal{A}})_{i \in I}$ be a function which assigns to every element $i$ of $I$ an $n_i$-ary operation $f_i^{\mathcal{A}}$ defined on $A$. Then the pair $\mathcal{A} = (A; (f_i^{\mathcal{A}})_{i \in I})$ is called an (*indexed*) *algebra*. The set $A$ is called the *base* set or *carrier set* or *universe* of the algebra, while $(f_i^{\mathcal{A}})_{i \in I}$ is called the *sequence of fundamental operations* of the algebra. The *type* of the algebra is the sequence $\tau := (n_i)_{i \in I}$ of the arities of the fundamental operations $f_i^{\mathcal{A}}$. We will use the name $Alg(\tau)$ for the class of all algebras of type $\tau$.

In our definition we have required that the base set $A$ of an algebra be a non-empty set. We can extend this definition to include an empty algebra, with the empty set as universe, as long as the type of the algebra does not include any nullary operations. We shall want to include the possibility of an empty coalgebra later as well.

Let us consider some examples of algebras of various types. An algebra $\mathcal{G} = (G; \cdot)$ of type $\tau = (2)$, that is, having one binary operation, is called a *groupoid*. The binary operation $f^{\mathcal{G}}$ is often denoted by $\cdot$ using infix notation, so we write $x \cdot y$ or even just $xy$ when the operation is clear from the context. If the binary operation is associative, $\mathcal{G}$ is called a *semigroup*. A semigroup $\mathcal{G} = (G; \cdot)$ is called a *semilattice* if its operation $\cdot$ also satisfies commutativity and idempotence. A semigroup with an additional nullary operation $e$ is called a *monoid* if $e$ is an identity element with respect to $\cdot$, meaning that for all elements $x$ in the base set the equations $x \cdot e = e \cdot x = x$ are satisfied. Monoids are thus algebras of type $\tau = (2, 0)$.

A *group* $\mathcal{G} = (G; \cdot)$ is a semigroup which satisfies the axiom

$$\forall a, b \in G \; \exists x, y \in G \; (a \cdot x = b \text{ and } y \cdot a = b) \qquad \text{(invertibility)}.$$

A group can also be regarded as an algebra $\mathcal{G} = (G; \cdot, ^{-1}, e)$ of type $(2, 1, 0)$ which satisfies the associative law and the axioms

$$\forall x \in G \; (x \cdot x^{-1} = x^{-1} \cdot x = e)$$

and

$$\forall x \in G \; (x \cdot e = x = e \cdot x).$$

An algebra $\mathcal{R} = (R; +, \cdot)$ of type $\tau = (2, 2)$ is called a *semiring* if both its binary operations are associative and it satisfies the two distributive laws

$$\forall x, y, z \in R \; (x \cdot (y + z) = x \cdot y + x \cdot z)$$

and

$$\forall x, y, z \in R \ ((x + y) \cdot z \ = \ x \cdot z + y \cdot z).$$

*Rings* are semirings in which the first operation $+$ is both commutative and invertible; a commutative ring $\mathcal{K} = (K; +, \cdot)$ is called a *field* if $(K \setminus \{0\}; \cdot)$ is a group, where the zero element $0$ is the neutral element with respect to addition.

Another important example of an algebra is a *lattice*. Lattices are of type $(2, 2)$, having two binary operations which are usually called meet and join, denoted by $\wedge$ and $\vee$ respectively. An algebra $\mathcal{L} = (L; \wedge, \vee)$ of type $(2, 2)$ is called a lattice if it satisfies the following equations:

| | | |
|---|---|---|
| $\forall x, y \in L$ | $(x \vee y = y \vee x),$ | |
| $\forall x, y \in L$ | $(x \wedge y = y \wedge x),$ | (commutativity) |
| $\forall x, y, z \in L$ | $(x \vee (y \vee z) = (x \vee y) \vee z),$ | |
| $\forall x, y, z \in L$ | $(x \wedge (y \wedge z) = (x \wedge y) \wedge z),$ | (associativity) |
| $\forall x \in L$ | $(x \vee x = x),$ | |
| $\forall x \in L$ | $(x \wedge x = x)$ | (idempotency), |
| $\forall x, y \in L$ | $(x \vee (x \wedge y) = x),$ | |
| $\forall x, y \in L$ | $(x \wedge (x \vee y) = x)$ | (absorption laws). |

If in addition the lattice satisfies the following *distributive laws*,

$$\forall x, y, z \in L \quad (x \wedge (y \vee z) = (x \wedge y) \vee (x \wedge z)),$$
$$\forall x, y, z \in L \quad (x \vee (y \wedge z) = (x \vee y) \wedge (x \vee z)),$$

then the lattice is said to be *distributive*.

Besides being examples of a kind of algebra, lattices are used in the study of all other kinds of algebras. This is because every algebra has some lattices associated with it; for instance, for any group we can form the lattice of subgroups of the group. Another important feature of lattices is that they involve an order relation. The following theorem relates lattices to partially ordered sets, which are sets with a relation $\leq$ which is reflexive, anti-symmetric and transitive.

**Theorem 1.1.2.** *Let $(L; \leq)$ be a partially ordered set in which for all $x, y \in L$ both the infimum $\bigwedge\{x, y\}$ and the supremum $\bigvee\{x, y\}$ exist. Then the infimum and supremum operations make $(L; \wedge, \vee)$ into a lattice. Conversely, every lattice defines a partially ordered set in which for all $x, y$ the infimum $\bigwedge\{x, y\}$ and the supremum $\bigvee\{x, y\}$ exist.*

**Proof:** Let $(L; \leq)$ be a partially ordered set, in which for any two-element set $\{x, y\} \subseteq L$ the infimum $\bigwedge\{x, y\}$ and the supremum $\bigvee\{x, y\}$ exist. We define $x \wedge y := \bigwedge\{x, y\}$ and $x \vee y := \bigvee\{x, y\}$. Then it is straightforward to verify that these operations make $(L; \wedge, \vee)$ a lattice.

Conversely, let $(L; \wedge, \vee)$ be a lattice. We can define an order relation on $L$ by setting

$$x \leq y : \Leftrightarrow x \wedge y = x.$$

Again it is easy to check that this gives a partial order relation on $L$, with $\bigwedge\{x, y\} = x \wedge y$ and $\bigvee\{x, y\} = x \vee y$. ∎

An easy example of a lattice is a *totally ordered set* or *chain*, i.e. a partially ordered set $(L; \leq)$ such that $x \leq y$ or $y \leq x$ for all $x, y \in L$. A lattice $\mathcal{L}$ in which for all sets $B \subseteq L$ the infimum $\bigwedge B$ and the supremum $\bigvee B$ exist is called a *complete lattice*. Obviously, any finite lattice is complete.

A lattice is called *bounded* if it has a least element and a greatest element. These special elements are usually denoted by 0 and 1 respectively, and we include them as selected elements in our type. That is, a *bounded lattice* $(L; \wedge, \vee, 0, 1)$ is an algebra of type $(2, 2, 0, 0)$ which is a lattice, with two additional nullary operations 0 and 1, which satisfy

$$\forall x \in L \ (x \wedge 0 = 0) \text{ and}$$
$$\forall x \in L \ (x \vee 1 = 1).$$

Finally, an algebra $\mathcal{B} = (B; \wedge, \vee, \neg, 0, 1)$ is called a *Boolean algebra*, if $(B; \wedge, \vee, 0, 1)$ is a bounded distributive lattice with an additional unary operation $\neg$ called complementation, which satisfies

$$\forall x \in B \ (x \wedge \neg x = 0)$$

and

$$\forall x \in B \ (x \vee \neg x = 1).$$

We mention two important examples of Boolean algebras. The first is the two-element Boolean algebra $(\{0, 1\}; \wedge, \vee, \neg, 0, 1)$, with binary operations of conjunction (as meet) and disjunction (as join) and unary operation of negation. By identifying 0 with False and 1 with True, we can use this Boolean algebra to model the usual binary or Boolean logic. The other example uses the power set of any base set $A$. The *power set algebra* $(\mathcal{P}(A); \cap, \cup, \sim, \emptyset, A)$ of $A$ uses intersection, union, complementation, the empty set and $A$ as operations.

Now we turn to the study of some algebraic constructions which produce new algebras from given ones. The first construction we discuss is the formation of subalgebras of a given algebra. Essentially we want an algebra $\mathcal{A}$ to be a subalgebra of an algebra $\mathcal{B}$ if $A$ is a subset of $B$ and the operations of $\mathcal{A}$ are those of $\mathcal{B}$ restricted to set $A$. To describe this formally, we define the *restriction* $f^B|A$ of $f^B$ to $A$, where $A \subseteq B$ and $f^B$ is an $n$-ary operation on $B$, by $(f^B|A)(a_1, \ldots, a_n) := f^A(a_1, \ldots, a_n)$ for all $a_1, \ldots, a_n \in A$.

**Definition 1.1.3.** Let $\mathcal{B} = (B; (f_i^B)_{i \in I})$ be an algebra of type $\tau$. Then an algebra $\mathcal{A}$ is called a *subalgebra* of $\mathcal{B}$, written as $\mathcal{A} \subseteq \mathcal{B}$, if the following conditions are satisfied:

(i) $\mathcal{A} = (A; (f_i^A)_{i \in I})$ is an algebra of type $\tau$;
(ii) $A \subseteq B$;
(iii) $\forall i \in I$, the graph of $f_i^A$ is a subset of the graph of $f_i^B$.

An equivalent version of the properties needed for one algebra to be a subalgebra of another one is given by the following Subalgebra Criterion.

**Lemma 1.1.4.** *(Subalgebra Criterion) Let $\mathcal{B} = (B; (f_i^B)_{i \in I})$ be an algebra of type $\tau$ and let $A \subseteq B$ be a subset of $B$ for which $f_i^A = f_i^B \mid A$ for all $i \in I$. Then $\mathcal{A} = (A; (f_i^A)_{i \in I})$ is a subalgebra of $\mathcal{B} = (B; (f_i^B)_{i \in I})$ iff $A$ is closed with respect to all the operations $f_i^B$ for $i \in I$; that is, if $f_i^B(A^{n_i}) \subseteq A$ for all $i \in I$.*

We say that an operation $f_i^B$ *preserves* the subset $A$ of $B$ when $f_i^B(A^{n_i}) \subseteq A$. It is obvious that the subalgebra relation is both *reflexive* and *transitive*, that is, for all algebras $\mathcal{A}$, $\mathcal{B}$ and $\mathcal{C}$, we have $\mathcal{A} \subseteq \mathcal{A}$ and if $\mathcal{A} \subseteq \mathcal{B} \subseteq \mathcal{C}$ then $\mathcal{A} \subseteq \mathcal{C}$.

Now suppose that $\{\mathcal{A}_j \mid j \in J\}$ is a family of subalgebras of a given algebra $\mathcal{B}$. It is easy to see that the intersection $A := \bigcap_{j \in J} A_j$ of the universes is the universe of a subalgebra of $\mathcal{B}$. This subalgebra is called the *intersection* of the family $\{\mathcal{A}_j \mid j \in J\}$, and we denote it by $\bigcap_{j \in J} \mathcal{A}_j$. This allows us to consider for any subset $X \subseteq B$ of the universe of an algebra $\mathcal{B}$ the subalgebra

$$\langle X \rangle_{\mathcal{B}} := \bigcap \{\mathcal{A} \mid \mathcal{A} \subseteq \mathcal{B} \text{ and } X \subseteq A\}$$

of $\mathcal{B}$ generated by $X$. The set $X$ is called a *generating system* of this new algebra. The process of subalgebra generation satisfies the following three *closure properties*.

**Theorem 1.1.5.** *Let $\mathcal{B}$ be an algebra. For all subsets $X$ and $Y$ of $B$, the following closure properties hold:*

(i) $X \subseteq \langle X \rangle_\mathcal{B}$,     *(extensivity)*;
(ii) $X \subseteq Y \Rightarrow \langle X \rangle_\mathcal{B} \subseteq \langle Y \rangle_\mathcal{B}$, *(monotonicity)*;
(iii) $\langle X \rangle_\mathcal{B} = \langle \langle X \rangle_\mathcal{B} \rangle_\mathcal{B}$,     *(idempotence)*.

We can use the fact that the intersection of two subalgebras of an algebra $\mathcal{B}$ is a subalgebra of $\mathcal{B}$ to define a lattice on the set $Sub(\mathcal{B})$ of all subalgebras of $\mathcal{B}$. We begin by setting the meet operation on $Sub(\mathcal{B})$ to be intersection. That is, we want to define the binary operation $\wedge$ on $Sub(\mathcal{B})$ by

$$\wedge : (A_1, A_2) \mapsto A_1 \wedge A_2 : = A_1 \cap A_2.$$

For our join operation on $Sub(\mathcal{B})$, we cannot use the usual union, since the union $A_1 \cup A_2$ of the universes of two subalgebras of $\mathcal{B}$ is in general not a universe of a subalgebra of $\mathcal{B}$. Instead we map the pair $(A_1, A_2)$ to the subalgebra of $\mathcal{B}$ which is generated by the union $A_1 \cup A_2$:

$$\vee : (A_1, A_2) \mapsto A_1 \vee A_2 := \langle A_1 \cup A_2 \rangle_\mathcal{B}.$$

Altogether we have the following result.

**Theorem 1.1.6.** *For every algebra $\mathcal{B}$, the algebra $(Sub(\mathcal{B}); \wedge, \vee)$ is a lattice, called the subalgebra lattice of $\mathcal{B}$.*

Our second important algebraic construction is the formation of *homomorphic images*.

**Definition 1.1.7.** Let $\mathcal{A} = (A; (f_i^A)_{i \in I})$ and $\mathcal{B} = (B; (f_i^B)_{i \in I})$ be algebras of the same type $\tau$. A function $h : A \to B$ is called a *homomorphism* $h : \mathcal{A} \to \mathcal{B}$ of $\mathcal{A}$ into $\mathcal{B}$ if for all $i \in I$ we have

$$h(f_i^A(a_1, \ldots, a_{n_i})) = f_i^B(h(a_1), \ldots, h(a_{n_i})),$$

for all $a_1, \ldots, a_{n_i} \in A$. In the special case that $n_i = 0$, this equation means that $h(f_i^A(\emptyset)) = f_i^B(\emptyset)$. Therefore the element designated by a nullary operation $f_i^A$ in $A$ must be mapped to the corresponding element $f_i^B$ in $B$.

If $h$ is surjective (onto), then $\mathcal{B}$ is called a *homomorphic image* of $\mathcal{A}$. If the function $h$ is bijective, that is both one-to-one (injective) and onto $B$ (surjective), then the homomorphism $h : \mathcal{A} \to \mathcal{B}$ is called an *isomorphism* from $\mathcal{A}$ onto $\mathcal{B}$. An injective homomorphism from $\mathcal{A}$ into $\mathcal{B}$ is also called an *embedding* of $\mathcal{A}$ into $\mathcal{B}$.

A homomorphism $h : \mathcal{A} \to \mathcal{A}$ of an algebra $\mathcal{A}$ into itself is called an *endomorphism* of $\mathcal{A}$, and an isomorphism $h : \mathcal{A} \to \mathcal{A}$ from $A$ onto $A$ is called an *automorphism* of $\mathcal{A}$. Clearly, the identity mapping $id_A$ on the set $A$ is always an automorphism of the algebra $\mathcal{A}$.

It is easy to see that the image $\mathcal{B}_1 = h(\mathcal{A}_1)$ of a subalgebra $\mathcal{A}_1$ of $\mathcal{A}$ under the homomorphism $h$ is a subalgebra of $\mathcal{B}$ and that the preimage $h^{-1}(\mathcal{B}') = \mathcal{A}'$ of a subalgebra $\mathcal{B}'$ of $h(\mathcal{A}) \subseteq \mathcal{B}$ is a subalgebra of $\mathcal{A}$.

Suppose that $X \subseteq A$ is a subset of the universe of an algebra $\mathcal{A}$ of type $\tau$. If $\langle X \rangle_\mathcal{A}$ is the subalgebra of $\mathcal{A}$ generated by $X$ and if $h : \mathcal{A} \to \mathcal{B}$ is a homomorphism, then

$$\langle h(X) \rangle_\mathcal{B} = h(\langle X \rangle_\mathcal{A}).$$

In particular, if $X$ is a generating system of $\mathcal{A}$ and if $h : \mathcal{A} \to \mathcal{B}$ is surjective, then $h(X)$ generates $\mathcal{B}$ since

$$h(\langle X \rangle_\mathcal{A}) = h(\mathcal{A}) = \mathcal{B} = \langle h(X) \rangle_\mathcal{B}.$$

Let $\mathcal{A}$, $\mathcal{B}$ and $\mathcal{C}$ be algebras of the same type, and let $h_1 : \mathcal{A} \to \mathcal{B}$ and $h_2 : \mathcal{B} \to \mathcal{C}$ be homomorphisms. Then $h_2 \circ h_1 : \mathcal{A} \to \mathcal{C}$ is also a homomorphism; and if both $h_1$ and $h_2$ are surjective, injective or bijective, then the composition has the same property.

Any homomorphism $h : \mathcal{A} \to \mathcal{B}$ maps from the algebra $\mathcal{A}$ to the algebra $\mathcal{B}$. We can also give a more "internal" description of the homomorphic image, inside $\mathcal{A}$, using the concepts of a kernel and a quotient algebra. Every function $h : A \to B$ from a set $A$ onto a set $B$ defines a partition of $A$, into classes of elements having the same image under $h$. Any partition of a set $A$ corresponds to an equivalence relation on $A$, by the rule that two elements of $A$ are related to each other if and only if they belong to the same block of the partition. Now let $A$ and $B$ be the universes of two algebras $\mathcal{A}$ and $\mathcal{B}$ of type $\tau$. If $h : \mathcal{A} \to \mathcal{B}$ is a surjective homomorphism, then $h$ can be compatible with an equivalence relation $\theta$ on $A$, in the following sense.

**Definition 1.1.8.** Let $A$ be a set, let $\theta \subseteq A \times A$ be an equivalence relation on $A$, and let $f$ be an $n$-ary operation from $O^n(A)$. Then $f$ is said to be *compatible* with $\theta$, or to *preserve* $\theta$, if for all $a_1, \ldots, a_n, b_1, \ldots, b_n \in A$,

$$(a_1, b_1) \in \theta, \quad \ldots, \quad (a_n, b_n) \in \theta$$

implies        $(f(a_1, \ldots, a_n), f(b_1, \ldots, b_n)) \in \theta.$

**Definition 1.1.9.** Let $\mathcal{A} = (A; (f_i^{\mathcal{A}})_{i \in I})$ be an algebra of type $\tau$. An equivalence relation $\theta$ on $A$ is called a *congruence relation* on $\mathcal{A}$ if all the fundamental operations $f_i^{\mathcal{A}}$ are compatible with $\theta$. We denote by $Con(\mathcal{A})$ the set of all congruence relations of the algebra $\mathcal{A}$. For every algebra $\mathcal{A} = (A; (f_i^{\mathcal{A}})_{i \in I})$ the trivial equivalence relations

$$\Delta_A := \{(a, a) \mid a \in A\} \text{ and } \nabla_A = A \times A$$

are congruence relations. An algebra which has no congruence relations except $\Delta_A$ and $\nabla_A$ is called *simple*.

The intersection of any family of congruence relations in $Con(\mathcal{A})$ is again a congruence relation on $\mathcal{A}$. This allows us to define a binary operation

$$\wedge : \quad Con(\mathcal{A}) \times Con(\mathcal{A}) \to Con(\mathcal{A}) \quad \text{by } (\theta_1, \theta_2) \mapsto \theta_1 \cap \theta_2.$$

Simple examples show that the union $\theta_1 \cup \theta_2$ of two equivalence relations $\theta_1$ and $\theta_2$ on the set $A$ need not be an equivalence relation on $A$. So to define a binary join operation on $Con(\mathcal{A})$, we first consider the congruence relation generated by a binary relation on $A$.

**Definition 1.1.10.** Let $\mathcal{A}$ be an algebra, and let $\theta$ be a binary relation on $A$. We define the congruence relation $\langle \theta \rangle_{Con(\mathcal{A})}$ on $\mathcal{A}$ generated by $\theta$ to be the intersection of all congruence relations $\theta'$ on $\mathcal{A}$ which contain $\theta$:

$$\langle \theta \rangle_{Con(\mathcal{A})} : \;\; = \;\; \bigcap \{\theta' \mid \theta' \in Con(\mathcal{A}) \text{ and } \theta \subseteq \theta'\}.$$

Again it can be seen that $\langle \theta \rangle_{Con(\mathcal{A})}$ has the three important properties of a closure operator:

$$\theta \subseteq \langle \theta \rangle_{Con(\mathcal{A})},$$
$$\theta_1 \subseteq \theta_2 \Rightarrow \langle \theta_1 \rangle_{Con(\mathcal{A})} \subseteq \langle \theta_2 \rangle_{Con(\mathcal{A})},$$
$$\langle \langle \theta \rangle_{Con(\mathcal{A})} \rangle_{Con(\mathcal{A})} = \langle \theta \rangle_{Con(\mathcal{A})}.$$

Now we define the second binary operation on $Con(\mathcal{A})$ by

$$\vee : \quad Con(\mathcal{A}) \times Con(\mathcal{A}) \to Con(\mathcal{A}) \quad \text{with } (\theta_1, \theta_2) \mapsto \langle \theta_1 \cup \theta_2 \rangle_{Con(\mathcal{A})}.$$

It is straightforward to verify that these operations of meet and join on $Con(\mathcal{A})$ satisfy all the lattice identities, giving us the following theorem.

**Theorem 1.1.11.** *For every algebra $\mathcal{A}$, the algebra $(Con(\mathcal{A}); \wedge, \vee)$ is a lattice, called the congruence lattice of $\mathcal{A}$.*

Congruences are an important kind of equivalence relation on an algebra for another reason. Given a congruence relation $\theta$ on $\mathcal{A}$, we can partition the set $A$ into blocks with respect to $\theta$, producing the *quotient set* $A/\theta$. Then we can define operations of type $\tau$ on this quotient set in a natural way: for each $i \in I$, we define an $n_i$-ary operation $f_i^{A/\theta}$ on the quotient set by

$$f_i^{A/\theta} : (A/\theta)^{n_i} \to A/\theta$$

with

$$([a_1]_\theta, \ldots, [a_{n_i}]_\theta) \mapsto f_i^{A/\theta}([a_1]_\theta, \ldots, [a_{n_i}]_\theta) := [f_i^A(a_1, \ldots, a_{n_i})]_\theta.$$

Of course, in order to define operations on equivalence classes like this we have to verify that our operations are well-defined, that is, that they are independent of the representatives chosen for the classes. But this requirement is exactly the condition that our equivalence relation $\theta$ on $A$ be a congruence on $\mathcal{A}$; that is, that each operation $f_i^A$ be compatible with $\theta$. Thus we obtain a new algebra $\mathcal{A}/\theta := (A/\theta; (f_i^{A/\theta})_{i \in I})$, called the *quotient algebra* (or *factor algebra*) of $\mathcal{A}$ modulo $\theta$.

There is a close connection between quotient algebras of an algebra $\mathcal{A}$ and homomorphic images of $\mathcal{A}$. In one direction, every quotient algebra $\mathcal{A}/\theta$ by a congruence $\theta$ is a homomorphic image of $\mathcal{A}$: we use the *natural homomorphism* defined by

$$nat(\theta) : \mathcal{A} \to \mathcal{A}/\theta \ \ with \ a \mapsto [a]_\theta \ for \ every \ a \in A.$$

It is easy to check that $nat(\theta)$ is in fact a surjective homomorphism. Conversely, we can show that every surjective homomorphism from $\mathcal{A}$ determines a congruence on $\mathcal{A}$, in the following way.

**Lemma 1.1.12.** *Let $\mathcal{A}$ and $\mathcal{B}$ be algebras of type $\tau$. The kernel*

$$Ker \ h := \{(a, b) \in A^2 \mid h(a) = h(b)\}$$

*of any homomorphism $h : \mathcal{A} \to \mathcal{B}$ is a congruence relation on $\mathcal{A}$.*

Now suppose that we have a homomorphism $h : \mathcal{A} \to \mathcal{B}$. We have seen that $Ker \ h$ is a congruence on $\mathcal{A}$, so we can form the quotient algebra $\mathcal{A}/Ker \ h$, along with the natural homomorphism $nat(Ker \ h) : \mathcal{A} \to \mathcal{A}/Ker \ h$ which maps the algebra $\mathcal{A}$ onto this quotient algebra. Now we have two homomorphic images of $\mathcal{A}$: the original $h(\mathcal{A})$ and the new quotient

$\mathcal{A}/Ker\ h$. How are these two homomorphic images related? The answer to this question is given by the well-known *Homomorphic Image Theorem*.

**Theorem 1.1.13.** (*Homomorphic Image Theorem*) *Let* $h : \mathcal{A} \to \mathcal{B}$ *be a surjective homomorphism. Then there exists a unique isomorphism $f$ from* $\mathcal{A}/Ker\ h$ *onto* $\mathcal{B}$ *with* $f \circ nat(Ker\ h) = h$. *That is, the diagram in* Figure 1.1 *commutes.*

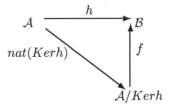

Fig. 1.1   Homomorphic Image Theorem

Later we shall need the sublattice $Con_{inv}(\mathcal{A})$ of all *fully invariant* congruence relations on $\mathcal{A}$.

**Definition 1.1.14.** Let $\mathcal{A}$ be an algebra of type $\tau$. A congruence relation $\theta \in Con(\mathcal{A})$ is called *fully invariant* if for all endomorphisms $\varphi : \mathcal{A} \to \mathcal{A}$ we have

$$(a,b) \in \theta \Rightarrow (\varphi(a), \varphi(b)) \in \theta \text{ for all } a, b \in A.$$

The congruence relations $\Delta_A$ and $\nabla_A$ are always fully invariant.

**Proposition 1.1.15.** *The set* $Con_{inv}(\mathcal{A})$ *of all fully invariant congruence relations of an algebra* $\mathcal{A}$ *forms a sublattice of* $Con(\mathcal{A})$.

## 1.2   Direct and Subdirect Products

The third important algebraic construction is the formation of product algebras. This construction, unlike subalgebras or homomorphic images, can produce algebras with larger cardinalities than those we started with. We begin by defining the *direct product* of a family of algebras.

**Definition 1.2.1.** Let $(\mathcal{A}_j)_{j \in J}$ be a family of algebras of type $\tau$. The *direct product* $\prod_{j \in J} \mathcal{A}_j$ of the $\mathcal{A}_j$ is defined as an algebra with the carrier set

$$P := \prod_{j \in J} A_j := \{(x_j)_{j \in J} \mid \forall j \in J\ (x_j \in A_j)\}$$

and the operations

$$(f_i^{\mathcal{P}}(\underline{a}_1, \ldots, \underline{a}_{n_i}))(j) \;\; = \;\; f_i^{\mathcal{A}_j}(\underline{a}_1(j), \ldots, \underline{a}_{n_i}(j)),$$

for $\underline{a}_1, \ldots, \underline{a}_{n_i}$ in $P$; that is,

$$f_i^{\mathcal{P}}((a_{1j})_{j\in J}, \ldots, (a_{n_ij})_{j\in J}) \;\; = \;\; (f_i^{\mathcal{A}_j}(a_{1j}, \ldots, a_{n_ij}))_{j\in J}.$$

If $\mathcal{A}_j = \mathcal{A}$ for all $j \in J$, then we usually write $\mathcal{A}^J$ instead of $\prod_{j\in J} \mathcal{A}_j$. If $J = \emptyset$, then $\mathcal{A}^{\emptyset}$ is defined to be the one-element (trivial) algebra of type $\tau$. If $J = \{1, \ldots, n\}$, then the direct product can be written as $\mathcal{A}_1 \times \cdots \times \mathcal{A}_n$.

The *projections* of the direct product $\prod_{j\in J} \mathcal{A}_j$ are the mappings

$$p_k : \prod_{j\in J} A_j \to A_k \text{ defined by } (a_j)_{j\in J} \mapsto a_k,$$

for $k \in J$.

It is easy to check that the projections of the direct product are in fact surjective homomorphisms.

We now consider a direct product of two factors. In this case we have two projection mappings, $p_1$ and $p_2$, each of which has a kernel which is a congruence relation on the product, since $p_1$ and $p_2$ are homomorphisms. To describe the special properties of these two kernels, we recall the definition of the product (composition) $\theta_1 \circ \theta_2$ of two binary relations $\theta_1, \theta_2$ on any set $A$:

$$\theta_1 \circ \theta_2 := \{(a,b) \mid \exists c \in A \, ((a,c) \in \theta_2 \wedge (c,b) \in \theta_1)\}.$$

Two binary relations $\theta_1, \theta_2$ on $A$ are called *permutable* if $\theta_1 \circ \theta_2 = \theta_2 \circ \theta_1$.

**Lemma 1.2.2.** *Let $\mathcal{A}_1, \mathcal{A}_2$ be two algebras of type $\tau$ and let $\mathcal{A}_1 \times \mathcal{A}_2$ be their direct product. Then:*

(i) *$ker\ p_1 \wedge ker\ p_2 = \Delta_{A_1 \times A_2}$;*
(ii) *$ker\ p_1 \circ ker\ p_2 = ker\ p_2 \circ ker\ p_1$;*
(iii) *$ker\ p_1 \vee ker\ p_2 = (A_1 \times A_2)^2 = \nabla_{A_1 \times A_2}$.*

This lemma motivates the following definition:

**Definition 1.2.3.** A congruence $\theta$ on $A$ is called a *factor congruence* if there is a congruence $\theta'$ on $\mathcal{A}$ such that

$$\theta \wedge \theta' \;\; = \;\; \Delta_A \qquad \text{and}$$
$$\theta \vee \theta' \;\; = \;\; \nabla_A.$$

The pair $(\theta, \theta')$ is called a pair of *factor congruences* on $\mathcal{A}$.

**Theorem 1.2.4.** *If $(\theta, \theta')$ is a pair of factor congruences on an algebra $\mathcal{A}$, then $\mathcal{A}$ is isomorphic to the direct product $\mathcal{A}/\theta \times \mathcal{A}/\theta'$, under the isomorphism given by $a \mapsto ([a]_\theta, [a]_{\theta'})$.*

**Definition 1.2.5.** An algebra $\mathcal{A}$ is called *directly indecomposable* if whenever $\mathcal{A} \cong \mathcal{B}_1 \times \mathcal{B}_2$, either $|B_1| = 1$ or $|B_2| = 1$.

The directly indecomposable algebras are in a sense "building blocks" for other algebras. For finite algebras, the following result can be shown by induction on the cardinality of $\mathcal{A}$.

**Theorem 1.2.6.** *Every finite algebra $\mathcal{A}$ is isomorphic to a direct product of directly indecomposable algebras.*

However, this theorem is not true for infinite algebras. It is well-known that there is up to isomorphism only one non-trivial directly indecomposable Boolean algebra, the two-element Boolean algebra $(\{0, 1\}; \wedge, \vee, \neg, 0, 1)$. But cardinality arguments show that a countably infinite Boolean algebra cannot be isomorphic to a direct product of directly indecomposable Boolean algebras. This means that directly indecomposable algebras are not the most general "building blocks" in the study of universal algebra. Therefore, we define another kind of product construction.

**Definition 1.2.7.** Let $(\mathcal{A}_j)_{j \in J}$ be a family of algebras of type $\tau$. A subalgebra $\mathcal{B} \subseteq \prod_{j \in J} \mathcal{A}_j$ of the direct product of the algebras $\mathcal{A}_j$ is called a *subdirect product* of the algebras $\mathcal{A}_j$, if for every projection mapping $p_k : \prod_{j \in J} \mathcal{A}_j \to \mathcal{A}_k$ we have

$$p_k(\mathcal{B}) = \mathcal{A}_k.$$

Examples of subdirect products are the diagonal $\Delta_A = \{(a, a) \mid a \in A\}$, as well as any direct product. The lattice given by the Hasse diagram in Figure 1.2 is the direct product of the two-element chain (lattice) $\mathcal{C}_2$ and the three-element chain $\mathcal{C}_3$. The sublattice $\mathcal{L} \subseteq \mathcal{C}_2 \times \mathcal{C}_3$ described by the Hasse diagram in Figure 1.3 is obviously a subdirect product of $\mathcal{C}_2$ and $\mathcal{C}_3$.

It can easily be checked that for any subdirect product $\mathcal{B} = \prod_{j \in J} \mathcal{A}_j$ of the family $(\mathcal{A}_j)_{j \in J}$, the projection mappings $p_k : \prod_{j \in J} \mathcal{A}_j \to \mathcal{A}_k$ satisfy the equation $\bigcap_{j \in J} ker(p_j \mid B) = \Delta_B$. In fact this property of the kernels of the

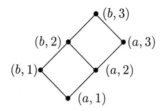

Fig. 1.2   Example of a Direct Product

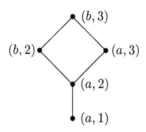

Fig. 1.3   Example of a Subdirect Product

projection mappings can be used to characterize subdirect products, in the sense that any set of congruences on an algebra with this property can be used to express the algebra as a subdirect product.

**Theorem 1.2.8.** *Let $\mathcal{A}$ be an algebra. Let $\{\theta_j \mid j \in J\}$ be a family of congruence relations on $\mathcal{A}$, which satisfy the equation $\bigcap_{j \in J} \theta_j = \Delta_A$. Then $\mathcal{A}$ is isomorphic to a subdirect product of the algebras $\mathcal{A}/\theta_j$, for $j \in J$. In particular, the mapping $\varphi(a) := ([a]_{\theta_j})_{j \in J}$ defines an embedding $\varphi : \mathcal{A} \to \prod_{j \in J} (\mathcal{A}/\theta_j)$, whose image $\varphi(\mathcal{A})$ is a subdirect product of the algebras $\mathcal{A}/\theta_j$.*

Algebras which cannot be expressed as a subdirect product of other smaller algebras, except in trivial ways, are called *subdirectly irreducible*.

**Definition 1.2.9.** An algebra $\mathcal{A}$ of type $\tau$ is called *subdirectly irreducible* if every family $\{\theta_j \mid j \in J\}$ of congruences on $\mathcal{A}$, none of which is equal to $\Delta_A$, has an intersection which is different from $\Delta_A$.

In this case, the conditions of Theorem 1.2.8 are not satisfied by any family of congruences on $\mathcal{A}$, and no representation of $\mathcal{A}$ as a subdirect product is possible.

It is easy to see that an algebra $\mathcal{A}$ is subdirectly irreducible if and only if $\Delta_A$ has exactly one upper neighbour or cover in the lattice $Con(\mathcal{A})$ of all congruence relations on $\mathcal{A}$. Then the congruence lattice has the form shown in the diagram of Figure 1.4.

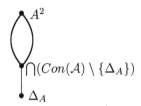

Fig. 1.4   Congruence Lattice of a Subdirect Irreducible Algebra

That subdirect products do act as the general "building blocks" in universal algebra is shown by the following theorem, due to G. Birkhoff ([9]).

**Theorem 1.2.10.** *Every algebra is isomorphic to a subdirect product of subdirectly irreducible algebras.*

## 1.3   Term Algebras, Identities and Free Algebras

In this section we turn from the study of constructions on algebras to another approach, the classification of algebras by logical expressions called identities. To do this we first need a formal language to use. This formal language is built out of variables from an $n$-element set $X_n = \{x_1, \ldots, x_n\}$ called an *alphabet*, for $n \geq 1$. We also need a set $\{f_i | i \in I\}$ of operation symbols, indexed by the set $I$; these symbols must be disjoint from the set $X_n$. To every operation symbol $f_i$ we assign a natural number $n_i \geq 1$, called the arity of $f_i$. The sequence $\tau = (n_i)_{i \in I}$ is called the *type* of the language. Now we define the terms of our type $\tau$, which will act as the "words" of our language.

**Definition 1.3.1.** Let $n \geq 1$. The *$n$-ary terms* of type $\tau$ are defined in the following inductive way:

(i) Every variable $x_j \in X_n$ is an $n$-ary term.
(ii) If $t_1, \ldots, t_{n_i}$ are $n$-ary terms and $f_i$ is an $n_i$-ary operation symbol, then $f_i(t_1, \ldots, t_{n_i})$ is an $n$-ary term.
(iii) The set $W_\tau(X_n) = W_\tau(x_1, \ldots, x_n)$ of all $n$-ary terms is the smallest set which contains $x_1, \ldots, x_n$ and is closed under finite application of (ii).

It follows immediately from the definition that every $n$-ary term is also $k$-ary, for $k > n$. This definition does not allow nullary terms, but we could add a fourth condition to the definition to make every nullary operation symbol of type $\tau$ into an $n$-ary term.

Our definition of terms is inductive, based on the number of occurrences of operation symbols in a term, and many proofs regarding terms and identities proceed on the basis of induction on the complexity of a term. Another commonly used measure of complexity of terms is the *depth* of a term. Both the depth and the operation-symbol-count can be defined inductively. Terms can be illustrated by tree diagrams, also called *semantic trees*. For example, the diagram of the term

$$t = f_2(f_1(f_2(f_2(x_1, x_2), f_1(x_3))), f_1(f_2(f_1(x_1), f_1(f_1(x_2)))))$$

is shown in Figure 1.5, where $f_2$ is a binary and $f_1$ is a unary operation symbol. For any term $t$, the depth of $t$ is the length of the longest path from the root to a leaf in the tree diagram for $t$.

Let $\tau$ be a fixed type. Let $X$ be the union of all the sets $X_n$ of variables, so $X = \{x_1, x_2, \ldots\}$. We denote by $W_\tau(X)$ the set of all terms of type $\tau$ over the countably infinite alphabet $X$:

$$W_\tau(X) = \bigcup_{n=1}^{\infty} W_\tau(X_n).$$

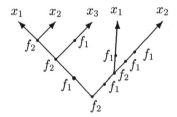

Fig. 1.5   Example of a Tree Diagram

This set $W_\tau(X)$ also forms the universe of an algebra of type $\tau$: for every $i \in I$ we define an $n_i$-ary operation $\bar{f}_i$ on $W_\tau(X)$, with

$$\bar{f}_i : W_\tau(X)^{n_i} \to W_\tau(X) \quad \text{by} \quad (t_1, \ldots, t_{n_i}) \mapsto f_i(t_1, \ldots, t_{n_i}).$$

**Definition 1.3.2.** The algebra $\mathcal{F}_\tau(X) := (W_\tau(X); (\bar{f}_i)_{i \in I})$ is called the *term algebra*, or the *absolutely free algebra*, or the *anarchic algebra* of type $\tau$ over the set $X$.

The term algebra $\mathcal{F}_\tau(X)$ is generated by the set $X$, and has what is known as the "absolute freeness" property. This means that for every algebra $\mathcal{A} \in Alg(\tau)$ and every mapping $f : X \to A$, there exists a unique homomorphism $\hat{f} : \mathcal{F}_\tau(X) \to \mathcal{A}$ which extends the mapping $f$ and such that $\hat{f} \circ \varphi = f$, where $\varphi : X \to \mathcal{F}_\tau(X)$ is the embedding of $X$ into $\mathcal{F}_\tau(X)$. This can be illustrated by Figure 1.6.

It should be clear that two absolutely free algebras $\mathcal{F}_\tau(X)$ and $\mathcal{F}_\tau(Y)$ are isomorphic if $X$ and $Y$ have the same cardinality. This means that we can use any countably infinite set of variables. In the case of the finite alphabet $X_n$, an analogous process leads to the algebra $\mathcal{F}_\tau(X_n)$ of all $n$-ary terms of type $\tau$, the absolutely free algebra of type $\tau$ on $n$ generators.

So far, terms are purely formal or syntactical expressions on our formal language of type $\tau$. In order to relate this syntax to semantics, that is, to formulate equations which are true or false in a given algebra $\mathcal{A}$, we need a way to attach a term operation on $\mathcal{A}$ to each term $t$ of type $\tau$.

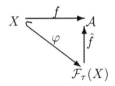

Fig. 1.6   Absolutely Free Algebra

We first evaluate the variables using elements of the concrete set $A$; then we interpret the operation symbols of the terms as concrete operations on this set. This process corresponds to the formation of a rational function from a polynomial $f(x) = \sum_{i=0}^{n} a_i x^i$ over a ring $\mathcal{R}$, by substituting for $x$ the elements of an overring $\mathcal{S} \supseteq \mathcal{R}$.

We need some notation: for $n \geq 1$ and $1 \leq j \leq n$, the mapping $e_j^{n,A}$ on $A$ defined by $e_j^{n,A}(a_1, \ldots, a_n) := a_j$ is called the *j-th projection mapping* on $A$.

**Definition 1.3.3.** Let $\mathcal{A}$ be an algebra of type $\tau$ and let $t$ be an $n$-ary term of type $\tau$ over $X_n$. Then $t$ induces an $n$-ary operation $t^\mathcal{A}$ on $\mathcal{A}$, called the *term operation induced by the term $t$* on the algebra $\mathcal{A}$, via the following steps:

1. If $t = x_j \in X_n$, then $t^\mathcal{A} = x_j^\mathcal{A} = e_j^{n,A}$.
2. If $t = f_i(t_1, \ldots, t_{n_i})$ is an $n$-ary term of type $\tau$, and $t_1^\mathcal{A}, \ldots, t_{n_i}^\mathcal{A}$

are the term operations which are induced by $t_1, \ldots, t_{n_i}$, then $t^A = f_i^A(t_1^A, \ldots, t_{n_i}^A)$.

We denote by $W_\tau(X_n)^A$ the set of all $n$-ary term operations of the algebra $A$, and by $W_\tau(X)^A$ the set of all (finitary) term operations on $A$.

Several properties of the fundamental operations of an algebra $A$ are also valid for term operations. For instance, for any homomorphism $h : A \to B$ and any $n$-ary term $t$ of the corresponding type, the following equality holds for all $a_1, \ldots, a_n$ in $A$:

$$h(t^A(a_1, \ldots, a_n)) = t^B(h(a_1), \ldots, h(a_n)).$$

Similarly, for a congruence relation $\theta$ of $A$ the implication from Definition 1.1.8 is valid for arbitrary $n$-ary term operations of $A$. The universes of subalgebras are preserved not only by all fundamental operations of $A$ but also by all its term operations.

**Definition 1.3.4.** Let $s$ and $t$ be terms from $W_\tau(X)$ and let $A$ be an algebra of type $\tau$. The equation $s \approx t$ is said to be an *identity* in the algebra $A$ of type $\tau$ if $s^A = t^A$; that is, if the term operations induced by $s$ and $t$ on the algebra $A$ are equal. In this case we also say that the equation $s \approx t$ is *satisfied* or *modelled* by the algebra $A$, and we write $A \models s \approx t$. If the equation $s \approx t$ is satisfied by every algebra $A$ of a class $K$ of algebras of the same type $\tau$, we write $K \models s \approx t$. Let $IdA$ be the set of all identities satisfied in the algebra $A$. If $K \subseteq Alg(\tau)$ is a class of algebras, then $IdK$ denotes the set of all equations which are identities in every algebra of $K$.

When two terms $s$ and $t$ satisfy the equation $s^A = t^A$ for some algebra $A$, it follows that for every mapping $f : X \to A$, we have $\hat{f}(s) = \hat{f}(t)$, where $\hat{f}$ is the uniquely determined extension of $f$ to the free algebra $\mathcal{F}_\tau(X)$. This means that $s \approx t$ is an identity in $A$ iff $(s, t) \in Ker\hat{f}$ for all mappings $f : X \to A$. Thus an identity $s \approx t$ holds in an algebra $A$ (or in a class $K$ of algebras), iff $(s, t)$ is in the intersection of the kernels of $\hat{f}$, for every map $f : X \to A$ (for every algebra $A$ in $K$).

For $K$ a class of algebras of type $\tau$, the set $IdK$ of all identities satisfied in every algebra from $K$ can be regarded as a binary relation on the set $W_\tau(X)$. This relation turns out to be a fully invariant congruence on the free algebra $\mathcal{F}_\tau(X)$.

**Proposition 1.3.5.** *Let $K \subseteq Alg(\tau)$ be a class of algebras of type $\tau$ and let $IdK$ be the set of all identities satisfied in every algebra $A \in K$. Then*

*IdK is a fully invariant congruence relation on the absolutely free algebra* $\mathcal{F}_\tau(X)$ *of type* $\tau$.

**Proof:** By definition, $IdK$ is an equivalence relation on the set $W_\tau(X)$. Now suppose that $s_1 \approx t_1$, ..., $s_{n_i} \approx t_{n_i}$ are in $IdK$. Then $s_1^{\mathcal{A}} = t_1^{\mathcal{A}}$, ..., $s_{n_i}^{\mathcal{A}} = t_{n_i}^{\mathcal{A}}$ for every algebra $\mathcal{A}$ in $K$. For any $n_i$-ary operation symbol $f_i$, we get $f_i^{\mathcal{A}}(s_1^{\mathcal{A}}, \ldots, s_{n_i}^{\mathcal{A}}) = f_i^{\mathcal{A}}(t_1^{\mathcal{A}}, \ldots, t_{n_i}^{\mathcal{A}})$. By the inductive step of the definition of a term operation induced by a term, this means that $[f_i(s_1, \ldots, s_{n_i})]^{\mathcal{A}} = [f_i(t_1, \ldots, t_{n_i})]^{\mathcal{A}}$, which gives us $f_i(s_1, \ldots, s_{n_i}) \approx f_i(t_1, \ldots, t_{n_i}) \in Id\mathcal{A}$. Thus $f_i(s_1, \ldots, s_{n_i}) \approx f_i(t_1, \ldots, t_{n_i}) \in IdK$, as required for a congruence.

To show that $IdK$ is fully invariant, we show that $(\varphi(s), \varphi(t))$ is in $IdK$ for every pair $(s, t)$ in $IdK$ and every endomorphism $\varphi$ of $\mathcal{F}_\tau(X)$. We use the property mentioned above that $IdK$ is equal to the intersection of the kernels of the homomorphisms $\hat{f}$, for all maps $f : X \to A$ and all algebras $\mathcal{A}$ in $K$. For any such $\mathcal{A}$ and map $f$, the map $\hat{f} \circ \varphi$ is also a homomorphism from $\mathcal{F}_\tau(X)$ into $\mathcal{A}$, and is the extension of some map $g$ from $X$ into $A$. Thus our pair $(s, t)$ from $IdK$ must also be in the kernel of this new homomorphism $\hat{f} \circ \varphi$. This means that the pair $(\varphi(s), \varphi(t))$ must be in $Ker\ \hat{f}$. Since this is true for all algebras $\mathcal{A}$ and maps $f : X \to A$, we must have $(\varphi(s), \varphi(t))$ in $IdK$. ∎

Now that we have a fully invariant congruence $IdK$ on the free algebra $\mathcal{F}_\tau(X)$, we can form a quotient algebra.

**Definition 1.3.6.** Let $X$ be a non-empty set of variables, and $K$ be a class of algebras of type $\tau$. The algebra $\mathcal{F}_K(X) := \mathcal{F}_\tau(X)/IdK$ is called the *K-free algebra over X* or the *free algebra relative to K* generated by $\overline{X} = X/IdK$.

Strictly speaking, $\mathcal{F}_K(X)$ is generated by the set $\overline{X} = X/IdK = \{[x]_{IdK} | x \in X\}$, and we should write $\mathcal{F}_K(\overline{X})$ rather than $\mathcal{F}_K(X)$. But for reasons of notational convenience this is not usually done, and we identify the equivalence classes $[x]_{IdK}$ with the terms $x$. This new algebra $\mathcal{F}_K(X)$ also satisfies a relative "freeness" property analogous to the absolutely freeness property:

**Theorem 1.3.7.** *For any algebra* $\mathcal{A} \in K \subseteq Alg(\tau)$ *and every mapping* $f : \overline{X} \to A$, *there exists a unique homomorphism* $\hat{f} : \mathcal{F}_K(X) \to \mathcal{A}$ *which extends* $f$.

## 1.4   The Galois Connection ($Id, Mod$)

We now have a relationship between algebras of type $\tau$ and pairs of terms of type $\tau$. The relation of satisfaction, in which an algebra is related to a pair $(s, t)$ of terms iff the algebra satisfies the identity $s \approx t$, gives a binary relation between the set $Alg(\tau)$ of all algebras of type $\tau$ and the set $W_\tau(X)^2$ of all pairs of terms of type $\tau$. We can also extend this relation to classes of algebras and sets of identities. For any subset $\Sigma \subseteq W_\tau(X)^2$ and any subclass $K \subseteq Alg(\tau)$ we consider

$$Mod\Sigma = \{\mathcal{A} \in Alg(\tau) \mid \forall s \approx t \in \Sigma,\ (\mathcal{A} \models s \approx t)\} \quad \text{and}$$

$$IdK = \{s \approx t \in W_\tau(X)^2 \mid \forall \mathcal{A} \in K,\ (\mathcal{A} \models s \approx t)\}.$$

This defines a pair $(Id, Mod)$ of mappings: $Mod$ maps from the power set of $W_\tau(X)^2$ into the power set of $Alg(\tau)$, while $Id$ maps back from the power set of $Alg(\tau)$ into the power set of $W_\tau(X)^2$. The next theorem gives some important properties of this pair of mappings.

**Theorem 1.4.1.**

(i)  *For all subsets $\Sigma$ and $\Sigma'$ of $W_\tau(X) \times W_\tau(X)$ and all subclasses $K$ and $K'$ of $Alg(\tau)$, we have*
$\Sigma \subseteq \Sigma' \Rightarrow Mod\Sigma \supseteq Mod\Sigma'$ *and* $K \subseteq K' \Rightarrow IdK \supseteq IdK'$.

(ii)  *For all subsets $\Sigma$ of $W_\tau(X) \times W_\tau(X)$ and all subclasses $K$ of $Alg(\tau)$, we have $\Sigma \subseteq IdMod\Sigma$ and $K \subseteq ModIdK$.*

(iii)  *The maps $IdMod$ and $ModId$ are closure operators on $W_\tau(X) \times W_\tau(X)$ and on $Alg(\tau)$, respectively.*

(iv)  *The sets closed under $ModId$ are exactly the sets of the form $Mod\Sigma$, for some $\Sigma \subseteq W_\tau(X) \times W_\tau(X)$, and the sets closed under $IdMod$ are exactly the sets of the form $IdK$, for some $K \subseteq Alg(\tau)$.*

A pair of mappings which satisfies conditions (i) and (ii) of this theorem is called a *Galois connection*.

**Definition 1.4.2.** A class $K$ of algebras of type $\tau$ is called an *equational class*, or is said to be *equationally definable*, if there is a set $\Sigma$ of equations such that $K = Mod\Sigma$. A set $\Sigma \subseteq W_\tau(X) \times W_\tau(X)$ is called an *equational theory* if there is a class $K \subseteq Alg(\tau)$ such that $\Sigma = IdK$.

From condition (iv) of Theorem 1.4.1, equational classes are exactly the fixed points with respect to the closure operator $ModId$, and dually equational theories are exactly the closed sets, or fixed points, with respect to the closure operator $IdMod$. The proof of the following theorem can be found in [29].

**Theorem 1.4.3.** *The collection of all equational classes of type $\tau$ forms a complete lattice $\mathcal{L}(\tau)$, and the collection of all equational theories of type $\tau$ forms a complete lattice $\mathcal{E}(\tau)$. These two lattices are dually isomorphic: there exists a bijection $\varphi : \mathcal{L}(\tau) \to \mathcal{E}(\tau)$, satisfying $\varphi(K_1 \vee K_2) = \varphi(K_1) \wedge \varphi(K_2)$ and $\varphi(K_1 \wedge K_2) = \varphi(K_1) \vee \varphi(K_2)$. (Note that $\varphi$ corresponds to $Id$ and $\varphi^{-1}$ is the mapping $Mod$.)*

The equational approach to algebra, using equations and identities, has led us to classes of algebras called equational classes. Now we return to the approach using the algebraic constructions of subalgebras, homomorphic images and products, to see what classes of algebras they determine.

**Definition 1.4.4.** We define the following operators on the class $Alg(\tau)$ of all algebras of type $\tau$: for any class $K \subseteq Alg(\tau)$,
$S(K)$ is the class of all subalgebras of algebras from $K$,
$H(K)$ is the class of all homomorphic images of algebras from $K$,
$P(K)$ is the class of all direct products of families of algebras from $K$,
$I(K)$ is the class of all algebras which are isomorphic to algebras from $K$,
$P_S(K)$ is the class of all subdirect products of families of algebras from $K$.

**Definition 1.4.5.** A class $K \subseteq Alg(\tau)$ is called a *variety* if $K$ is closed under the operators $H$, $S$ and $P$; that is, if $H(K) \subseteq K$, $S(K) \subseteq K$ and $P(K) \subseteq K$.

For any class $K$ of algebras of type $\tau$, it can be shown that the class $V(K) := HSP(K)$ is the least variety which contains $K$. When $K = \{\mathcal{A}\}$, a singleton class, we usually write $V(\mathcal{A}) := HSP(\{\mathcal{A}\})$.

*Birkhoff's Theorem* is the main theorem of equational theory; it tells us that the two approaches to universal algebra, by algebraic constructions and by equations and identities, lead to the same classes of algebras. (For a proof see for instance [12].)

**Theorem 1.4.6.** (*Birkhoff's Main Theorem of Equational Theory*) *A class $K$ of algebras of type $\tau$ is equationally definable if and only if it is a variety.*

There are two important ways to describe completely all the algebras of a given variety. The first method uses the subdirectly irreducible algebras of the variety, and the fact that every algebra of a variety $V$ is isomorphic to a subdirect product of subdirectly irreducible algebras from $V$.

The second method uses the relatively free algebras $\mathcal{F}_V(X_n)$ for natural numbers $n \geq 1$, and the fact that

$$V = HSP(\{\mathcal{F}_V(X_n) \mid n \in \mathbb{N}, n \geq 1\}).$$

Here the question arises whether the free algebras $\mathcal{F}_V(X_n)$ are finite. A variety $V$ is said to be *locally finite* if every finitely generated algebra in $V$ is finite. This is the case iff all the free algebras $\mathcal{F}_V(X_n)$ are finite, for $n \geq 1$.

We saw above that the collection of all equational classes of algebras of a fixed type $\tau$ forms a complete lattice. By Birkhoff's Theorem this is the same as the lattice $\mathcal{L}(\tau)$ of all varieties of type $\tau$. This is an algebraic lattice, with greatest element the variety $Alg(\tau)$ consisting of all algebras of type $\tau$ and least element the trivial variety $T$ consisting exactly of all one-element algebras of type $\tau$. Clearly, $Alg(\tau) = Mod\{x \approx x\}$ and $T = Mod\{x \approx y\}$. Dually, the greatest element in the lattice $\mathcal{E}(\tau)$ of all equational theories of type $\tau$ is the equational theory generated by $\{x \approx y\}$, while the least element is the equational theory generated by $\{x \approx x\}$.

A subclass $W$ of a variety $V$ which is also a variety is called a *subvariety* of $V$. A non-trivial variety $V$ is called *minimal* or *equationally complete* if it has no proper subvarieties other than the trivial variety. It can be shown that every non-trivial $V \in \mathcal{L}(\tau)$ contains a minimal subvariety (see for instance [29]). If $V$ is a given variety of type $\tau$, then the collection of all subvarieties of $V$ forms a complete lattice $\mathcal{L}(V) := \{W \mid W \in \mathcal{L}(\tau)$ and $W \subseteq V\}$. This lattice is called the *subvariety lattice* of $V$.

Equational theories were defined above as sets $\Sigma$ of equations for which $IdMod\Sigma = \Sigma$. By Theorem 1.4.1(iv), a set $\Sigma$ is an equational theory of type $\tau$ iff there is a class $K$ of algebras of type $\tau$ with $\Sigma = IdK$. We also saw in Proposition 1.3.5 that sets of equations of the form $IdK$ are fully invariant congruence relations on the absolutely free algebra $\mathcal{F}_\tau(X)$.

The properties of a fully invariant congruence relation give us a system of equational logic, a formal way to derive from a given set of equations satisfied as identities in a class of algebras new equations which are also satisfied as identities. We write $\Sigma \vdash s \approx t$, read as "$\Sigma$ yields $s \approx t$," if there is a formal deduction of $s \approx t$ starting with identities in $\Sigma$, using the following five *rules of consequence* or *derivation* or *deduction rules*:

(1) $\emptyset \vdash s \approx s$.
(2) $\{s \approx t\} \vdash t \approx s$.

(3) $\{t_1 \approx t_2, t_2 \approx t_3\} \vdash t_1 \approx t_3.$

(4) $\{t_j \approx t'_j \mid 1 \leq j \leq n_i\} \vdash f_i(t_1, \ldots, t_{n_i}) \approx f_i(t'_1, \ldots, t'_{n_i}),$ for every operation symbol $f_i$ and $i \in I.$ (This is called the *replacement rule*.)

(5) Let $s, t, r \in W_\tau(X)$ and let $\tilde{s}, \tilde{t}$ be the terms obtained from $s, t$ by replacing every occurrence of a given variable $x \in X$ by $r$. Then $\{s \approx t\} \vdash \tilde{s} \approx \tilde{t}.$ (This is called the *substitution rule*.)

These five rules correspond directly to the properties of a fully invariant congruence relation: the first three are the properties of an equivalence relation, the fourth describes the congruence property and the fifth the fully invariant property. Again, equational theories $\Sigma$ are precisely sets of equations which are closed with respect to finite application of these five rules.

Let $s \approx t$ be an identity and $\Sigma$ a set of identities of type $\tau$. We write $\Sigma \models s \approx t$ to mean that $s \approx t$ is satisfied as an identity in every algebra $\mathcal{A}$ of type $\tau$ in which all equations from $\Sigma$ are satisfied as identities. The connection between "$\vdash$" and "$\models$" is given by the Completeness and Consistency Theorem of equational logic: For $\Sigma \subseteq W_\tau(X) \times W_\tau(X)$ and $s \approx t \in W_\tau(X) \times W_\tau(X)$, we have

$$\Sigma \models s \approx t \quad \Leftrightarrow \quad \Sigma \vdash s \approx t.$$

The "$\Rightarrow$"-direction of this theorem means that any equation which is "true" is derivable, and is usually called the *completeness* property. The "$\Leftarrow$"-direction is called *consistency*, since it means that every derivable equation is "true".

## 1.5 Multi-based Algebras and Clones

In many mathematical systems which we want to model algebraically, it is natural to consider more than one set of objects, and operations on them, rather than the single universe of an algebra. In this section we present some examples and a bit of the theory of *many-sorted algebras* or *heterogeneous* or *multi-based algebras*.

A ring can be regarded as an algebra $\mathcal{R} = (R; +, \cdot, -, 0)$ of type $(2, 2, 1, 0)$ where $(R; +, -, 0)$ is an abelian (commutative) group (with addition as the binary operation and 0 as an identity element) and $(R; \cdot)$ is a semigroup, and where the two *distributive laws* are satisfied.

Let $\mathcal{R} = (R; +, \cdot, -, 0)$ be a ring such that $R$ is infinite. As an example we consider an algebra with an infinite number of operation symbols, one for each element of the ring $\mathcal{R}$. An algebra $(M; +, -, 0, R)$ of type $(2, 1, 0, (1)_{r \in R})$ is called an $\mathcal{R}$-*module* (or a *module over* $\mathcal{R}$) if $(M; +, -, 0)$ is an abelian group and the following identities are satisfied, for all $r$ and $s$ in $R$ and $x$ and $y$ in $M$:

(M1)        $r(x + y) = r(x) + r(y),$
(M2)        $(r + s)(x) = r(x) + s(x),$
(M3)        $(r \cdot s)(x) = r(s(x)).$

If $\mathcal{R}$ also has an identity element $1$, then the following additional identity is required for an $\mathcal{R}$-module:

(M4)        $1(x) = x.$

In the special case that $\mathcal{R}$ is a field, an $\mathcal{R}$-module is usually called an $\mathcal{R}$-*vector space*. The elements of the module are then called *vectors*, while the field elements are called *scalars*.

This example shows that modules and vector spaces can be viewed as algebras with infinitely many operations. There is another way to describe modules and vector spaces algebraically, which avoids this infinitary type, and which is more often used. We have so far considered only what are called *homogeneous* or *one-sorted* algebras, with a single base set of objects and some operations defined on this set. Now we define *heterogeneous* or *many-sorted* or *multi-based* algebras, where there are two or more base sets of objects and operations are allowed on more than one kind of object.

For vector spaces, for example, we can use both the set $V$ of all vectors and the universe of the field $\mathcal{R}$ as two different base sets in such a heterogeneous algebra. Then we have two operations, $+ : V \times V \to V$ for the usual vector addition and $\cdot : R \times V \to V$ for the scalar multiplication of a scalar from $R$ times a vector from $V$. The structure $(V, R; +, \cdot)$ is thus an example of a heterogeneous algebra.

The algebraic theory of many-sorted algebras can be developed in a completely analogous way to the development for (one-sorted) algebras. For instance, the universes of subalgebras of many-sorted algebras are sequences of subsets of the sequence of universes of the given algebra, which are closed under application of the heterogeneous operations. Homomorphisms are sequences of mappings which are compatible with the heterogeneous operations. For more details see for instance [10].

*Clones* are an important example of many-sorted algebras. Let $A$ be any set, and let $O^n(A)$ be the set of all $n$-ary operations defined on $A$. We recall the notation from Section 1.1 that $O(A) = \bigcup_{n \geq 1} O^n(A)$ for the set of all finitary operations on $A$.

**Definition 1.5.1.** Let $O^n(A), n \geq 1$, be the set of all $n$-ary operations defined on the set $A$. For each $n, m \geq 1$, we define an operation

$$S_m^{n,A} : O^n(A) \times (O^m(A))^n \to O^m(A),$$

by
$$S_m^{n,A}(f^A, g_1^A, \ldots, g_n^A)(a_1, \ldots, a_m)$$
$$:= f^A(g_1^A(a_1, \ldots, a_m), \ldots, g_n^A(a_1, \ldots, a_m))$$
for all $a_1, \ldots, a_m \in A$.

The operations $S_m^{n,A}$ are called superposition operations on $O(A)$, and $O(A)$ is closed under these operations.

A *clone of operations* is a subset of $O(A)$ which is closed under all the operations $S_m^{n,A}$ for $m, n \in \mathbb{N}^+ := \mathbb{N} \setminus \{0\}$ and contains all the projections $e_j^{n,A}$ (as defined in Section 1.3). $O(A)$ itself is called the *full clone* of operations defined on $A$.

Using the sets $O^n(A)$, for $n \geq 1$, as the universes, we can view the full clone on $A$ as a many-sorted algebra

$$Clone\,A = (\ (O^n(A));\ S_m^{n,A},\ e_j^n)_{n,m \geq 1, 1 \leq j \leq n}.$$

**Definition 1.5.2.** Let $C \subseteq O(A)$ be a set of operations on a set $A$. Then the *clone generated by* $C$, denoted by $\langle C \rangle$, is the smallest subset of $O(A)$ which contains $C$, is closed under composition, and contains all the projections $e_j^{n,A} : A^n \to A$ for arbitrary $n \geq 1$ and $1 \leq j \leq n$.

For any type $\tau$, we can also define superposition operations $S_m^n$ on the sets $W_\tau(X_n)$ of terms of type $\tau$. Here

$$S_m^n : W_\tau(X_n) \times (W_\tau(X_m))^n \to W_\tau(X_m).$$

This gives us a clone

$$((W_\tau(X_n));\ S_m^n, e_j^n)_{n,m \geq 1, 1 \leq j \leq n}$$

called the *term clone* of type $\tau$. The clone of all term operations on an algebra $\mathcal{A}$ can be defined similarly.

The concept of a term clone can be used for (one-sorted) algebras of terms as well. If we fix $n \geq 1$, we can consider the set $W_\tau(X_n)$ of $n$-ary terms, with the superposition operation $S_n^n$. It is also common to write $t(t_1, \ldots, t_n)$ instead of $S^n(t, t_1, \ldots, t_n)$. Selecting the variable terms $x_1, \ldots, x_n$ for the nullary operations, we form the algebra

$$n\text{-}clone \; \tau := (W_\tau(X_n); \; S^n, x_1, \ldots, x_n),$$

called the $n$-clone of type $\tau$. This algebra is an example of what is called a *unitary Menger algebra of rank n*.

## 1.6    Exercises for Chapter 1

1. Prove that the composition of homomorphisms (where defined) is a homomorphism.

2. Prove that the inverse of a bijective homomorphism is a homomorphism.

3. Let $\mathcal{A}$ and $\mathcal{B}$ be algebras with $h$ a homomorphism from $\mathcal{A}$ to $\mathcal{B}$.
a) Prove that if $\mathcal{A}_1$ a subalgebra of $\mathcal{A}$ then $h(\mathcal{A}_1)$ is the universe of a subalgebra of $\mathcal{B}$.
b) Prove that if $\mathcal{B}_1$ is a subalgebra of $\mathcal{B}$ then the set $h^{-1}(\mathcal{B}_1)$ is the universe of a subalgebra of $\mathcal{A}$.

4. Prove that any equational class is a variety. That is, prove that if a family of algebras satisfies an identity $s \approx t$, then so does any homomorphic image or subalgebra of an algebra in the family and so does any product of algebras in the family.

5. Prove that the set $Con_{inv}(\mathcal{A})$ of all fully invariant congruence relations of an algebra $\mathcal{A}$ forms a sublattice of the lattice $Con(\mathcal{A})$.

6. Prove Lemma 1.2.2.

7. Prove Theorem 1.2.4.

8. Let $B$ be the variety of bands, that is, a variety of algebras of type $\tau = (2)$ satisfying the associative and the idempotent identity. Determine all elements of the free algebra $F_B(X_2)$ over the two-element alphabet

$X_2 = \{x_1, x_2\}$.

9. Determine all elements of the free algebra $F_V(X_2)$ for the following varieties of type $\tau = (2)$:

(i) the variety of *right zero semigroups* $RZ = Mod\{x_1 x_2 \approx x_2\}$,
(ii) the variety of *left zero semigroups* $LZ = Mod\{x_1 x_2 \approx x_1\}$,
(iii) the variety of *zero semigroups* $Z = Mod\{x_1 x_2 \approx x_3 x_4\}$,
(iv) the variety of *semilattices* $SL = Mod\{x_1(x_2 x_3) \approx (x_1 x_2)x_3, x_1 x_2 \approx x_2 x_1, x_1 x_1 \approx x_1\}$, and
(v) the variety of *rectangular bands* $RB = Mod\{x_1(x_2 x_3) \approx (x_1 x_2)x_3 \approx x_1 x_3, x_1 x_1 \approx x_1\}$.

10. Define congruence relations for many-sorted algebras and prove that the collection of all congruence relations of a many-sorted algebra $\mathcal{A}$ forms a complete lattice.

11. Define many-sorted subalgebras for many-sorted algebras and prove that the collection of all many-sorted subalgebras of a many-sorted algebra $\mathcal{A}$ forms a complete lattice.

12. Formulate and prove the homomorphism theorem for many-sorted algebras.

13. Define a binary operation $+$ on the set $W_\tau(X_n)$ by $t_1 + t_2 := S^n(t_1, t_2, \ldots, t_2)$ and prove that $+$ is associative.

14. Let $\mathcal{A}$ and $\mathcal{B}$ be any algebras, with $h : \mathcal{A} \to \mathcal{B}$ a homomorphism.
a) Prove that for any $n$-ary term $t$, the following equality holds for all $a_1, \ldots, a_n$ in $A$:

$$h(t^A(a_1, \ldots, a_n)) = t^B(h(a_1), \ldots, h(a_n)).$$

15. Prove Theorem 1.3.7.

# Chapter 2

# State-based Systems

Universal coalgebra can be viewed as the theory of *state-based systems*. Before we formally define what a coalgebra is, we present a number of examples of state-based systems, and of mappings between such systems. Then our definitions of coalgebras and homomorphisms of coalgebras will arise as abstractions of properties of these various examples.

What is a state-based system? Any state-based system has inputs, outputs and inner states. The behaviour of such a system depends not only on the inputs but also on the inner states, in the sense that identical inputs can produce different outputs in different inner states. In general, state-based systems meet the following criteria:

(i) The behaviour of the system depends on inner states which are not visible to the user of the system.
(ii) The system can interact with its environment.
(iii) This interaction is based on operations.

In specifying a state-based system, our only interest lies in the input-output behaviour of the system; the states are taken to be a fixed part of the implementation. Once a working system is obtained, one may try to minimize the set of inner states by removing any unnecessary states, in particular, states which show no difference with respect to the input-output behaviour of the system. This lack of difference will be expressed by the important concept of *bisimulation* (also called *bisimilarity*). In general, two inner states $s$ and $s'$ are called bisimilar, and we write $s \sim s'$, if $s$ and $s'$ cannot be distinguished with respect to their input-output behaviour. This relationship $\sim$ is reflexive and symmetric, and in some sense to be explained later, compatible with the behaviour of the system.

## 2.1   Black Boxes

*Black boxes* form a particularly simple class of state-based systems having a display and two buttons labelled $h$ and $t$. When button $h$ is pressed, the display shows an element $d$ of some fixed set $D$ of data (for instance a number). The button $t$ changes the inner state of the black box, in such a way that when $h$ is pressed after $t$ the black box shows a (possibly new) element $d' \in D$ on its display. Let $S$ be the set of inner states of the black box. Then the black box can be described by a pair of mappings

$$h : S \to D,$$

$$t : S \to S.$$

**Definition 2.1.1.** Let $D$ be a set of data elements. A black box over $D$ is a triple $(S; h, t)$ where $S$ is a set of elements called states and $h : S \to D$ and $t : S \to S$ are mappings.

Given some inner state $s$ of a black box, we can observe an infinite stream of data elements:

$$(h(s), h(t(s)), h(t(t(s))), \ldots).$$

For this reason black boxes are also called *stream automata*.

A black box can be regarded as an algebraic structure in several ways. Using both the set $D$ of data and the set $S$ of states as base sets, we can form a many-sorted algebra $(S, D; h, t)$. To consider a black box instead as a coalgebra, we want to combine our two mappings $h$ and $t$ into one mapping.

We will describe this combining process first for arbitrary sets and functions. Let $A$, $B$ and $C$ be sets, and let $f : C \to A$ and $g : C \to B$ be mappings. We define a mapping $f \otimes g : C \to A \times B$, called the *tensor product* of $f$ and $g$, by setting $(f \otimes g)(c) := (f(c), g(c))$ for all $c \in C$. Let $p_1 : A \times B \to A$ and $p_2 : A \times B \to B$ be the usual first and second projection mappings on a product $A \times B$. For any functions $h : A \to B$ and $k : C \to D$ we define $h \times k := (h \circ p_1) \otimes (k \circ p_2) : A \times C \to B \times D$ by the rule that $(h \times k)(a, c) := (h(a), k(c))$ for any pair $(a, c) \in A \times C$. We remark that the mapping $f \otimes g$ has the following uniqueness property, known as the universal property for the product: if $f : C \to A$ and $g : C \to B$ are mappings, then $f \otimes g$ is the uniquely determined mapping satisfying $p_1 \circ (f \otimes g) = f$ and $p_2 \circ (f \otimes g) = g$.

Returning now to black boxes, we can combine our mappings $h : S \to D$ and $t : S \to S$ into the mapping $h \otimes t : S \to D \times S$. Then our black box (over the fixed set $D$ of data) can be regarded as a pair $(S; h \otimes t)$. Notice that our single mapping now goes from the base set $S$ to the product $D \times S$.

Now we consider the question of bisimulation of states in a black box. Two states which produce different outputs from $h$ can certainly be distinguished. Two states can also be distinguished if a series of identical inputs gives different outputs. Therefore $s \sim s'$ should mean that $h(s) = h(s')$ and $t(s) \sim t(s')$. This can be formulated as the following *rule of inference*:

$$\frac{s \sim s'}{h(s) = h(s') \text{ and } t(s) \sim t(s')} \; .$$

As a rule of inference this means that $s \sim s'$ implies $h(s) = h(s')$ and $t(s) \sim t(s')$. A *bisimulation* on a black box $(S; h \otimes t)$ is then any relation $\sim \subseteq S \times S$ which satisfies this rule of inference.

As an example, we consider a black box with an eight-element set of states, $s_1$ to $s_8$. We represent the action of this black box by a transition mapping, shown below, in which each state $s_j$ is shown by its output, $(h(s_j))$ and an arrow from state $s_j$ to state $s_k$ means that $t(s_j) = s_k$.

$$(33) \longrightarrow (17) \longleftrightarrow (42) \longleftarrow (17) \longleftarrow (42) \longleftarrow (33) \quad (42) \longleftrightarrow (17)$$

Then two states with different labels in the picture have different outputs, and can be distinguished. Among the states with the same labels, the two states labelled by 33 are also distinguishable: pressing $t$ and then $h$ gives 17 in one case and 42 in the other. Starting from the first state $s_1$ which is labelled by 33 we obtain the data sequence $33, 17, 42, 17, 42, \ldots$. The state $s_6$ which is also labelled by 33 produces the sequence $33, 42, 17, 42, 17, \ldots$. The three states $s_3$, $s_5$ and $s_7$ which are labelled by 42 however are not distinguishable, since they all produce the same data sequence $42, 17, 42, 17, \ldots$.

## 2.2 Data Streams

A *stream of data* is an infinite list of data. A stream $s$ consists of its head, the first element in the list, and then the remainder of the list, which is called the tail of $s$. An infinite stream from a set $D$ of data can then be considered as a mapping $\tau : \omega \to D$, where $\omega$ is the ordered set of natural numbers and where $\tau(k)$ is the $k$-th element of the stream, for $k \in \omega$.

In Unix the stream given by an infinite sequence of **y**'s is called **yes**:

$$\textbf{yes} := [\textbf{y}, \textbf{y}, \textbf{y}, \ldots].$$

The command " remove * | yes " deletes all files. For every file $f$ the program asks: *Do you really want to delete $f$ [y/n]?* The answer is **y** as often as necessary.

Streams are closely connected to black boxes. A stream of elements from a set $D$ of data together with two mappings $hd$ and $tl$ which pick out the head and tail of the stream can be considered as a special black box in which the set of states is the set $D^\omega$ of all mappings from $\omega$ to $D$:

$$
\begin{aligned}
hd &: D^\omega \to D \\
tl &: D^\omega \to D^\omega.
\end{aligned}
$$

For any $\tau \in D^\omega$ the black-box mappings $h$ and $t$ are given by $h(\tau) := hd(\tau) := \tau(0)$ and $t(\tau) := tl(\tau)$, with $tl(\tau)(k) := \tau(k+1)$. Writing $\tau$ as an infinite list $(\tau(0), \tau(1), \ldots)$, we have $hd(\tau) = \tau(0)$ as the head and $tl(\tau) = (\tau(1), \tau(2), \ldots)$ as the tail of the list. We may then consider a stream as a system $(D, D^\omega; hd \otimes tl)$ with $hd \otimes tl : D^\omega \to D \times D^\omega$; or if we fix the set $D$ of data, we may use the pair $(D^\omega; hd \otimes tl)$.

Streams have the property that any two different states can be distinguished. This can be written as a rule of inference in the form

$$
\frac{s \sim s'}{s = s'}.
$$

The bisimilarity relations on our set $D^\omega$ determined by this rule of inference are then any subsets of the diagonal relation on $D^\omega$. This rule of inference is also called the principle of *coinduction*, and will be discussed further later in Chapter 4 on coalgebras.

## 2.3    Data Types

Data types from computer science are usually considered algebraically as many-sorted algebras. For instance, for the data type *stack*, the universes of the many-sorted algebra *Stack* are the sets $D$ and *Stack*, and there are two operations given by

$$
\begin{aligned}
emptystack &: \quad \emptyset \to Stack, \\
push &: \quad D \times Stack \to Stack.
\end{aligned}
$$

Any stack is either the empty stack or has the form

$$
push(d_1, (push(d_2, (\cdots (push(d_n, emptystack) \cdots )))))
$$

for some natural number $n$. The algebra *Stack* also has two partial operations *pop* and *top*:

$$pop \quad : \quad Stack \rightarrow Stack,$$
$$top \quad : \quad Stack \rightarrow D.$$

From the coalgebraic point of view, we want to combine the two operations *emptystack* and *push* into one operation on the base set *Stack*. First, we use the one-element set $1 := \{*\}$ and the notation $+$ for the disjoint union of two sets to define a mapping

$$k : 1 + (D \times Stack) \rightarrow Stack,$$

where $k(*) := emptystack$ and $k(d, s) := push(d, s)$. This mapping $k$ is bijective, and has an inverse mapping

$$l : Stack \rightarrow 1 + (D \times Stack)$$

given by

$$l(s) = \begin{cases} * & \text{if } s = emptystack \\ (d, s') & \text{if } s = push(d, s'). \end{cases}$$

This total operation $l$ combines the partial operations *top* and *pop*.

Now we can consider <u>*Stack*</u> as a pair $(Stack; l)$, where $l$ is a mapping from *Stack* to $1 + (D \times Stack)$. This is similar to the process we carried out to express any black box in coalgebra form. We can also interpret $(Stack; l)$ as a black box. Two states cannot be distinguished, and we write $s \sim s'$, if either $l(s) = * = l(s')$ or $top(s) = top(s')$ and $pop(s) \sim pop(s')$. This gives the following rule of inference:

$$\frac{s \sim s'}{(emptystack(s) \wedge emptystack(s')) \vee (top(s) = top(s') \wedge pop(s) \sim pop(s'))}.$$

The elements of the set *Stack* can be viewed as states. Then starting from an arbitrary state $s_0$, repeated application of mapping $l$ produces a sequence of states $s_0, s_1, s_2, \ldots$, with $s_{i+1} = pop(s_i)$. In parallel there is a sequence of outputs $top(s_0), top(s_1), \ldots$. These sequences will stop at step $i$ if $l(s_i) = *$, that is, if the stack is empty. Thus what is produced starting from state $s_0$ is a sequence, either finite or infinite, of elements from $D$.

Conversely, from the set $D^\infty := D^* + D^\omega$ of all finite or infinite sequences of elements from $D$, a system of the type previously described can be built up: we set

$$l(\sigma) = \begin{cases} * & \text{if } \sigma = \varepsilon \text{ (empty word)} \\ (hd(\sigma), tl(\sigma)) & \text{otherwise.} \end{cases}$$

Here again we obtain the coinduction rule of inference,

$$\frac{s \sim s'}{s = s'}.$$

## 2.4   Automata

In this section we consider various kinds of automata. We begin with an *automaton without output*, also called an *acceptor* or a *recognizer*. This is a many-sorted algebra $\mathcal{H} = (I, S; \delta)$, where $I$ and $S$ are non-empty sets, called the sets of inputs and states, respectively, and $\delta : S \times I \to S$ is a mapping called the state transition mapping. An *automaton (with output)* is a quintuple $\mathcal{A} = (I, S, O; \delta, \gamma)$, where $(I, S; \delta)$ is an acceptor, $O$ (or sometimes $D$) is a non-empty set called the set of outputs, and $\gamma : S \times I \to O$ is a mapping called the output mapping. For any state $s \in S$ and any input $e \in I$, the value $\delta(s, e)$ is a state $s'$ and the value $\gamma(s, e)$ is the output which results when the input $e$ is read when the machine is in the state $s$. When all the sets $I$, $S$ and $O$ are finite, then the automaton is said to be finite; otherwise it is said to be infinite. In the finite case we will sometimes also write $I_n, S_m, O_l$, to indicate that these sets contain $n, m$ and $l$ elements, respectively.

What we have defined so far is a *deterministic* automaton. Since $\delta$ and $\gamma$ are mappings, they each have exactly one image for each state-input pair $(s, e)$, and the behaviour of the automaton is completely determined. It is also possible to allow $\delta$ and $\gamma$ to be relations which are not mappings, that is, to take on more than one value or to be undefined for some input pairs. In this case the automaton is called *non-deterministic*.

We mention also that there are two different types of automata, depending on the nature of the output function $\gamma$. What we have defined here so far is the Mealy type of automata, where $\gamma : S \times I \to O$ is a binary function. The Moore type of automata uses instead a unary output function $\gamma : S \to O$, in which case the output does not depend on the input element but only on the current state. We shall mostly discuss Mealy automata, but occasionally use Moore automata instead.

Another variation on the basic definition is that sometimes one state $s_0 \in I$ is selected as an *initial state*, and in this case we write $(I, S; \delta, s_0)$ or $(I, S, O; \delta, \gamma, s_0)$ for the automaton.

In the finite case, an automaton may be fully described by means of tables for $\delta$ and $\gamma$, or by a directed graph. The vertices of the graph correspond to the states, and there is an edge labelled by $e_j; d_k$ going from vertex $s_2$ to vertex $s_1$ when $\delta(s_2, e_j) = s_1$ and $\gamma(s_2, e_j) = d_k$. We illustrate this with the graph shown in Figure 2.1, which corresponds to the automaton with $S = \{s_1, s_2, s_3\}$, $I = \{e_1, e_2, e_3\}$, and $O = \{d_1, d_2\}$, with the tables shown below for $\delta$ and $\gamma$.

| $\delta$ | $e_1$ | $e_2$ | $e_3$ |
|----------|-------|-------|-------|
| $s_1$ | $s_1$ | $s_1$ | $s_3$ |
| $s_2$ | $s_2$ | $s_1$ | $s_3$ |
| $s_3$ | $s_3$ | $s_2$ | $s_1$ |

| $\gamma$ | $e_1$ | $e_2$ | $e_3$ |
|----------|-------|-------|-------|
| $s_1$ | $d_2$ | $d_1$ | $d_2$ |
| $s_2$ | $d_2$ | $d_1$ | $d_2$ |
| $s_3$ | $d_1$ | $d_1$ | $d_2$ |

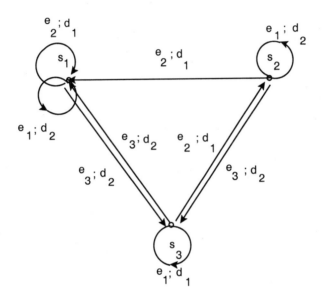

Fig. 2.1   Graph of an Automaton

In general, the behaviour of a deterministic automaton on a sequence of input elements can be described as follows. Let $s_0$ be the initial state, and let $e_1, \ldots, e_n$ be a sequence of inputs. Then a corresponding sequence of states is reached: from $s_0$ we have $s_1 = \delta(s_0, e_1)$, $s_2 = \delta(\delta(s_0, e_1), e_2)$, $\ldots$, $s_n = \delta(s_{n-1}, e_n)$. The previous example shows that there can be equal elements in the sequence. The sequence of output elements is given by $d_1 = \gamma(s_0, e_1), \ldots, d_n = \gamma(s_{n-1}, e_n)$.

It is clear that two states $s$ and $s'$ in an automaton cannot be distinguished if they produce the same output and the same state for each input. Bisimulation for automata is thus defined by the following rule of inference:

$$\frac{s \sim s'}{\forall e \in I \ (\gamma(s, e) = \gamma(s', e), \delta(s, e) \sim \delta(s', e))}.$$

That is, a bisimulation on an automaton is any relation $\sim \subseteq S \times S$ which satisfies this rule of inference.

Finite automata are used to recognize languages, or sets of words. In order to discuss the languages accepted by finite automata, we first recall some definitions and notation. Since we shall be concerned only with finite automata, we shall consider a finite set $I_n = \{e_1, \ldots, e_n\}$ which will be called a finite alphabet of size $n \geq 1$. We denote by $I_n^*$ the universe set of the free monoid generated by $I_n$. It is customary to denote the binary operation of this monoid by juxtaposition, and using the associativity of this operation in a monoid allows us to omit all brackets from terms. Thus any monoid term from $I_n^*$ may be expressed as a *word* composed of the letters from $I_n$. Note that the letters may be repeated; for instance, each of $e_1e_2$, $e_2e_1e_1$ and $e_1e_2e_3e_3$ is a word on the alphabet $I_3 = \{e_1, e_2, e_3\}$. Formally, any word $w$ may be expressed as $w = e_{i_1}e_{i_2}\cdots e_{i_m}$, for some $m \geq 0$ and $e_{i_1}, \ldots, e_{i_m} \in I_n$. The number $m$ is called the *length* of the word $w$, and is denoted by $|w|$. The special case that $m = 0$ corresponds to the *empty word*, denoted by $\varepsilon$. The set $I_n^+$ is defined to be the set of all non-empty words on the alphabet $I_n$. This set is the universe of the free semigroup generated by the set $I_n$. Of course, $I_n^+$ and $I_n^*$ are (the universes of) a semigroup and a monoid respectively, with the binary operation of juxtaposition or *concatenation* of words.

Languages are sets of words. More precisely, a *language over the alphabet* $I_n$ is any subset of the universe of the monoid $I_n^*$. The power set $P(I_n^*)$ is then the family of all possible languages on $I_n$. There are several operations defined on sets of languages. One of these is the usual set-theoretic union. The *product* of two languages, usually denoted by juxtaposition, is the operation which maps any two languages $U$ and $V$ to the language

$$UV := \{uv \mid u \in U, v \in V\}.$$

We have $U(VW) = (UV)W$ for all languages $U$, $V$ and $W$ on $I_n$, so this product is associative. It also has the properties that $U\emptyset = \emptyset U = \emptyset$ and $U\{\varepsilon\} = \{\varepsilon\}U = U$ for every language $U$.

The product operation can be inductively extended to powers of languages. For any language $U$, we define powers $U^m$ for all $m \geq 0$, by
(i) $U^0 = \{\varepsilon\}$,    and
(ii) $U^m = U^{m-1}U$, for $m \geq 0$.

Then we define $U^* = \bigcup_{m \in \mathbb{N}} U^m$ and $U^+ = \bigcup_{m \geq 1} U^m$. A word $w \in I_n^*$ belongs to $U^*$ if and only if it is the empty word or it can be expressed in the form $u_1u_2 \cdots u_m$ for some $m \geq 1$ and some words $u_1, \ldots, u_m \in U$.

Then $I_n^m$ is the set of all words of length $m$ on the alphabet $I_n$, and $I_n^* = \bigcup_{m \in \mathbb{N}} I_n^m$. The unary operation taking a language $U$ to the language $U^*$ is called *iteration*. The three operations discussed thus far, the union, product and iteration of languages, are called the *regular language operations*.

**Definition 2.4.1.** The set $RegI_n$ of all regular languages over an alphabet $I_n$ is the smallest set $R$ such that

(i) $\emptyset \in R$ and $\{x\} \in R$ for each $x \in I_n$, and
(ii) for any $U$ and $V$ in $R$, all of $U \cup V$, $UV$ and $U^*$ are in $R$.

It is immediate from this definition that any finite language is regular. The set $RegI_n$ is the smallest set of languages over $I_n$ which contains all the finite languages and is closed under the three regular language operations.

We have defined three operations on the set $\mathcal{P}(I_n^*)$. To consider this set as the base of an algebra, we can use the binary symbol $+$ for the union operation, the binary symbol $\cdot$ (or simply juxtaposition) for the product operation, and a unary symbol $*$ for the iteration operation. We also need nullary operation symbols to pick out $\emptyset$ and the empty word $\varepsilon$. This gives an algebra of type $(2, 2, 1, 0, 0)$. The terms of this algebra are called the *regular expressions* over the alphabet $I_n$. In fact terms of this type correspond precisely to regular languages over $I_n$.

Now we return to our automaton or acceptor $\mathcal{A} = (I, S, O; \delta, \gamma)$. We will use the notation for words and languages to consider first bisimulations on $\mathcal{A}$ and then the language accepted by $\mathcal{A}$. For the next definition and lemma we use the Moore model of an automaton, that is, we use a unary output mapping $\gamma : S \to O$. Our state transition mapping $\delta$, which is a mapping from $S \times I_n$ to $S$, can be extended to a mapping $\delta^* : S \times I_n^* \to S$ in the following inductive way:

(i) $\delta^*(s, \varepsilon) = s$, for all $s \in S$,
(ii) $\delta^*(s, ew) = \delta^*(\delta(s, e), w)$, for all $w \in I^*$ and all $e \in I_n$.

As we saw above, bisimulation for automata is defined by the rule of inference

$$\frac{s \sim s'}{\gamma(s) = \gamma(s'), \forall e \in I_n \ (\delta(s, e) \sim \delta(s', e))}.$$

We now introduce a new congruence defined on any automaton, which we shall show characterizes the largest bisimulation possible on that automaton.

**Definition 2.4.2.** Let $\mathcal{A} = (I, S, O; \delta, \gamma)$ be an automaton. The *Nerode* or *syntactic congruence* on $\mathcal{A}$ is the relation $\sim_N$ defined on $S$ as follows: for any two states $s$ and $s'$ of $S$, we have $s \sim_N s' :\Leftrightarrow \forall w \in I^* \ (\gamma(\delta^*(s,w)) = \gamma(\delta^*(s',w)))$.

This means that
$$s \sim_N s'$$
$$\Leftrightarrow \quad \forall w \in I^* \ ((\delta^*(s,w), \delta^*(s',w)) \in Ker\gamma)$$
$$\Leftrightarrow \quad \forall w \in I^* \ ((s,w), (s',w)) \in Ker(\gamma \circ \delta^*).$$

**Lemma 2.4.3.** *Let $\mathcal{A} = (I, S, O; \delta, \gamma)$ be a many-sorted algebra representing an automaton. The Nerode congruence is the greatest bisimulation and the greatest congruence $\theta$ on $\mathcal{A}$ to satisfy $\theta \subseteq Ker\gamma$.*

**Proof:** It is clear from the definition that $\sim_N$ is an equivalence relation on $S$. To see that it is a congruence, we let $s$ and $s'$ be states with $s \sim_N s'$, and let $e \in I$ and $w \in I^*$. Then $\gamma(s) = \gamma(\delta^*(s, \varepsilon)) = \gamma(\delta^*(s', \varepsilon)) = \gamma(s')$. Moreover, for any $e \in I$ and $w \in I^*$ we have
$$\gamma(\delta^*(\delta(s,e), w))$$
$$= \quad \gamma(\delta^*(s, ew))$$
$$= \quad \gamma(\delta^*(s', ew))$$
$$= \quad \gamma(\delta^*(\delta(s',e), w)),$$
which shows that $\delta(s,e) \sim_N \delta(s',e)$. Therefore $\sim_N$ is a congruence on $S$.

Now let $\sim$ be an arbitrary bisimulation on $S$. We want to show that $\sim \,\subseteq\, \sim_N$. We shall prove first the claim that for all words $w \in I^*$ and all states $s$ and $s'$ in $S$, $\quad s \sim s' \Rightarrow \gamma(\delta^*(s,w)) = \gamma(\delta^*(s',w))$.

We prove the claim by induction on the complexity of the word $w$. For $w = \varepsilon$ the claim follows directly from the definition of a bisimulation. Inductively, suppose that $w = ev$ for some letter $e$ and some word $v$ for which $\gamma(\delta^*(s,v)) = \gamma(\delta^*(s',v))$. Then
$$\gamma(\delta^*(s, ev))$$
$$= \quad \gamma(\delta^*(\delta(s,e), v))$$
$$= \quad \gamma(\delta^*(\delta(s',e), v)) \text{ (since it follows from } s \sim s' \text{ that}$$
$$\gamma(\delta^*(\delta(s,e), v)) = \gamma(\delta^*(\delta(s',e), v)))$$
$$= \quad \gamma(\delta^*(s', ev)).$$
Altogether we have $\sim \,\subseteq\, \sim_N$.

For each bisimulation $\sim$ we have $\sim \,\subseteq\, Ker\gamma$, and we want to show now that for any $\theta \subseteq Ker\gamma$ which is a congruence of the many-sorted algebra $\mathcal{A} = (I, S, O; \delta, \gamma)$ we must have $\theta \subseteq \,\sim_N$. Let $(s, s')$ be a pair in $\theta$. Since $\theta$ is a congruence we have $\gamma(s) = \gamma(s')$ and also $\delta^*(s, \varepsilon) = \delta^*(s', \varepsilon)$.

Inductively, suppose that $w = eu$ for some letter $e$ and some word $u$ for which $\delta^*(s,u) = \delta^*(s',u)$. Then $(\delta(s,e), \delta(s',e)) \in \theta$ and $\delta^*(\delta(s,e),u) = \delta^*(\delta(s',e),u)$. It follows that

$$\delta^*(s,eu) = \delta^*(\delta(s,e),u) = \delta^*(\delta(s',e),u) = \delta^*(s',eu).$$

This shows that $\theta \subseteq \sim_N$. ∎

Finally we describe the language accepted or recognized by an automaton. Given a set $S$ of states of a finite deterministic automaton or acceptor $\mathcal{A}$, we fix a state $s_0$ called the *initial state* and a set $F \subseteq S$ of *final* or *accepting states*. A word $w$ is said to be accepted by the automaton if $\delta^*(s_0,w) \in F$. Let $L(\mathcal{A})$ be the language consisting of all words which are accepted by $\mathcal{A}$. A language $L$ is called *recognizable* if there is a finite deterministic automaton (or acceptor) $\mathcal{A}$ such that $L = L(\mathcal{A})$. Kleene's famous theorem tells us that a language is recognizable if and only if it is regular.

Let $L$ be a language over the alphabet $I$. For all $e \in I$ we define the *derivation* of $L$ to $e$ by

$$L_e := \{w \in I^* \mid ew \in L\}.$$

We can assign to each state $s$ of an automaton (acceptor) $\mathcal{A}$ the language which consists of all words which go from $s$ into an accepting state:

$$L(\mathcal{A},s) := \{w \in I^* \mid \delta^*(s,w) \in F\}.$$

**Lemma 2.4.4.** *Let $S$ be the set of states of an automaton $\mathcal{A}$ over the alphabet $I$ and let $F \subseteq S$ be the set of final states of $\mathcal{A}$. For any states $s$ and $s'$ in $S$ and any letter $e \in I$,*

$$\delta(s,e) = s' \iff L(\mathcal{A},s)_e = L(\mathcal{A},s').$$

**Proof:** Assume that $\delta(s,e) = s'$. Then

$$
\begin{aligned}
&L(\mathcal{A},s)_e \\
=\ & \{w \in I^* \mid ew \in L(\mathcal{A},s)\} \\
=\ & \{w \in I^* \mid \delta^*(s,ew) \in F\} \\
=\ & \{w \in I^* \mid \delta^*(\delta(s,e),w) \in F\} \\
=\ & \{w \in I^* \mid \delta^*(s',w) \in F\} \\
=\ & L(\mathcal{A},s').
\end{aligned}
$$

If conversely $L(\mathcal{A},s)_e = L(\mathcal{A},s')$, then

$$\{w \in I^* \mid \delta^*(\delta(s,e),w) \in F\} = \{w \in I^* \mid \delta^*(s',w) \in F\}.$$

Then it follows that $\delta(s, e) = s'$.                  ∎

We conclude this section with a look at how automata can be expressed from the coalgebraic point of view, that is, how to combine the two automata mappings into a single mapping. We have mentioned already that an automaton can be regarded as a many-sorted algebra $(I, O, S; \delta, \gamma)$. If we agree to fix the sets $I$ of inputs and $O$ of outputs, we need consider only the triple $(S; \delta, \gamma)$. As for other state-based machines, we can then combine the two mappings $\delta$ and $\gamma$ into one mapping. We use $\alpha_S : S \rightarrow O \times S^I$ for Moore automata and $\alpha_S : S \rightarrow (O \times S)^I$, for Mealy automata. Then any state $s$ is mapped to a pair consisting of an output element and a mapping, the latter of which maps each input to a state or to a mapping which maps each pair consisting of an output element and a state to a mapping from $I$ to $O \times S$. The system $(S; \alpha_S)$ will then be an example of a coalgebra.

Acceptors are also useful, since they can be used to check whether or not a given word belongs to a given language. Let $s_0 \in S$ be chosen as an initial state and let $F \subseteq S$ be a set of final states. This set of final states can be encoded by an output mapping $\gamma : S \rightarrow \{0, 1\}$, defined by

$$\gamma(s) = \begin{cases} 1 & \text{if } s \in F \\ 0 & \text{otherwise.} \end{cases}$$

Therefore, any acceptor can be regarded as an automaton.

## 2.5   Automata with Error Conditions

It is possible to extend the concept of an automaton to incorporate the case of a failure to perform a certain transition. As an example, consider a soft-drink vending machine as a deterministic automaton. The inputs are coins, while the outputs are the drinks served. This system can be modelled by an automaton when no failures occur: if the machine is in state $s$, an input gives us both a new state $s'$ and an output, depending on the current state and the input. Now we want to extend the workings of the machine, to give a signal if for instance the machine is out of change or merchandise. We fix an additional set $E$ of errors, such as $E = \{$ "out of change", "out of merchandise"$\}$. If the automaton encounters one of these situations it should produce an error condition $e \in E$. Assuming that $S$ and $E$ are disjoint sets, we replace $\delta$ with the mapping $next : S \times I \rightarrow S \cup E$. The disjointness needed in the codomain $S \cup E$ here can be expressed by the concept of a coproduct. Coproducts will be defined more generally in Chapter 3, but here we discuss the special case of the coproduct of two sets.

Let $A, B$ be sets. The *coproduct* of $A$ and $B$ is a set $A + B$ defined by

$$A + B := \{(a, 0) \mid a \in A\} \cup \{(b, 1) \mid b \in B\}.$$

Let $C$ be another set and let $f : A \to C$ and $g : B \to C$ be mappings. We define a mapping $[f, g] : A + B \to C$ by

$$[f, g](x, i) = \begin{cases} f(x) & \text{if } i = 0 \\ g(x) & \text{if } i = 1. \end{cases}$$

The canonical injections $\iota_r : B \to A + B$ and $\iota_l : A \to A + B$ of the coproduct are defined by $\iota_r(b) = (b, 1)$ and $\iota_l(a) = (a, 0)$. For any mappings $f_1 : A \to B$ and $g_1 : C \to D$, we define

$$f_1 + g_1 := [\iota_r \circ f_1, \iota_l \circ g_1] : A + C \to B + D,$$

which means that $(f_1 + g_1)(a, 0) = (f_1(a), 0)$ and $(f_1 + g_1)(c, 1) = (g_1(c), 1)$.

**Lemma 2.5.1.** (Universal property of the coproduct) Let $f : A \to C$ and $g : B \to C$ be any mappings. Then the mapping $[f, g]$ is the unique mapping satisfying $[f, g] \circ \iota_l = f$ and $[f, g] \circ \iota_r = g$.

We leave the proof of this Lemma as an exercise.

Recall that our extended automaton has an output mapping and a state transition mapping with error condition. We define *out* and *next* as mappings $out : S \times I \to D + E$ and $next : S \times I \to S + E$. As in previous sections, we can combine these two mappings into one. We use the tensor product $out \otimes next$, which maps $S \times I \to (D + E) \times (S + E)$. This will allow us to view the automaton with error conditions as a coalgebra.

## 2.6 Object-oriented Programming

In object-oriented programming, a *class* is a set of data elements called objects. All the objects of a given class have a common structure, given by a list of attributes and methods. A user can access objects only by methods which belong to *public*. As an example we show the implementation of a bank account in the programming language Java:

**Class Bank account {**
**private int balance;**
**account(){ balance = 0; }**

**public void trans(int n){balance += n; } public show(int n){return balance; }**
**}**

When a new account is opened, the integer variable **balance** will be initialized to 0. The variable **balance** has **private** as an attribute, so that the user has no direct access to this part of the class. The public method **show** allows the variable **balance** to be read, and **trans** allows transfers into the account. The user need not know how to implement **trans** and **show**, but can check that they are working by the equation

$$x \cdot \textbf{trans}(n_1) \cdot \textbf{trans}(n_2) \cdot \textbf{show}() \;=\; x \cdot \textbf{trans}(n_1 + n_2) \cdot \textbf{show}(),$$

where $x$ is a bank account. This equation means that two subsequent transfers of amounts $n_1$ and with $n_2$ give the same result as one transfer of the amount $n_1 + n_2$. The account $x$ may contain other information to which the user has no access, so the request $x \cdot \textbf{trans}(n_1) \cdot \textbf{trans}(n_2) = x \cdot \textbf{trans}(n_1 + n_2)$ will not always be satisfied.

A mathematical model of this system may be made using the mappings $show : S \to \mathbb{Z}$ and $trans : S \times \mathbb{Z} \to S$. As before, these mappings can then be combined into one mapping from $S$ to some set formed from $S$.

## 2.7 Kripke Structures

Many information systems consist of multiple components, packages of programs which are connected and which must work together. But sometimes the interaction of such components is not deterministic, and instead of a mapping to specify the state changes and outputs a set of constraints is used. These constraints are given by a relation $R$ which is not right-unique: that is, it can contain pairs $(a, b)$ and $(a, c)$ with $b \neq c$. Such situations can be modelled by non-deterministic automata, which are known in logic and theoretical computer science as *Kripke structures*. An important property of a Kripke structure then is that the successor state is determined not by a transition mapping but by a transition relation.

**Definition 2.7.1.** Let $\Phi$ be a non-empty set. A Kripke-structure over $\Phi$ is a triple $(S; R, v)$ consisting of a set $S$ of states, a binary relation $R \subseteq S \times S$ and a mapping $v : \Phi \to \mathcal{P}(S)$, where $\mathcal{P}(S)$ is the power set of $S$. The pair $(S; R)$ is also called a *Kripke frame*.

In many applications $\Phi$ consists of a set of elementary (atomic) propositions. We think of $v$ as a valuation mapping, taking $p$ to the set $v(p)$ of all states where the proposition $p \in \Phi$ is satisfied. Along with the mapping $v : \Phi \to \mathcal{P}(S)$ we shall also use a related mapping $prop : S \to \mathcal{P}(\Phi)$, which assigns to each state $s$ the set of propositions true in that state.

A change from a state $s$ into a state $s'$ in a Kripke structure corresponds to having the pair $(s, s')$ in the relation $R$. We call this a *transition*, and denote it by $s \xrightarrow{R} s'$.

Like deterministic automata, Kripke structures can be described by directed graphs, in which the states are used as vertices and the transitions as directed edges. The atomic propositions from $v(s)$ are the labellings of the states. The following pictures describe two Kripke structures which will be used later on.

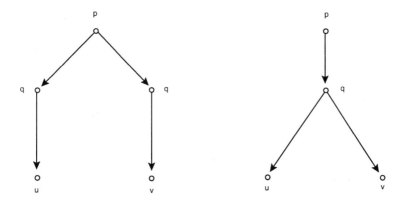

Fig. 2.2   Examples of Kripke Structures

To describe a Kripke structure by means of a single operation on $S$, we have to combine the transition relation $R \subseteq S \times S$ and the valuation $v : \Phi \to \mathcal{P}(S)$ into one mapping with domain $S$. As a first step, we consider how to convey the information in the relation $R$ in the form of a mapping.

Let $R \subseteq A \times B$ be any relation. We can define a partial mapping $f : A \to \mathcal{P}(B)$ by $f(a) = C \subseteq B$ with $C := \{c \in B \mid (a, c) \in R\}$; that is, we assign to each element $a$ the set of all elements which are related to it by $R$. Conversely, any mapping $f : A \to \mathcal{P}(B)$ determines a relation $R_f \subseteq A \times B$, by $R_f := \{(a, c) \mid c \in C \text{ such that } f(a) = C\}$. This gives a one-to-one correspondence between relations $R \subseteq A \times B$ and partial mappings $f : A \to \mathcal{P}(B)$.

Using this correspondence on our Kripke structure, we change from the transition relation $R \subseteq S \times S$ to the induced mapping $next : S \to \mathcal{P}(S)$. In addition, we have a mapping $prop : S \to \mathcal{P}(\Phi)$ which is defined by $s \mapsto \{a \in \Phi \mid s \in v(a)\}$. If these mappings both have domain $S$, then we can combine them using the tensor product and get $next \otimes prop : S \to \mathcal{P}(S) \times \mathcal{P}(\Phi)$. Thus our Kripke structure $(S; R, v)$ can be regarded as a pair $(S; next \otimes prop)$.

For bisimulation in a Kripke structure, we note first that two states can be distinguished if different atomic propositions are valid in these states. This means that we want

$$\frac{x \sim y}{v(x) = v(y)}$$

as a rule of inference. If two states $x$ and $y$ cannot be distinguished, it must be the case that for each transition starting from $x$ there must be a transition starting from $y$. This gives the rules of inference

$$\frac{x \sim y \wedge x \to x'}{\exists y' \; (y \to y' \wedge x' \sim y')}$$

and

$$\frac{x \sim y \wedge y \to y'}{\exists x' \; (x \to x' \wedge x' \sim y')}$$

Any relation $\sim$ on $S$ which satisfies these rules of inference is called a bisimulation on the Kripke structure.

## 2.8    The Concept of a Coalgebra

We have now surveyed a number of examples of state-based systems: what do all these examples have in common? In each example our structure had a set $S$ of states and one or more transition mappings (or relations) which assigned to each state a new state and/or an output. In each case, we condensed the information from the mappings or relations into a single mapping. This mapping had the form $\alpha : S \to F(S)$, where $F(S)$ was the result of some set-theoretical construction which provided a combination of states and outputs. Let us briefly review the various systems considered.

*Black box*

For a black box with data set $D = \mathbb{N}$, we combined the two mappings $h : S \to \mathbb{N}$ and $t : S \to S$ into a mapping $\alpha : S \to \mathbb{N} \times S$, defined by $\alpha(s) := (h \otimes t)(s) = (a, s')$.

*Bank account*

In the bank account program, the two mappings **show** : $S \to \mathbb{Z}$ and **trans** : $S \times \mathbb{Z} \to S$ can be combined into one mapping $\alpha : S \to \mathbb{Z} \times S^{\mathbb{Z}}$. This mapping takes each state $s \in S$ to a pair consisting of an integer and a mapping from $\mathbb{Z}$ to $S$.

*Automaton*

For a finite automaton (of Moore type) the two mappings $\gamma : S \to D$ and $\delta : S \times I \to S$ can be combined into a mapping $\alpha : S \to D \times S^{I}$. This mapping $\alpha$ maps each state $s$ to a pair consisting of an output element and a mapping which maps each input element to a state.

*Acceptor*

For an acceptor with the set $F \subseteq S$ of finale states, we can combine $F$ and the mapping $\delta : S \times I \to S$ into one mapping $\alpha : S \to \{0,1\} \times S^{I}$. Here $\alpha$ maps each state $s$ to a pair consisting of 0 or 1 (not accepted or accepted) and a mapping from $I$ to $S$.

*Kripke structure*

We have seen that in a Kripke structure $(S; R, v)$, the relation $R \subseteq S \times S$ can be replaced by a mapping $next : S \to \mathcal{P}(S)$. To obtain an everywhere defined mapping $next$ we have to assume that $R$ uses each state from $S$ as a first component. Combining this mapping with the mapping $prop : S \to \mathcal{P}(\Phi)$ which is defined by $s \mapsto \{a \in \Phi \mid s \in v(a)\}$, the Kripke structure can be regarded as a pair $(S; next \otimes prop)$, where $next \otimes prop : S \to \mathcal{P}(S) \times \mathcal{P}(\Phi)$.

In all of these examples we replaced our original structure by one of the form $(A; \alpha_A)$, where $\alpha_A : A \to F(A)$ is a mapping from a set $A$ to the result of a set-theoretical construction $F(A)$. We shall see that the category-theoretical concept of a functor explains more precisely what we mean by a set-theoretical construction $F(A)$; this will be discussed in more detail in Section 2.9 and Chapter 3. For now let us clarify the difference between an algebra and a coalgebra. An algebra of type $\tau$ is a set $A$ together with a sequence $(f_i^A)_{i \in I}$ of $n_i$-ary operations $f_i^A : A^{n_i} \to A$, indexed by an index set $I$. The type $\tau$ is the sequence $(n_i)_{i \in I}$ of the arities of the $f_i^A$'s. All of the operations could be combined into one mapping $f : \bigcup_{i \in I} A^{n_i} \to A$ from the disjoint union of the sets $A^{n_i}$ into $A$. Conversely, from each such mapping one can get back the $n_i$-ary operations $f_i^A$. Using $F(X) := \bigcup_{i \in I} X^{n_i}$ as our set-theoretical construction, the algebra can be regarded as a pair $(A; \alpha^A)$ with $\alpha^A : F(A) \to A$.

The important feature here is that in an algebra, operations are mappings from some set-theoretical construction $F(A)$ on $A$ to $A$ itself. But in the state-based machines described in this chapter, we want our mappings to go in the other direction, from $A$ to some set $F(A)$ made out of $A$. This is the basic idea of an $F$-coalgebra, or for short a coalgebra.

**Definition 2.8.1.** An $F$-*coalgebra*, or simply a *coalgebra*, is a system $(A; \alpha_A)$, consisting of a set $A$ and a mapping $\alpha_A : A \to F(A)$ for some set-theoretical construction $F(A)$.

## 2.9 Homomorphisms of State-based Systems

We saw in Chapter 1 that a homomorphism of algebras is a structure-preserving mapping between two algebras of the same type. Now we want to consider such structure-preserving maps for pairs of state-based systems of various kinds, in order to motivate a definition of a coalgebra homomorphism. We begin by discussing mappings between the various kinds of state-based systems considered so far.

*Black boxes and streams*

For a stream of data on a set $D$, we defined the operations $hd : D^\omega \to D$ and $tl : D^\omega \to D^\omega$, which correspond to the operations of a black box on $D$. This gave us a many-sorted algebra $(D, D^\omega; hd, tl)$. If we fix the data set $D$, we can use the triple $(D^\omega; hd, tl)$. A coalgebraic representation of this structure was given by the pair $(D^\omega; hd \otimes tl)$, with the mapping $hd \otimes tl : D^\omega \to D \times D^\omega$. Note here that the set of states is the set $S := D^\omega$. Now we define a homomorphism between such black boxes.

**Definition 2.9.1.** Let $(S; hd, tl)$ and $(S'; hd', tl')$ be black boxes. A mapping $f : S \to S'$ is called a *homomorphism of black boxes* if for every $s \in S$,

(i) $hd(s) = hd'(f(s))$, and

(ii) $f(tl(s)) = tl'(f(s))$.

Now we consider what this definition means for the coalgebra representation of a black box. If conditions (i) and (ii) are satisfied, then $(hd(s), (f \circ tl)(s)) = ((hd' \circ f)(s), (tl' \circ f)(s))$ for any $s \in S$. This condition can be written in the form

$$(1_D \times f) \circ (hd \otimes tl) = (hd' \otimes tl') \circ f, \qquad (*)$$

where $1_D : D \to D$ is the identity mapping on $D$. This equation can be expressed by the commutativity of the diagram in Figure 2.3.

**Proposition 2.9.2.** *The mapping $f : S \to S'$ is a homomorphism of black boxes if and only if $f$ satisfies condition $(*)$ above.*

**Proof:** We have seen above that any black box homomorphism $f$ must satisfy condition $(*)$. For the converse, we recall the universal property of the cartesian product. For any mappings $f : C \to A$ and $g : C \to B$, the mapping $f \otimes g : C \to A \times B$ is the unique mapping satisfying $p_1 \circ (f \otimes g) = f$ and $p_2 \circ (f \otimes g) = g$, where $p_1$ and $p_2$ are the canonical projections. Suppose that condition $(*)$ is satisfied by a mapping $f$. Then for every $s \in S$ we have

$$
\begin{aligned}
& (1_D \times f) \circ (hd \otimes tl)(s) \\
=\ & (1_D \times f)(hd(s), tl(s)) \\
=\ & ((1_D \circ hd)(s), (f \circ tl)(s)) \\
=\ & (1_D(hd(s)), f(tl(s))) \\
=\ & (hd(s), f(tl(s))).
\end{aligned}
$$

Moreover, we have

$$
\begin{aligned}
& ((hd' \otimes tl') \circ f)(s) \\
=\ & hd'(f(s)) \otimes tl'(f(s)) \\
=\ & (hd'(f(s)), tl'(f(s))).
\end{aligned}
$$

Therefore for every $s \in S$ we get $(hd(s), f(tl(s))) = (hd'(f(s)), tl'(f(s)))$ iff condition $(*)$ is satisfied. Now applying the first and second projections on both sides of $(*)$ we get $hd(s) = hd'(f(s))$ and $f(tl(s)) = tl'(f(s))$, which are the equations from the definition of a black box homomorphism. ∎

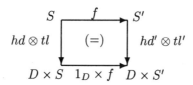

Fig. 2.3    Homomorphism of Black Boxes

As usual for an operation $f : A \to A$ the power $f^n$ is inductively defined by $f^0 = id_A, f^n = f \circ f^{n-1}$ for $n > 0$. Any black box produces for each state $s$ the sequence $(hd(s), (hd \circ tl)(s), (hd \circ tl^2)(s), (hd \circ tl^3)(s), \ldots)$ of data elements. We can describe the behaviour of such a sequence under a black box homomorphism.

**Lemma 2.9.3.** *Let $(S; hd, tl)$ and $(S'; hd', tl')$ be black boxes and let $f$ :*

$S \to S'$ be a homomorphism. Then for every $s \in S$,

$$(hd \circ tl^n)(s) = (hd' \circ (tl')^n)(f(s)).$$

**Proof:** We show first by induction on $n$ that $f \circ tl^n = (tl')^n \circ f$, if $f : S \to S'$ is a black box homomorphism. For $n = 0$ this is immediate, and for $n = 1$ it is the condition (ii) in the definition of a homomorphism. Inductively, let $n > 1$ and let $s \in S$. Then

$$
\begin{aligned}
&(f \circ tl^n)(s) \\
=\ &(f \circ tl \circ tl^{n-1})(s) \\
=\ &(tl' \circ f \circ tl^{n-1})(s) \\
=\ &(tl' \circ (tl')^{n-1} \circ f)(s) \qquad \text{by induction} \\
=\ &((tl')^n \circ f)(s).
\end{aligned}
$$

From this result we get

$$
\begin{aligned}
&(hd' \circ (tl')^n \circ f)(s) \\
=\ &(hd' \circ f \circ tl^n)(s) \\
=\ &(hd \circ tl^n)(s), \qquad \text{by the definition of homomorphism.} \quad\blacksquare
\end{aligned}
$$

### Automata

We shall consider automata over a fixed input alphabet $I$ and a fixed output alphabet $D$. This allows us to represent a finite deterministic automaton by a triple $(S; \delta, \gamma)$, where $\delta : S \times I \to S$ is the state transition mapping and $\gamma : S \times I \to D$ is the output mapping. In automata theory the following definition of a homomorphism is used.

**Definition 2.9.4.** Let $(S; \delta, \gamma)$ and $(S'; \delta', \gamma')$ be automata. A mapping $f : S \to S'$ is a *homomorphism* of automata if for every $s \in S$ and every $e \in I$,

(i) $\gamma(s, e) = \gamma'(f(s), e)$,     and
(ii) $(f \circ \delta)(s, e) = \delta'(f(s), e)$.

Condition (i) of this definition means that the states $s$ and $f(s)$ produce the same output for every input $e$. Since $\gamma(s, e) \in D$ and $f : S \to S'$, the function $f$ occurs only on one side, making it different from the usual (algebraic) definition of a homomorphism. Condition (ii) means that $f$ is compatible with the mappings $\delta$ and $\delta'$.

We introduce the following notation. Let $f : A \to B$ be a mapping and let $C$ be any set. Let $A^C$ denote the set of all mappings from $C$ to $A$. Then $f^C : A^C \to B^C$ is the mapping which assigns to each $h \in A^C$ the mapping $f^C(h) := f \circ h$ in $B^C$.

**Lemma 2.9.5.** *Let $(S; \delta, \gamma)$ and $(S'; \delta', \gamma')$ be finite deterministic automata. A mapping $f : S \to S'$ is a homomorphism of automata if and only if the diagram in* Figure 2.4 *commutes.*

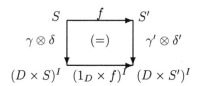

Fig. 2.4   Homomorphisms of Automata

**Proof:** In our notation, we have $((1_D \times f)^I \circ (\gamma \otimes \delta))(s) = (1_D \times f)^I \circ ((\gamma \otimes \delta)(s)) = ((1_D \times f) \circ (\gamma \otimes \delta))(s)$. Then we see that

$$f \text{ is an automata homomorphism}$$

$$\Leftrightarrow \quad (\gamma(s, e), (f \circ \delta)(s, e)) = (\gamma'(f(s), e), \delta'(f(s), e))$$

$$\Leftrightarrow \quad ((1_D \times f) \circ (\gamma \otimes \delta))(s, e) = (\gamma' \otimes \delta')(f(s), e)$$

$$\Leftrightarrow \quad (1_D \times f)((\gamma \otimes \delta)(s)(e)) = ((\gamma' \otimes \delta') \circ f)(s)(e)$$

$$\Leftrightarrow \quad ((1_D \times f)^I \circ (\gamma \otimes \delta))(s)(e) = ((\gamma' \otimes \delta') \circ f)(s)(e)$$

for all states $s$ and inputs $e$,

which is equivalent to the commutativity of the diagram.   ∎

Let us now consider the structural connections between homomorphisms of automata and homomorphisms of black boxes. A black box as a coalgebra uses the mapping $\alpha_S : S \to F(S)$, where $F$ is the set-valued mapping with $F(X) = D \times X$ for all sets $X$. A Mealy automaton regarded as a coalgebra uses the mapping $\alpha_S : S \to T(S)$, where the set-valued mapping $T$ is defined by $T(X) = (D \times X)^I$ for all sets $X$. The two mappings $F$ and $T$ can also be applied to mappings, in the following way. Let $f : S \to S'$. Then $F(f) = 1_D \times f : D \times S \to D \times S'$ in the first case, and $T(f) = (1_D \times f)^I : (D \times S)^I \to (D \times S')^I$ in the second case. Thus if $(S; \alpha_S)$ and $(S'; \alpha_{S'})$ are coalgebraic representations of either black boxes or automata, the mapping $f : S \to S'$ is a homomorphism if and only if $F(f) \circ \alpha_S = \alpha_{S'} \circ f$. This is equivalent to having the diagram in Figure 2.5 commute.

If we want homomorphisms to be mappings which preserve the structure, we need the following minimal requirements to hold:

(i) the identity mapping must be a homomorphism;   and

(ii) the composition of two homomorphisms must be a homomorphism.

In the coalgebra setting, we want to be able to extend our set-valued mapping $F$ to mappings too. That is, we require that for any mapping $f : A \to B$ the mapping $F(f) : F(A) \to F(B)$ exists. Assuming this property, let us consider what other properties must hold in order for the two homomorphism conditions (i) and (ii) to be fulfilled. For any coalgebra $(A; \alpha_A)$, the identity mapping $1_A$ on $A$ is a homomorphism if the rectangle shown in Figure 2.6 is commutative.

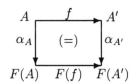

Fig. 2.5   Coalgebra Homomorphism

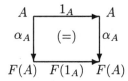

Fig. 2.6   Identity Homomorphism

This forces the condition that $\alpha_A \circ 1_A = F(1_A) \circ \alpha_A$. This will hold if $F$ preserves the identity mapping $1_A$, in the sense that $F(1_A) = 1_{F(A)}$.

For the composition condition, let $(A; \alpha_A)$, $(A'; \alpha_{A'})$ and $(A''; \alpha_{A''})$ be $F$-coalgebras and let $f : (A; \alpha_A) \to (A'; \alpha_{A'})$ and $g : (A'; \alpha_{A'}) \to (A''; \alpha_{A''})$ be homomorphisms. The fact that $f$ and $g$ are homomorphisms means that the two smaller rectangles in the diagram of Figure 2.7 commute.

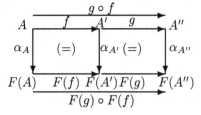

Fig. 2.7   Composition of Homomorphisms

The mapping $g \circ f$ will be a homomorphism if the larger rectangle also commutes, that is, if $\alpha_{A''} \circ (g \circ f) = F(g \circ f) \circ \alpha_A$. This is certainly the case if $F(g \circ f) = F(g) \circ F(f)$.

Our discussion of the conditions needed for $F$ leads us to the following definition.

**Definition 2.9.6.** A *functor* on sets is an operation $F$ on sets and mappings such that

(i) $F(A)$ is a set whenever $A$ is a set.
(ii) $F(f)$ is a mapping whenever $f$ is a mapping and
  (a) $F(f) : F(A) \to F(B)$ if $f : A \to B$.
  (b) $F(1_A) = 1_{F(A)}$.
  (c) $F(f \circ g) = F(f) \circ F(g)$ whenever $f : A \to B$ and $g : B \to C$ are composable mappings.

Using category theory notation to be introduced in Chapter 3, we will write $F : \mathbf{Set} \to \mathbf{Set}$ for a functor $F$ on sets. Here $\mathbf{Set}$ denotes the category consisting of sets as objects and mappings between sets as morphisms. More information on functors and the category $\mathbf{Set}$ will be given in Chapter 3. For now we can define coalgebra homomorphisms, and record our conclusion that they have the properties we wanted.

**Definition 2.9.7.** Let $F : \mathbf{Set} \to \mathbf{Set}$ be a functor and $(A; \alpha_A)$ and $(A'; \alpha_{A'})$ be $F$-coalgebras. A mapping $f : A \to B$ is a *coalgebra homomorphism* from $(A; \alpha_A)$ to $(A'; \alpha_{A'})$ if $F(f) \circ \alpha_A = \alpha_{A'} \circ f$; that is, if the diagram in Figure 2.8 commutes.

**Proposition 2.9.8.** *Let* $F : \mathbf{Set} \to \mathbf{Set}$ *be a functor. Then for any F-coalgebras* $(A; \alpha_A)$, $(A'; \alpha_{A'})$ *and* $(A''; \alpha_{A''})$,

(i) $1_A : (A; \alpha_A) \to (A; \alpha_A)$ *is a coalgebra homomorphism.*
(ii) *If* $f : (A; \alpha_A) \to (A'; \alpha_{A'})$ *and* $g : (A'; \alpha_{A'}) \to (A''; \alpha_{A''})$ *are coalgebra homomorphisms, then* $g \circ f : (A; \alpha_A) \to (A''; \alpha_{A''})$ *is also a coalgebra homomorphism.*

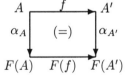

Fig. 2.8   Coalgebra Homomorphism

*Kripke structures*

For Kripke structures we go in the opposite direction: we start with our general definition of a coalgebra homomorphism and look at what that definition means for Kripke structures. We defined a Kripke structure over a set $\Phi$ of elementary propositions as a triple $(S; R, v)$, where $S$ is a set of states, $R \subseteq S \times S$ is a binary relation and $v : \Phi \to \mathcal{P}(S)$ is a mapping called the valuation mapping. We recall that along with $v : \Phi \to \mathcal{P}(S)$ we can also consider the mapping $prop : S \to \mathcal{P}(\Phi)$; these two mappings $v$ and $prop$ are related to each other by the condition that $s \in v(p) \Leftrightarrow p \in prop(s)$ for all $s \in S$ and $p \in \Phi$. Kripke structures are in one-to-one correspondence with $F$-coalgebras, for a set-valued functor $F$ with $F(X) := \mathcal{P}(X) \times \mathcal{P}(\Phi)$. This correspondence associates to our triple $(S; R, v)$ the $F$-coalgebra $(S; \alpha_S)$ with $\alpha_S : S \to F(S)$.

To consider homomorphisms of Kripke structures we need to extend the set-valued mapping $F$ to a functor, which means that we have to define the action of $F$ on mappings. We begin with the action of $F(f)$ on the formation of power sets. We want to define $F(f) : \mathcal{P}(A) \to \mathcal{P}(B)$ for $f : A \to B$. We define $\mathcal{P}(f)(X) := \{f(x) \mid x \in X\} = f(X)$ for any $X \subseteq A$. Now for $f : A \to B$ we define $F(f) : \mathcal{P}(A) \times \mathcal{P}(\Phi) \to \mathcal{P}(B) \times \mathcal{P}(\Phi)$ by $\mathcal{P}(f) \times 1_{\mathcal{P}(\Phi)}$ with $(\mathcal{P}(f) \times 1_{\mathcal{P}(\Phi)})(X, P) := (\mathcal{P}(f)(X), 1_{\mathcal{P}(\Phi)}(P))$ for $X \subseteq A$ and $P \subseteq \Phi$. Therefore for any $s \in S$ we have $\alpha_S(s) = (\{s' \in S \mid (s, s') \in R\}, \{p \in \Phi \mid s \in v(p)\})$. By definition of a coalgebra homomorphism the diagram in Figure 2.9 commutes.

Now we can express what the coalgebra homomorphism conditions mean for Kripke structures.

**Lemma 2.9.9.** *Let $(S; R, v)$ and $(S', R', v')$ be Kripke structures over $\Phi$, with corresponding coalgebra representations $(S; \alpha_S)$ and $(S', \alpha_{S'})$. A mapping $f : S \to S'$ is a coalgebra homomorphism $f : (S; \alpha_S) \to (S'; \alpha_{S'})$ if and only if the following conditions are satisfied:*

(i) *For all $s \in S$ and all $p \in \Phi$,    $s \in v(p)$  iff  $f(s) \in v'(p)$.*
(ii) *For all $s_0, s_1 \in S$,    if $(s_0, s_1) \in R$, then $(f(s_0), f(s_1)) \in R'$.*
(iii) *For all $s_0 \in S$ and $s' \in S'$, if $(f(s_0), s_1') \in R'$, then there is an element $s_1 \in S$ such that $f(s_1) = s_1'$ and $(s_0, s_1) \in R$.*

**Proof:** The mapping $f$ is a coalgebra homomorphism if and only if the diagram in Figure 2.9 commutes. Therefore $(\alpha_{S'} \circ f)(s) = ((\mathcal{P}(f) \times 1_{\mathcal{P}(\Phi)}) \circ \alpha_S)(s)$ for all $s \in S$. This means that $\alpha_{S'}(f(s)) = ((\alpha_{S'})_1(f(s)), (\alpha_{S'})_2(f(s))) = (\mathcal{P}(f) \times 1_{\mathcal{P}(\Phi)})(\alpha_S(s)) = (\mathcal{P}(f) \times 1_{\mathcal{P}(\Phi)})((\alpha_S)_1(s), (\alpha_S)_2(s)) = (\mathcal{P}(f)((\alpha_S)_1(s)), 1_{\mathcal{P}}(\Phi)((\alpha_S)_2(s)))$. From this equation it follows that (i') $(\alpha_S)_2(s) = (\alpha_{S'})_2(f(s))$ for all $s \in S$,

(ii') $\mathcal{P}(f) \circ (\alpha_S)_1 \subseteq (\alpha_{S'})_1 \circ f$,

(iii') $\mathcal{P}(f) \circ (\alpha_S)_1 \supseteq (\alpha_{S'})_1 \circ f$.

Now we claim that these three equations are equivalent to (i), (ii) and (iii) respectively. For the equivalence of (i) and (i') we have

$$\forall s \in S \; \forall p \in \Phi \; (s \in v(p) \Leftrightarrow f(s) \in v'(p))$$

$\Leftrightarrow \quad \forall s \in S \; \forall p \in \Phi \; (p \in (\alpha_S)_2(s) \Leftrightarrow p \in (\alpha_{S'})_2(f(s)))$

$\Leftrightarrow \quad$ (i')

The other equivalences can be shown similarly, and we leave them as exercises. ∎

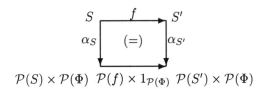

$$\mathcal{P}(S) \times \mathcal{P}(\Phi) \quad \mathcal{P}(f) \times 1_{\mathcal{P}(\Phi)} \quad \mathcal{P}(S') \times \mathcal{P}(\Phi)$$

Fig. 2.9   Homomorphisms of Kripke Structures

## 2.10   Behavioural Equivalence

In this section we return to the question of bisimulation of states. The goal is to identify any two states of a machine which cannot be distinguished from the outside. Such states will be called *behaviourally equivalent*. Such equivalence is one of the main concerns of the theory of state-based systems, for the following reasons:

(i) If two states cannot be distinguished from each other from outside the system, then we can substitute one for the other without affecting the behaviour of the system.

(ii) We want to ensure that any assertions we make about a system are based only on its visible behaviour, and do not constrain its internal implementation.

As an example, consider the automaton given by the directed graph in Figure 2.10. The second integer labelling an edge is the output element. For this automaton we have $I = D = \{0, 1\}$, and $\gamma(z_0, 0) = 1, \gamma(z_2, 0) = 1$, $\gamma(z_0, 1) = 0$ and $\gamma(z_2, 1) = 0$; and $\delta(z_0, 0) = z_2, \delta(z_0, 1) = z_1, \delta(z_2, 0) = z_2$

and $\delta(z_2, 1) = z_1$. It is clear that states $z_0$ and $z_2$ are equivalent.

In general, two states of an automaton cannot be distinguished if they produce the same output and the same next state for every input. Bisimilarity is then defined by the following rule of inference:

$$\frac{s \sim s'}{\forall e \in I \ (\gamma(s, e) = \gamma(s', e), \delta(s, e) \sim \delta(s', e))}.$$

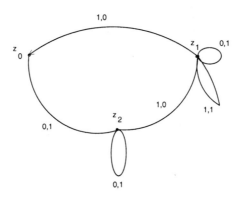

Fig. 2.10   Example of an Automaton

### Black Boxes

We saw in Section 2.1 that in a black box system, two states with different outputs can immediately be distinguished. Two states can also be distinguished if a series of identical inputs gives different outputs. Therefore $s \sim s'$ means that $h(s) = h(s')$ and $t(s) = t(s')$. This can be formulated as the rule of inference

$$\frac{s \sim s'}{h(s) = h(s') \wedge t(s) = t(s')}.$$

### Streams

A stream was considered as a black box with mappings $hd : D^\omega \to D, tl : D^\omega \to D^\omega$. The behaviour of the stream associated with a state $s$ is the infinite sequence $(hd(s), (hd \circ tl)(s), (hd \circ tl^2)(s), \ldots)$. Two states are called behaviourally equivalent if they have the same behaviour, that is, if they produce the same sequence of data. Now we consider the set $D^\omega := \{(d_0, d_1, d_2, \ldots) \mid d_i \in D \text{ for all } i \geq 0\}$ of all possible behaviours. We define two operations $\zeta_1 : D^\omega \to D$ and $\zeta_2 : D^\omega \to D^\omega$ by $\zeta_1(d_0, d_1, d_2 \ldots,) = d_0$ and $\zeta_2(d_0, d_1, d_2, \ldots) = (d_1, d_2, \ldots)$. Then $\zeta_1$ produces the head and

$\zeta_2$ the tail of any sequence. With these mappings we obtain an algebra $(D^\omega; \zeta_1, \zeta_2)$.

**Lemma 2.10.1.** *Let $(S; hd, tl)$ be a black box. Then the mapping bhvr :* $S \to D^\omega$ *defined by $bhvr(s) = (hd(s), (hd \circ tl)(s), (hd \circ tl^2)(s), \ldots)$ is a homomorphism from $(S; hd, tl)$ to $(D^\omega; \zeta_1, \zeta_2)$.*

**Proof**: From our definition of black box homomorphisms, we have to show that

(i) $hd(s) = \zeta_1(bhvr(s))$ and
(ii) $bhvr(tl(s)) = \zeta_2(bhvr(s))$.

We have $\zeta_1(hd(s), (hd \circ tl)(s), (hd \circ tl^2)(s), \ldots) = hd(s)$, so that (i) is satisfied. For (ii) we have $\zeta_2(bhvr(s)) = \zeta_2(hd(s), (hd \circ tl)(s), (hd \circ tl^2)(s), \ldots) = ((hd \circ tl)(s), (hd \circ tl^2)(s), \ldots) = (hd(s), (hd \circ tl)(s), \ldots)(tl(s)) = bhvr(tl(s))$. ∎

**Remark**: It can also be shown that the mapping *bhvr* is in fact the only homomorphism from $(S; hd, tl)$ to $(D^\omega; \zeta_1, \zeta_2)$. We leave this as an exercise for the reader.

We have shown that two states $s$ and $s'$ of a machine are behaviourally equivalent if $bhvr(s) = bhvr(s')$. Moreover, there is a morphism, *bhvr*, which identifies these two states. This idea is used in the following more general definition of behavioural equivalence of states in two (possibly different) coalgebras.

**Definition 2.10.2.** Let $(A; \alpha_A)$ and $(B; \alpha_B)$ be $F$-coalgebras for some functor $F$. A pair $(a, b) \in A \times B$ is said to be behaviourally equivalent, written as $a \sim b$, if there is an $F$-coalgebra $(C; \alpha_C)$ and a pair of coalgebra homomorphisms $f : (A; \alpha_A) \to (C, \alpha_C)$ and $g : (B; \alpha_B) \to (C; \alpha_C)$ such that $f(a) = g(b)$. (See Figure 2.11.)

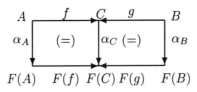

Fig. 2.11   Behavioural Equivalence

This defines a relation $R \subseteq A \times B$, between the universes of the two coalgebras, called the behavioural equivalence relation from $(A; \alpha_A)$ to

$(B, \alpha_B)$. In the special case of a behavioural equivalence $R$ from one coalgebra $(A; \alpha_A)$ to itself, this relation is easily seen to be symmetric, and the diagram in Figure 2.12 shows that it is also reflexive.

The following result shows that for streams, this definition of behavioural equivalence coincides with equality under the *bhvr* homomorphism.

**Proposition 2.10.3.** *Let* $(S; hd, tl)$ *and* $(S'; hd', tl')$ *be systems corresponding to streams. Then for any pair* $(s, s') \in S \times S'$ *we have* $s \sim s'$ *if and only if* $bhvr(s) = bhvr(s')$.

**Proof:** If $s \sim s'$, then there exists an algebra $(S''; hd'', tl'')$ and two homomorphisms $f : (S; hd, tl) \to (S''; hd'', tl'')$ and $g : (S'; hd', tl') \to (S''; hd'', tl'')$ such that $f(s) = g(s')$. By Lemma 2.9.3, $f$ satisfies the condition that $(hd \circ tl^n)(s) = (hd' \circ (tl')^n)(f(s))$. Therefore,

$$bhvr(s)$$
$$= \quad (hd(s), (hd \circ tl)(s), \ldots)$$
$$= \quad (hd'(f(s)), (hd' \circ tl')(f(s)), \ldots)$$
$$= \quad bhvr(f(s))$$

and in a similar way we get $bhvr(s') = bhvr(g(s'))$. But then we have $bhvr(f(s)) = bhvr(g(s'))$ and $bhvr(s) = bhvr(s')$.

For the converse, we have already proved that $bhvr : (S; hd, tl) \to (D^\omega; \zeta_1, \zeta_2)$ and $bhvr : (S'; hd', tl') \to (D^\omega; \zeta_1, \zeta_2)$ are homomorphisms with $bhvr(s) = bhvr(s')$. Therefore $s \sim s'$. ∎

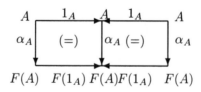

Fig. 2.12   Reflexivity of Behavioural Equivalence

*Finite Deterministic Automata*

We consider here a finite deterministic automaton $(I, S, D; \gamma, \delta)$, having input alphabet $I$, output alphabet $D$, set of states $S$, and state transition mapping $\delta : S \times I \to S$. For the output mapping we use the Moore automata form $\gamma : S \to D$. We want to show that the inference rule

$$\frac{s \sim s'}{\gamma(s) = \gamma(s'), \forall e \in I(\delta(s, e) \sim \delta(s', e))}$$

which we used in Section 2.4 for bisimulation meets the requirements for behavioural equivalence.

The behaviour of a deterministic automaton on a sequence $e_1, \ldots, e_n$ of input elements can be described as follows. Starting with the initial state $s_0$, the following states are produced in succession: $s_1 = \delta(s_0, e_1)$, $s_2 = \delta(\delta(s_0, e_1), e_2)$, ..., $s_n = \delta(s_{n-1}, e_n)$. The corresponding sequence of output elements is given by $d_0 = \gamma(s_0)$, ..., $d_n = \gamma(s_n)$. We combine the two mappings, to define for every state $s \in S$ a mapping $t(s) : I^* \to D^*$ by $t(s)(\varepsilon) := \varepsilon$ for the empty word $\varepsilon$ and $t(s)(e_1 e_2) := \gamma(s, e_1) t(\delta(s, e_1))(e_2)$ for $e_1, e_2 \in I^*$. We will show that two states $s$ and $s'$ are behaviourally equivalent if and only if $t(s) = t(s')$. This means that two states are behaviourally equivalent if and only if they are stable under some mapping.

**Lemma 2.10.4.** *Let $(S; \delta, \gamma)$ and $(S'; \delta', \gamma')$ be automata and let $s \sim s'$ for states $s \in S$ and $s' \in S'$. Then $t(s) = t(s')$.*

**Proof:** If $s \sim s'$, then there exist an automaton $(S''; \delta'', \gamma'')$ and two homomorphisms $f : (S; \delta, \gamma) \to (S''; \delta'', \gamma'')$ and $g : (S'; \delta', \gamma') \to (S''; \delta'', \gamma'')$ such that $f(s) = g(s)$. If we are able to prove that for any morphism $f$ we have $t(s) = t(f(s))$, then we have:

$$s \sim s' \Rightarrow f(s) = g(s') \Rightarrow t(f(s)) = t(g(s')) \Rightarrow t(s) = t(s').$$

To verify the equation $t(s) = t(f(s))$, we show by induction on the complexity of words that for all $i \in I^*$ the equation $t(s)(i) = t(f(s))(i)$ is satisfied. For the base case $i = \varepsilon$, we have $t(s)(\varepsilon) = \varepsilon = t(f(s))(\varepsilon)$. Inductively, for $i = i_0 i_1$ for some $i_0, i_1 \in I^*$, we have

$$f(\delta(s, i_0)) = \delta'(f(s), i_0)$$
$$\Rightarrow \quad t(\delta'(f(s), i_0)) = t(f(\delta(s, i_0))) = t(\delta(s, i_0)),$$

by hypothesis. This gives

$$
\begin{aligned}
&t(s)(i_0 i_1) \\
=\ &\gamma(s, i_0) t(\delta(s, i_0))(i_1) \\
=\ &\gamma'(f(s), i_0) t(\delta'(f(s), i_0))(i_1) \\
=\ &t(f(s))(i_0 i_1),
\end{aligned}
$$

as claimed. ∎

We introduce the following notation. Recall that for a word $w \in I^*$ the *length* $l(w)$ of $w$ is defined inductively by $l(\varepsilon) = 0$ and $l(e) = 1$ if $e \in I$, and $l(ew') = 1 + l(w')$. A *prefix* of $w$ is defined as a word $w'$ for which $w' w_1 = w$ for some word $w_1 \in I^*$. A word $w'$ is a *subword* of $w$ if there are words $w_1, w_2 \in I^*$ such that $w = w_1 w' w_2$.

The following proposition is easy to prove, and the proof is left as an exercise for the reader.

**Proposition 2.10.5.**

(i) *For every state $s$ in a finite automaton, the mapping $t(s) : I^* \to D^*$ is length-preserving; i.e. $l(w) = l(t(s)(w))$ for all $w \in I^*$.*
(ii) *For every state $s$ in a finite automaton, the mapping $t(s)$ is prefix-closed, i.e. if $w'$ is a prefix of $w$, then $t(s)(w')$ is a prefix of $t(s)(w)$.*

Now we define the set

$$\mathcal{M} := \{b : I^* \to D^* \mid b \text{ is length-preserving and prefix-closed}\}.$$

**Lemma 2.10.6.** *Let $(S; \delta, \gamma)$ and $(S'; \delta', \gamma')$ be finite deterministic automata and let $s \in S$ and $s' \in S'$ such that $t(s) = t(s')$. Then $s \sim s'$.*

**Proof:** We define the mapping $\gamma_M : \mathcal{M} \times I^* \to D^*$ by $\gamma(b, i) := b(i)$ for any word $i \in I^*$, and another mapping $\delta_M : \mathcal{M} \times I^* \to \mathcal{M}$ using a partial mapping $tl : D^* \to D^*$ which produces the tail by deleting the first element from a finite sequence. Let $\lambda_j$ be a prefix. We define $\delta_M(b, i) := \lambda_j(tl(b(ij)))$, and consider the automaton $(\mathcal{M}; \gamma_M, \delta_M)$ with $\lambda_j(tl(b(ij)))(i_1) = tl(b(i_1))$.
<u>Claim:</u> $t(b) = b$ for all $b \in \mathcal{M}$, that is, $t(b)(i) = b(i)$ for all $i \in I^*$.
We will prove the claim by induction on $i$. For the base case, if $i = \varepsilon$, we have $t(b)(i) = t(b)(\varepsilon) = \varepsilon$ by definition of $t$. But we also have $b(\varepsilon) = \varepsilon$ since $b$ is length-preserving.
Now suppose that $i = i_0 i_1$ for some $i_0, i_1 \in I^*$. Then

$$
\begin{aligned}
t(b)(i_0 i_1) &= \gamma_M(b, i_0) t(\delta_M(b, i_0))(i_1) && \text{by definition of } t \\
&= b(i_0) t(\lambda_j(tl(b(i_0 j))))(i_1) && \text{by definition of } \gamma_M \text{ and } \delta_M \\
&= b(i_0) \lambda_j(tl(b(i_0 j)))(i_1) && \text{by hypothesis} \\
&= b(i_0) tl(b(i_0 i_1)), && \text{since } i_0 \text{ is a prefix of } i_0 i_1 \\
& && \text{and thus } b(i_0) \text{ is a prefix of } b(i_0 i_1) \\
&= b(i_0 i_1).
\end{aligned}
$$

We now show that given any automaton $(S; \gamma, \delta)$, the mapping $t : s \mapsto t(s)$ is a homomorphism. Let $i \in I$ and $s \in S$. Then

$$
\begin{aligned}
&\gamma_M(t(s), i) \\
&= t(s)(i) && \text{by definition of } \gamma_M \\
&= \gamma_M(s, i) t(\delta_M(s, i))(\varepsilon) && \text{by definition of } t(s) \\
&= \gamma_M(s, i), && \text{since } t(s)(\varepsilon) = \varepsilon \text{ and } \gamma_M(s, i)\varepsilon = \gamma_M(s, i).
\end{aligned}
$$

Therefore the first condition for an automata homomorphism is satisfied. Next we have

$$\delta_M(t(s), i)$$
$$= \quad \lambda_j(tl(t(s)(ij)))$$
$$= \quad \lambda_j(tl(\gamma_M(s,i))t(\delta_M(s,i)(j))) \qquad \text{by definition of } t$$
$$= \quad \lambda_j(t(\delta_M(s,i))(j)) \qquad \text{since } \gamma_M(s,i) \text{ is removed by } tl$$
$$= \quad t(\delta_M(s,i)).$$

We have shown that the mappings $f$ and $g$ given by $f : s \mapsto t(s)$ and $g : s' \mapsto t(s')$ are coalgebra homomorphisms. Since $t(s) = t(s')$, we get $f(s) = g(s')$, and therefore $s \sim s'$. $\blacksquare$

### Kripke Structures

In this section we extend our definition of bisimulation to states of two possibly different Kripke structures, and show that two states are behaviourally equivalent iff they are bisimilar under this extended definition.

**Definition 2.10.7.** Let $(S; R, v)$ and $(S'; R', v')$ be two Kripke structures over a set $\Phi$ of elementary propositions. A bisimulation from $(S; R, v)$ to $(S'; R', v')$ is a binary relation $B \subseteq S \times S'$ which satisfies the following three conditions:

(i) For all $(s, s') \in B$ and for all $p \in \Phi$ we have $s \in v(p)$ iff $s' \in v'(p)$.
(ii) Whenever $(s_0, s_1) \in R$ and $(s_0, s'_0) \in B$, there exists an element $s'_1 \in S'$ such that $(s'_0, s'_1) \in R'$ and $(s_1, s'_1) \in B$.
(iii) Whenever $(s'_0, s'_1) \in R'$ and $(s_0, s'_0) \in B$, there exists an element $s_1 \in S$ such that $(s_0, s_1) \in R$ and $(s_1, s'_1) \in B$.

Two states $s \in S$ and $s' \in S'$ are called bisimilar if there is a bisimulation $B \subseteq S \times S'$ with $(s, s') \in B$.

The proof of the following lemma is left as an exercise for the reader.

**Lemma 2.10.8.** Let $(S; R, v)$, $(S'; R', v')$ and $(S''; R'', v'')$ be Kripke structures over the same set of elementary propositions. Then

(i) $\Delta_S = \{(s, s) \mid s \in S\}$ is a bisimulation on $(S; R, v)$.
(ii) If $B \subseteq S \times S'$ is a bisimulation from $(S; R, v)$ to $(S'; R', v')$, then $B^{op} := \{(s', s) \mid (s, s') \in B\}$ is a bisimulation from $(S'; R', v')$ to $(S; R, v)$.
(iii) If $B \subseteq S \times S'$ and $B' \subseteq S' \times S''$ are bisimulations, then $B' \circ B \subseteq S \times S''$ is a bisimulation from $(S; R, v)$ to $(S''; R'', v'')$.

The proof of the next lemma is also left as an exercise. We recall that the *graph* of a mapping $f : A \to B$ is defined by $G(f) := \{(a, f(a)) \mid a \in A\}$.

**Lemma 2.10.9.** *Let $(S; R, v)$ and $(S'; R', v')$ be Kripke structures over the same set of elementary propositions, with associated coalgebra representations $(S; \sigma)$ and $(S'; \sigma')$. For any coalgebra homomorphism $f : S \to S'$, the graph $G(f)$ of $f$ is a bisimulation from $(S; R, v)$ to $(S'; R', v')$.*

We are now ready to compare bisimulation with behavioural equivalence.

**Proposition 2.10.10.** *Let $(S; R, v)$ and $(S'; R', v')$ be Kripke structures over the same set of elementary propositions, and let $s \in S$ and $s' \in S'$ be behaviourally equivalent (as states of the associated coalgebras). Then $s$ and $s'$ are bisimilar.*

**Proof:** If $f$ and $g$ are coalgebra morphisms identifying $s$ and $s'$, then we know that $G(f)$ and $G(g)$ are bisimulations such that $(s, f(s)) \in G(f)$ and $(s', g(s')) \in G(g)$. Then $(g(s'), s') \in G(g)^{op}$, and $G(g)^{op}$ and $G(g)^{op} \circ G(g)$ are also bisimulations. Since $f(s) = g(s')$, $(s, f(s)) \in G(f)$ and $(g(s'), s') \in G(g)^{op}$, we have $(s, s') \in G(g)^{op} \circ G(f)$. ∎

This finishes the proof that any behavioural equivalence of Kripke systems is a bisimulation. It can be proved (see [64]) that the converse is also true, making the following proposition true.

**Proposition 2.10.11.** *Let $\Phi$ be a set. Let $F(X) = \mathcal{P}(X) \times \mathcal{P}(\Phi)$ and let $(S; \sigma)$ and $(S'; \sigma')$ be F-coalgebras. Let $B \subseteq S'$ be a bisimulation of the associated Kripke structures over $\Phi$. Then there is an F-coalgebra structure $\beta$ on $B$ and there are homomorphisms $\pi_1$ and $\pi_2$ such that $\pi_1(s) = \pi_2(s')$ whenever $(s, s') \in B$.*

## 2.11 Exercises for Chapter 2

1. Prove Lemma 2.5.1: for any mappings $f : A \to C$ and $g : B \to C$, the mapping $[f, g]$ is the unique mapping satisfying $[f, g] \circ \iota_l = f$ and $[f, g] \circ \iota_r = g$.

2. Prove Proposition 2.9.8: Let $F : \mathbf{Set} \to \mathbf{Set}$ be a functor, and let $(A; \alpha_A)$, $(A'; \alpha_{A'})$ and $(A''; \alpha_{A''})$ be any F-coalgebras.

(i) $1_A : (A; \alpha_A) \to (A; \alpha_A)$ is a coalgebra homomorphism,

(ii) if $f : (A; \alpha_A) \to (A'; \alpha_{A'})$ and $g : (A'; \alpha_{A'}) \to (A''; \alpha_{A''})$ are coalgebra homomorphisms, then $g \circ f : (A; \alpha_A) \to (A''; \alpha_{A''})$ is also a coalgebra homomorphism.

3. Complete the proof of Lemma 2.9.9, by proving the equivalences (ii) $\Leftrightarrow$ (ii') and (iii) $\Leftrightarrow$ (iii').

4. Prove that the mapping $bhvr$ of Lemma 2.10.1 is in fact the only homomorphism from $(S; hd, tl)$ to $(D^\omega; \zeta_1, \zeta_2)$.

5. Prove Proposition 2.10.5:
a) For every state $s$ in a finite automaton, the mapping $t(s) : I^* \to D^*$ is length-preserving; i.e. $l(w) = l(t(s)(w))$ for all $w \in I^*$.
b) For every state $s$ in a finite automaton, the mapping $t(s)$ is prefix-closed, i.e. if $w'$ is a prefix of $w$, then $t(s)(w')$ is a prefix of $t(s)(w)$.

6. Find the language over $\{0, 1\}$ which is accepted by the deterministic acceptor in Figure 2.13.

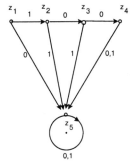

Fig. 2.13    Deterministic Acceptor

7. Prove Lemma 2.10.8: Let $(S; R, v)$, $(S'; R', v')$ and $(S''; R'', v'')$ be Kripke structures over the same set of elementary propositions. Then
a) $\Delta_S = \{(s, s) \mid s \in S\}$ is a bisimulation on $(S; R, v)$.
b) If $B \subseteq S \times S'$ is a bisimulation from $(S; R, v)$ to $(S'; R', v')$, then $B^{op} = \{(s', s) \mid (s, s') \in B\}$ is a bisimulation from $(S'; R', v')$ to $(S; R, v)$.
c) If $B \subseteq S \times S'$ and $B' \subseteq S' \times S''$ are bisimulations, then $B' \circ B \subseteq S \times S''$ is a bisimulation from $(S; R, v)$ to $(S''; R'', v'')$.

8. Prove Lemma 2.10.9: Let $(S; R, v)$ and $(S'; R', v')$ be Kripke struc-

tures over the same set of elementary propositions, with associated coalgebra representations $(S; \sigma)$ and $(S'; \sigma')$. For any coalgebra homomorphism $f : S \to S'$, the graph $G(f)$ of $f$ is a bisimulation from $(S; R, v)$ to $(S'; R', v')$.

9. Consider the language $PAL$ over the alphabet $\Sigma := \{a, b\}$ defined by $PAL := \{\varepsilon\} \cup \{x \mid x$ is a word over $\Sigma$ such that $reverse(x) = x\}$. Here $reverse(x)$ is the result of spelling the word $x$ backwards.
a) Is the language $PAL$ closed under concatenation ?
b) Prove that if $x$ is in $PAL$, then so is $x^n$ for any integer $n \geq 1$.
c) Prove: If $z^n$ is in $PAL$ for some $n > 0$, then $z$ is in $PAL$.

10. Prove the universal property of the coproduct.

# Chapter 3

# Basic Concepts from Category Theory

We have seen in Chapter 2 that in order to describe state-based systems as coalgebras we need the idea of a set-valued functor $F$. This is a special case of a more general concept, that of a functor on categories. This chapter provides a short, self-contained introduction to some elementary topics in category theory, including categories, functors and natural transformations. Some well-known constructions of set theory are also generalized to an arbitrary category. For more information on category theory we refer the reader to [11], [76] and [60].

## 3.1 The Concept of a Category

A category $\mathbf{C}$ consists of a class $\mathcal{O}$ of *objects* and a class $\mathcal{M}$ of *morphisms* between these objects. Each morphism $f$ has exactly one object $A$ as its *source* and exactly one object $B$ as its *target*. We will write $f : A \to B$ or $A \xrightarrow{f} B$ for a morphism from source $A$ to target $B$, and $Hom(A, B)$ for the set of all morphisms from $A$ to $B$. The objects of a category are classes which are not necessarily sets. (Every set is a class, but not conversely.) For a set-theoretical foundation one can use for instance the Set Theory of *von Neumann-Bernays-Gödel*. The concept of a class is then taken as a basic concept, and sets are those classes which are elements of a class. There exists a universal class which contains all sets as elements. For details see for instance [47] or [62].

A category in which the objects are sets with additional structure (such as operations, partial operations or relations) is called a *concrete category*. The following properties have to be satisfied by the objects and morphisms of such a category:

(i)   If $A$ is the base set of an object, then $id_A : A \rightarrow A$ defined by $id_A(x) = x$
      for all $x \in A$ is a morphism.

(ii)  The class of morphisms is closed under composition: if $A$, $B$ and $C$
      are objects and $f : A \rightarrow B$ and $g : B \rightarrow C$ are morphisms, then
      $g \circ f : A \rightarrow C$ is a morphism.

(iii) For any objects $A$ and $B$ and any morphism $f : A \rightarrow B$ the equations
      $f \circ id_A = f$ and $id_B \circ f = f$ are satisfied.

(iv)  The composition of morphisms is associative.

**Example 3.1.1.**

1. The objects of the category **S**et are sets and the morphisms are the
   usual mappings between sets.
2. The objects of the category **P**ar are sets and the morphisms are the
   partial mappings between sets.
3. The objects of the category **T**op are the topological spaces and the
   morphisms are the continuous mappings between them.
4. The objects of the category **G**roup are groups and morphisms are homomorphisms between them.
5. The objects of the category **A**lg$(\tau)$ are the algebras of type $\tau$ and the
   morphisms are (algebra) homomorphisms between them.
6. More generally, each variety of algebras can be regarded as a category,
   where the objects are the algebras of the variety and the morphisms
   are the homomorphisms between them.
7. The objects of the category **D**ia are points in a plane and morphisms
   are arrows between points.

## 3.2   Special Morphisms

In this section we introduce some special properties of morphisms, analogous to the invertibility properties of ordinary mappings.

**Definition 3.2.1.** A morphism $f : A \rightarrow B$ is called
a) *left-invertible* or a *coretraction* if there exists a morphism $h : B \rightarrow A$
such that $h \circ f = id_A$. In this case the morphism $h$ is called a *left inverse*
of $f$.
b) *right-invertible* or a *retraction* if there exists a morphism $g : B \rightarrow A$ such
that $f \circ g = id_B$. In this case $g$ is called a *right inverse* of $f$.

c) an *isomorphism* if there exists a morphism $g : B \to A$ such that $g \circ f = id_A$ and $f \circ g = id_B$. Such a morphism $g$ is called an *inverse* of $f$.

**Remark 3.2.2.**

1. If the morphism $f : A \to B$ is both left-invertible and right-invertible, then it is an isomorphism. It is a simple exercise to prove that if $f$ has both a left inverse and a right inverse, then these two inverses must be equal.

2. If $f : A \to B$ is an isomorphism, then there exists exactly one morphism $g$ with $g \circ f = id_A$ and $f \circ g = id_B$. This mapping $g$ is denoted by $f^{-1}$, and is called the inverse of $f$.

3. If there is an isomorphism between the objects $A$ and $B$, then $A$ and $B$ are called isomorphic and we write $A \cong B$.

**Definition 3.2.3.** A morphism $f : A \to B$ is said to be a *monomorphism* (or a *mono* for short) if

$$\forall g_1, g_2 : C \to A \quad (f \circ g_1 = f \circ g_2 \Rightarrow g_1 = g_2),$$

and an *epimorphism* (or an *epi*) if

$$\forall h_1, h_2 : B \to C \quad (h_1 \circ f = h_2 \circ f \Rightarrow h_1 = h_2).$$

**Lemma 3.2.4.** *Any left-invertible morphism is a monomorphism, and dually any right-invertible morphism is an epimorphism.*

**Proof:** If $f$ is a left-invertible morphism, then there is a morphism $h : B \to A$ such that $h \circ f = id_A$. Then for any morphisms $g_1$ and $g_2$ from $C$ to $A$,

$$
\begin{aligned}
& f \circ g_1 = f \circ g_2 \\
\Rightarrow \quad & (h \circ f) \circ g_1 = (h \circ f) \circ g_2 \\
\Rightarrow \quad & id_A \circ g_1 = id_A \circ g_2 \\
\Rightarrow \quad & g_1 = g_2.
\end{aligned}
$$

The proof for a right-invertible morphism is similar. ∎

Although right invertible implies epi, the converse of this lemma is not true, as the following counterexample shows. Let $\mathbb{Z}$ and $\mathbb{Q}$ be the rings of all integers and all rational numbers, respectively. Let $\mathcal{R}$ be an arbitrary ring. We denote by $\iota$ the embedding of $\mathbb{Z}$ into $\mathbb{Q}$, and by $\iota^{-1}$ the inverse mapping. Then $\iota^{-1} \circ \iota = id_{\mathbb{Z}}$, so that $\iota$ is left-invertible. The injection $\iota$ is not right-invertible, but nevertheless it is an epimorphism, as the following argument shows. For any homomorphisms $\varphi_1$ and $\varphi_2$ from $\mathbb{Q}$ to $\mathcal{R}$, we have

$$(\varphi_1 \circ \iota)(g) = (\varphi_2 \circ \iota)(g) \Rightarrow \varphi_1(\iota(g)) = \varphi_2(\iota(g)).$$

Since every homomorphism $\varphi : \mathbb{Q} \to \mathcal{R}$ is uniquely determined by the images of the integers, we get $\varphi_1 = \varphi_2$.

## 3.3  Terminal and Initial Objects

Terminal and initial objects of a category are objects with certain special properties.

**Definition 3.3.1.** An object $T$ of a category $\mathbf{C}$ is called *terminal* if for every object $A$ of $\mathbf{C}$ there is exactly one morphism $\tau_A : A \to T$.

If there are terminal objects in the category $\mathbf{C}$, then such objects are uniquely determined up to isomorphism. To see this, suppose that $T$ and $Q$ are both terminal. Then there is precisely one morphism $\tau_Q : Q \to T$ and precisely one morphism $\sigma_T : T \to Q$. But this gives two morphisms from $T$ to $T$, both $id_T$ and $\tau_Q \circ \sigma_T$, which force $\tau_Q \circ \sigma_T = id_T$. Similarly we can get $\sigma_T \circ \tau_Q = id_Q$, from which we conclude that $T \cong Q$.

**Example 3.3.2.** In the category $\mathbf{S}et$ every one-element set is terminal. In the category $\mathbf{G}roup$ every one-element group is terminal, and in $\mathbf{A}lg(\tau)$ the one-element algebras of type $\tau$ are terminal.

**Definition 3.3.3.** An object $I$ of a category $\mathbf{C}$ is called *initial* if for every object $A$ of $\mathbf{C}$ there is exactly one morphism $I \to A$.

The initial objects of a category are also uniquely determined up to isomorphism. The proof of this fact is very similar to the proof for terminal objects, and we leave it as an exercise.

**Example 3.3.4.** In the category $\mathbf{S}et$ the empty set $\emptyset$ is initial. In the category $\mathbf{G}roup$ every one-element group is initial.

## 3.4  Sums and Products

Now we look at two constructions defined on a family $(A_i)_{i \in I}$ of objects in a category.

**Definition 3.4.1.** Let $(A_i)_{i \in I}$ be a family of objects in a category **C**. An object $S$ together with a family $(e_i : A_i \to S)_{i \in I}$ of morphisms is called the *sum* of the $A_i$'s if for every object $Q$ from **C** with morphisms $(q_i : A_i \to Q)_{i \in I}$ there is exactly one morphism $s : S \to Q$ with $q_i = s \circ e_i$ for all $i \in I$. If such a sum exists, we write $S := \sum_{i \in I} A_i$ and call the morphisms $(e_i)_{i \in I}$ *injections* of the sum.

The diagram in Figure 3.1 illustrates the sum construction. The sum of a family of objects is also called their *coproduct*.

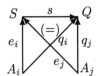

Fig. 3.1   Coproduct

**Example 3.4.2.** Consider the category **S**et of sets and set mappings. We have already seen in Section 2.5 a version of the coproduct of two sets $A_1$ and $A_2$. Using two indices such as 1 and 2, or 0 and 1, we take $S$ to be the disjoint sum

$$A_1 + A_2 = \{(1,a) \mid a \in A_1\} \cup \{(2,b) \mid b \in A_2\},$$

and as morphisms the mappings $e_1 : A_1 \to A_1 + A_2$ and $e_2 : A_2 \to A_1 + A_2$ defined by $e_1(a) = (1,a)$ and $e_2(a) = (2,b)$. To see that this fits our categorical definition of the sum, suppose that $Q$ is any set with mappings $q_i : A_i \to Q$ for $i = 1, 2$. Then there exists precisely one mapping $s : A_1 + A_2 \to Q$ with $q_i = s \circ e_i$, defined by $s(i,a) := q_i(a)$. For any $a \in A_1 \cup A_2$, we have $q_i(a) = (s \circ e_i)(a) = s(e_i(a)) = s((i,a))$, so $q_i = s \circ e_i$ for $i = 1, 2$.

The sum of a family $(A_i)_{i \in I}$ of objects of a category **C** is uniquely determined up to isomorphism. To see this, suppose that $S$ and $S_1$ are both sums of the family $(A_i)_{i \in I}$ of objects. Then there are morphisms $s_1 : S \to S_1$ and $s_2 : S_1 \to S$, and $s_2 \circ s_1 : S \to S$ satisfies $q_i = (s_2 \circ s_1) \circ e_i$ for $i \in I$. The identity morphism $id_S$ also satisfies $q_i = id_S \circ e_i$ for $i \in I$, and from the definition of a sum we obtain $s_2 \circ s_1 = id_S$. In a similar way we get $s_1 \circ s_2 = id_{S_1}$, making $S \cong S_1$.

The injections $(e_i)_{i \in I}$ associated with a sum must be epimorphisms. That is, for all morphisms $h_1, h_2 : S \to C$ and all $i \in I$ we have

$$h_1 \circ e_i = h_2 \circ e_i \quad \Rightarrow \quad h_1 = h_2.$$

The product of a family of objects is defined in the following dual way.

**Definition 3.4.3.** Let $(A_i)_{i \in I}$ be a family of objects in a category **C**. An object $P$ together with a family $(p_i : P \to A_i)_{i \in I}$ of morphisms is called the *product* of the $A_i$'s if for every object $Q$ from **C** with morphisms $(q_i : Q \to A_i)_{i \in I}$ there is exactly one morphism $s : Q \to P$ with $q_i = p_i \circ s$ for all $i \in I$. If the product exists, we write $P := \prod_{i \in I} A_i$, and the morphisms $(p_i)_{i \in I}$ are called *projections*.

The diagram in Figure 3.2 illustrates the product construction.

Fig. 3.2    Product

In the category **Set** of sets, the usual cartesian product of a family of sets is their product, with the usual projection mappings as morphisms. In the category **A**$lg(\tau)$ the usual direct product of a family $(\mathcal{A})_{i \in I}$ of algebras is a product.

## 3.5    Equalizers and Coequalizers

**Definition 3.5.1.** Let $(f_i : A \to B)_{i \in I}$ be a family of morphisms. A morphism $g : B \to C$ is called a *coequalizer* of the $f_i$'s if the following conditions are satisfied:

(i)  For all $i, j \in I$,   $g \circ f_i = g \circ f_j$;
(ii) For all objects $Q$ and all morphisms $q : B \to Q$ with $q \circ f_i = q \circ f_j$ for all $i, j \in I$, there exists a unique morphism $h : C \to Q$ such that $q = h \circ g$.

The diagram in Figure 3.3 illustrates this definition.

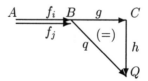

Fig. 3.3    Coequalizer

**Proposition 3.5.2.** *Let $f_i : A \to B$ be morphisms for $i = 1, 2$. Every coequalizer for $f_1$ and $f_2$ is an epimorphism.*

**Proof:** Let $g : B \to C$ be a coequalizer for $f_1$ and $f_2$, and assume that $\varphi$ and $\psi$ are morphisms satisfying $\varphi \circ g = \psi \circ g$. Then

$$
\begin{aligned}
g \circ f_1 &= g \circ f_2 \\
\Rightarrow \quad \varphi \circ (g \circ f_1) &= \varphi \circ (g \circ f_2) \\
\Rightarrow \quad (\varphi \circ g) \circ f_1 &= (\varphi \circ g) \circ f_2
\end{aligned}
$$

and

$$
\begin{aligned}
\varphi \circ g &= \psi \circ g \\
\Rightarrow \quad (\varphi \circ g) \circ f_1 &= (\psi \circ g) \circ f_1 \\
\Rightarrow \quad (\varphi \circ g) \circ f_1 &= (\varphi \circ g) \circ f_2.
\end{aligned}
$$

This shows that $\xi := \varphi \circ g = \psi \circ g$ satisfies $\xi \circ f_1 = \xi \circ f_2$. By part (ii) of the definition of a coequalizer, the equation $h \circ g = \xi$ has only one solution $h$. So $\varphi = \psi$ and $g$ is an epimorphism. ∎

**Example 3.5.3.** Let $f_1, f_2 : A \to B$ be mappings in the category **Set**. Let $\sim$ be the smallest equivalence relation on $B$ such that $f_1(a) \sim f_2(a)$. Then the natural mapping $nat \sim:\ B \to B/\sim$ with $a \mapsto [a]_\sim$ for all $a \in A$ is a coequalizer of $f_1$ and $f_2$. This is because $f_1(a) \sim f_2(a)$ implies $nat \sim\ \circ\ f_1$ $= nat \sim\ \circ f_2$ since $(nat \sim\ \circ\ f_1)(a) = nat \sim (f_1(a)) = [f_1(a)]_\sim = [f_2(a)]_\sim$ $= nat \sim (f_2(a)) = (nat \sim\ \circ\ f_2)(a)$ for all $a \in A$. If $q \circ f_1 = q \circ f_2$ for $q : B \to Q$, then there is exactly one mapping $h : B/\sim\ \to Q$ such that $h \circ nat \sim =\ q$. The diagram in Figure 3.4 illustrates this example.

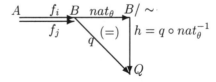

Fig. 3.4 Coequalizer in **Set**

If $s_1 \sim s_2$ and $s_1$ and $s_2$ do not have the form $f_1(a)$ and $f_2(a)$ for some $a \in A$, then the pair $(s_1, s_2)$ can be generated by a sequence of pairs of the form $(f_1(a_i), f_2(a_i))$ and we can repeat the previous argument.

An equalizer of a family of morphisms is defined dually.

**Definition 3.5.4.** Let $(f_i : A \to B)_{i \in I}$ be a family of morphisms in a category **C**. A morphism $g : C \to A$ is called an *equalizer* of the $f_i$'s if the following conditions are satisfied:

(i) For all $i, j \in I$,    $f_i \circ g = f_j \circ g$;
(ii) For all objects $Q$ and all morphisms $q : Q \to A$ with $f_i \circ q = f_j \circ q$ for all $i, j \in I$, there exists a unique morphism $h : C \to Q$ satisfying $g = q \circ h$.

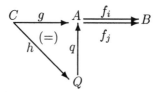

Fig. 3.5    Equalizer in **Set**

**Proposition 3.5.5.** *Every equalizer of two morphisms $f_1$ and $f_2$ is a monomorphism. Every equalizer which is an epimorphism is an isomorphism, and every left invertible morphism is an equalizer.*

**Proof:** The proof of the first proposition is similar to that of Proposition 3.5.2. If the equalizer $g$ of $f_1$ and $f_2$ is an epimorphism, then $f_1 = f_2$, and $id_A$ is also an equalizer of $f_1$ and $f_2$, making $g$ an isomorphism. If $q : Q \to A$ is a left invertible morphism with left inverse $t$, then one can check that $q$ is the equalizer of $id_A$ and $q \circ t$.      ∎

In the category **G***roup* of all groups, any subgroup with its inclusion mapping is an equalizer.

## 3.6    Pushouts and Pullbacks

**Definition 3.6.1.** Let $(f_i : A \to B_i)_{i \in I}$ be a family of morphisms in a category **C**. An object $P$ together with a family $(p_i : B_i \to P)_{i \in I}$ of morphisms is called a *pushout* of the $f_i$'s if

(i) For all $i, j \in I$,    $p_i \circ f_i = p_j \circ f_j$;
(ii) For all objects $Q$ and all morphisms $(q_i : B_i \to Q)_{i \in I}$ with $q_i \circ f_i = q_j \circ f_j$ for all $i, j \in I$, there exists a unique morphism $h : P \to Q$ satisfying $h \circ p_i = q_i$ for all $i \in I$.

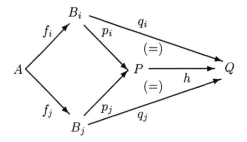

Fig. 3.6   Pushout

In fact the existence of sums and coequalizers in a category turns out to guarantee the existence of pushouts.

**Theorem 3.6.2.** *Let* **C** *be a category in which arbitrary sums and coequalizers exist. Then arbitrary pushouts also exist. Let* $(f_i : A \to B_i)_{i \in I}$ *be a family of morphisms. For the canonical injections* $e_i : B_i \to \sum_{i \in I} B_i$, *let the morphism* $g : \sum_{i \in I} B_i \to P$ *be the coequalizer of the family* $(e_i \circ f_i)_{i \in I}$. *Then the object* $P = \sum_{i \in I} B_i$ *with the morphisms* $(g \circ e_i)_{i \in I}$ *form the pushout of the family* $(f_i)_{i \in I}$.

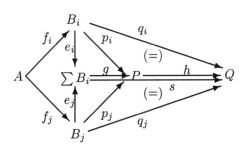

Fig. 3.7   Pushout in **Set**

**Proof:** Let $f_i : A \to B_i$ for $i \in I$. We have $g \circ (e_i \circ f_i) = g \circ (e_j \circ f_j)$ for all $i, j \in I$ since $g$ is the coequalizer of the family $(e_i \circ f_i)_{i \in I}$. By associativity then $(g \circ e_i) \circ f_i = (g \circ e_j) \circ f_j$. By the definition of the sum, for any object $Q$ together with a family $(q_i : B_i \to Q)_{i \in I}$ of morphisms with $q_i \circ f_i = q_j \circ f_j$ for all $i, j \in I$, there exists a morphism $s : \sum_{i \in I} B_i \to Q$ with $q_i = s \circ e_i$ for all $i \in I$. Then it follows that $(s \circ e_i) \circ f_i = (s \circ e_j) \circ f_j$ and $s \circ (e_i \circ f_i)$

$= s \circ (e_j \circ f_j)$ for all $i, j \in I$. By the definition of a coequalizer there exists precisely one morphism $h : P \to Q$ such that $s = h \circ g$. This gives $q_i = s \circ e_i = (h \circ g) \circ e_i$ for all $i \in I$. Altogether we get

(i) $\forall i, j \in I \ ((g \circ e_i) \circ f_i = (g \circ e_j) \circ f_j)$, and
(ii) $\forall Q \forall i \in I \ (q_i : B_i \to Q$ such that $q_i \circ f_i = q_j \circ f_j) \exists! h : P \to Q \ (h \circ (g \circ e_i) = q_i)$.

We note that since $s = h \circ g$ we get $q_i = s \circ e_i = h \circ (g \circ e_i)$ for all $i \in I$. By the definition of a coequalizer the morphism $h$ is uniquely determined. ∎

**Proposition 3.6.3.** *Let $f_1 : A \to B_1$ and $f_2 : A \to B_2$ be morphisms and let the object $P$ together with the morphisms $p_1 : B_1 \to P$ and $p_2 : B_2 \to P$ be the pushout of $f_1$ and $f_2$. If $f_1$ is an epimorphism, then $p_1$ is an epimorphism.*

**Proof:** Let $f_1$ be an epimorphism, and let morphisms $w_1$ and $w_2$ from $P$ to $Q$ satisfy $w_1 \circ p_1 = w_2 \circ p_1$. Then we have $w_1 \circ p_1 \circ f_1 = w_2 \circ p_1 \circ f_1$. From the definition of a pushout and the fact that $f_1$ is an epimorphism we obtain $w_1 = w_2$; that is, $p_1$ is an epimorphism. ∎

**Definition 3.6.4.** Let $(f_i : A_i \to B)_{i \in I}$ be a family of morphisms. An object $P$ together with a family $(p_i : P \to A_i)_{i \in I}$ of morphisms is called the *pullback* of the $f_i$'s if

(i) For all $i, j \in I$, $\quad f_i \circ p_i = f_j \circ p_j$;
(ii) For all objects $Q$ and all morphisms $q_i : Q \to A_i$ with $f_i \circ q_i = f_j \circ q_j$ for all $i, j \in I$, there exists a unique morphism $h : Q \to P$ such that $p_i \circ h = q_i$ for all $i \in I$.

Pullbacks are the dual of pushouts, and the dual results to Theorem 3.6.2 and Proposition 3.6.3 also hold. We leave the proofs as an exercise.

**Theorem 3.6.5.** *Let **C** be a category in which arbitrary products and equalizers exist. Then arbitrary pullbacks also exist. Let $(f_i)_{i \in I}$ be a family of morphisms. For the canonical projections $p_i : \prod_{i \in I} B_i \to B_i$, let the morphism $g : P \to \prod_{i \in I} B_i$ be the coequalizer of the family $(f_i \circ p_i)_{i \in I}$. Then $\prod_{i \in I} B_i$ with the morphisms $(p_i \circ g)_{i \in I}$ is the pullback of the family $(f_i)_{i \in I}$.*

A category is said to be *cocomplete* if arbitrary coequalizers and sums exist, and to be *finite cocomplete* if arbitrary coequalizers and arbitrary finite sums exist. Dually, a category is called *complete* if arbitrary equalizers and products exist and *finite complete* if arbitrary equalizers and arbitrary finite products exist. Sums, coequalizers and pushouts are special types of a more general concept called a *colimit* and products, equalizers and pullbacks are special kinds of *limits*. More detail on such constructions may be found in [76].

## 3.7 The Category Set

In this section we relate the basic properties of injectivity and surjectivity of set mappings to the monomorphisms and epimorphisms in the category Set, and consider colimits in Set.

**Theorem 3.7.1.**

(i) *A morphism in the category* Set *is a monomorphism iff it is an injective mapping.*

(ii) *A morphism in the category* Set *is an epimorphism iff it is a surjective mapping.*

**Proof:** (i) Let $f : A \to B$ be an injective mapping. Then for any morphisms $g_1$ and $g_2$,

$$f \circ g_1 \ = \ f \circ g_2$$
$$\Rightarrow \quad \forall c \in C \quad ((f \circ g_1)(c) \ = \ (f \circ g_2)(c))$$
$$\Rightarrow \quad \forall c \in C \quad (f(g_1(c)) \ = \ f(g_2(c)))$$
$$\Rightarrow \quad \forall c \in C \quad (g_1(c) \ = \ g_2(c)), \quad \text{since } f \text{ is injective}$$
$$\Rightarrow \quad g_1 \ = \ g_2,$$

showing that $f$ is a monomorphism.

Conversely, Let $f : A \to B$ be a monomorphism and let $f(a_1) = f(a_2)$ for $a_1, a_2 \in A$. Then for $i = 1, 2$ define mappings $g_i : C \to A$ with $g_i(c) = a_i$ for any element $c \in C$. Then

$$f(a_1) \ = \ f(a_2)$$
$$\Rightarrow \quad f(g_1(c)) \ = \ f(g_2(c))$$
$$\Rightarrow \quad (f \circ g_1)(c) \ = \ (f \circ g_2)(c)$$
$$\Rightarrow \quad g_1(c) \ = \ g_2(c)$$
$$\Rightarrow \quad a_1 \ = \ a_2.$$

(ii) First let $f : A \to B$ be surjective, and suppose that mappings $g_1$ and $g_2$ from $B$ to $C$ satisfy $g_1 \circ f = g_2 \circ f$. Then

$$g_1 \circ f \;=\; g_2 \circ f$$
$$\Rightarrow \quad \forall a \in A \;\; ((g_1 \circ f)(a) \;=\; (g_2 \circ f)(a))$$
$$\Rightarrow \quad \forall a \in A \;\; (g_1(f(a)) \;=\; g_2(f(a))).$$

Since $f$ is surjective, for every element $b \in B$ there exists an element $a \in A$ such that $b = f(a)$. Therefore we obtain $g_1(b) = g_2(b)$ for all $b \in B$, making $g_1 = g_2$.

Conversely, suppose that $f$ is not surjective. Then there is an element $b \in B$ such that $f(x) \neq b$ for all $x \in A$. We choose the two-element set $P := \{1,2\}$ and mappings $g, h : B \to P$ defined by $h(y) = 1$ for all $y \in B$ and

$$g(y) = \begin{cases} 1, & \text{if } y \neq b \\ 2, & \text{if } y = b. \end{cases}$$

Then $g \neq h$, but $(g \circ f)(x) = (h \circ f)(x)$ for all $x \in A$. Therefore $f$ is not an epimorphism. ∎

It is easy to see that injective mappings $f : A \to B$ are left-invertible as long as $A \neq \emptyset$. If $x_0 \in A$ is a fixed element, then define

$$f_{x_0}^-(y) := \begin{cases} x, & \text{if } f(x) = y \\ x_0 & \text{if } y \notin f(A). \end{cases}$$

Then we can check that $f_{x_0}^-$ is well-defined. If $y_1 = y_2$ and $f(x_1) = y_1$ and $f(x_2) = y_2$, then $x_1 = x_2$ since $f$ is injective. If $y_1 = y_2$ and $y_1(y_2) \notin f(x)$, then $f_{x_0}^-(y_1) = x_0 = f_{x_0}^-(y_2)$. Moreover, we have $(f_{x_0}^- \circ f)(x) = f_{x_0}^-(f(x)) = x$ for all $x \in A$, i.e. $f_{x_0}^- \circ f = id_A$.

A natural question then is whether every surjective mapping is right invertible. That is, if $f : A \to B$ is surjective, is there a mapping $g : B \to A$ such that $f \circ g = id_B$? In general, each $b \in B$ can have more than one preimage in $A$. The question is whether we can choose precisely one element from each class of elements from $A$ which have the same image in $B$ under $f$. To do so, we need the *Axiom of Choice*.

**Axiom of Choice**: *If $\mathcal{M}$ is a system of pairwise disjoint sets, then there is a set $P$ which contains exactly one element from each set of the system. The mapping which selects these elements is said to be a choice function on the system $\mathcal{M}$.*

It turns out that the Axiom of Choice is actually equivalent to the proposition that every surjective mapping is right invertible. We can see this as follows. Every surjective mapping $f : A \to B$ defines a family

$(f^{-1}(\{y\}))_{y \in B}$ of non-empty subsets of $A$. Every choice function for this family is obviously right-inverse to $f$. Conversely every family $(X_i)_{i \in I}$ of non-empty sets defines a surjective mapping $f : \bigcup_{i \in I} X_i \to I$. A function which is right inverse to $f$ is a choice function.

A frequently used proposition in category theory is the *Diagram Lemma*. It is the set-theoretical basis of the General Homomorphism Theorem in universal algebra (see Chapter 1 or [29], p. 55). We recall from Section 1.1 that the *kernel* of a mapping $f : A \to B$ is the relation on $A$ given by

$$Kerf = \{(a,b) \mid f(a) = f(b)\}.$$

Clearly, $Kerf$ is an equivalence relation on $A$.

**Lemma 3.7.2. (The Diagram Lemma)**
*Let $f : X \to Y$ be a surjective mapping and let $g : X \to Z$ be an arbitrary mapping. Then there is a mapping $h : Y \to Z$ with $h \circ f = g$ iff $Kerf \subseteq Kerg$; and this mapping $h$ is uniquely determined.*

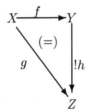

Fig. 3.8   Diagram Lemma

**Proof:** If there is a mapping $h : Y \to Z$ such that $h \circ f = g$, then it is straightforward to see that $Kerf \subseteq Ker(h \circ f) = Kerg$. For the converse direction, we consider $h := \{(f(x), g(x)) \mid x \in X\}$. Since $Kerf \subseteq Kerg$, if $f(x_1) = f(x_2)$ then $h(f(x_1)) = g(x_1) = g(x_2) = h(f(x_2))$, so that $h : Y \to Z$ is well-defined and satisfies $h \circ f = g$. Moreover, if there is a second mapping $h'$ satisfying $h' \circ f = g = h \circ f$, then the fact that $f$ is surjective and an epimorphism forces $h = h'$. ∎

As a consequence we get the so-called *E-M-Square Lemma*:

**Lemma 3.7.3.** *Let $e : X \to Y$, $g : X \to Z$, $f : Y \to U$ and $m : Z \to U$ be mappings, and let $e$ be an epimorphism and $m$ a monomorphism. If the equation $f \circ e = m \circ g$ is satisfied, then the corresponding "$E - M$-square"* has a diagonal, meaning that there is a mapping $d : Y \rightarrow Z$ such that $d \circ e = g$ and $m \circ d = f$.

The diagram in Figure 3.9 illustrates this lemma.

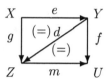

Fig. 3.9   E-M-Square Lemma

**Proof:** If $(a, b) \in Ker\ e$, then $e(a) = e(b)$ and so $(f \circ e)(a) = (f \circ e)(b)$. Using the equation $f \circ e = m \circ g$ gives $(m \circ g)(a) = (m \circ g)(b)$ and then $g(a) = g(b)$, since $m$ is injective. This shows that

$$Ker\ e \subseteq Ker(f \circ e) = Ker(m \circ g) \subseteq Ker g.$$

By the Diagram Lemma there exists exactly one mapping $d : Y \rightarrow Z$ with $d \circ e = g$. By composition with $m$ from the left we obtain $m \circ d \circ e = m \circ g$. Applying the equation $f \circ e = m \circ g$ and using the fact that $e$ is an epimorphism gives $m \circ d = f$.          ∎

Now we consider different kinds of colimits in the category **Set**.

*Sums*

If $(X_i)_{i \in I}$ is a family of sets, then the sum of the $X_i$'s is the disjoint union $\sum_{i \in I} X_i := \bigcup_{i \in I} \{(i, x) \mid x \in X_i\}$, with the injections defined by $e_i(x) := (i, x)$ for all $x \in X_i$. To verify this, let $Q$ be a set with mappings $(q_i : X_i \rightarrow Q)_{i \in I}$. Then there is exactly one mapping $\sigma : \sum_{i \in I} X_i \rightarrow Q$ with $(\sigma \circ e_i)(x) = \sigma(e_i(x)) = \sigma((i, x)) = q_i(x)$ for all $x \in X_i$. Uniqueness of $\sigma$ follows from the fact that the $e_i$'s are injective and therefore monomorphisms.

*Coequalizers*

For coequalizers in **Set** we have the following result.

**Lemma 3.7.4.** *Let $(f_i)_{i \in I} : X \rightarrow Y$ be a family of mappings and let $\theta$ be the equivalence relation which is generated by the set*

$$R := \{(f_i(x), f_j(x)) \mid x \in X, i, j \in I\}.$$

*Then the natural mapping $nat_\theta : Y \to Y/\theta$ defined by $nat_\theta(y) := [y]_\theta$ for all $y \in Y$ is the coequalizer of the $f_i$'s.*

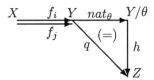

Fig. 3.10    Coequalizer in **Set**

**Proof:** For all elements $i, j$ from the index set $I$ and for all $x \in X$, we have $(nat_\theta \circ f_i)(x) = [f_i(x)]_\theta = [f_j(x)]_\theta = (nat_\theta \circ f_j)(x)$. This is exactly the condition (i) from the definition of a coequalizer.

If $Z$ is a set together with a morphism $q : Y \to Z$ such that $q \circ f_i = q \circ f_j$ for all $i, j \in I$, then

$$(a, b) \in \theta$$
$$\Rightarrow \quad [a]_\theta = [b]_\theta$$
$$\Rightarrow \quad [f_i(x)]_\theta = [f_j(x)]_\theta.$$

Moreover,

$$(q \circ f_i)(x) = (q \circ f_j)(x)$$
$$\Rightarrow \quad (a, b) \in Kerq,$$

and therefore $\theta \subseteq Kerq$.

If $(s_1, s_2) \in \theta$ and $(s_1, s_2)$ does not have the form $(f_i(x), f_j(x))$ for some $x \in X$, then $(s_1, s_2)$ is given by a sequence of those elements and we can proceed as before.

By the Diagram Lemma there is exactly one mapping $h : Y/\theta \to Z$ with $h \circ nat\theta = q$. ∎

*Pushouts*

By Theorem 3.6.2 pushouts can be constructed using sums and coequalizers. Thus the pushout is the coequalizer of the family $(f_i : X \to Y_i)_{i \in I}$ of mappings, i.e. the family $(nat_\theta \circ e_i)_{i \in I}$, where $e_i : Y_i \to \sum_{i \in I} Y_i$ are the canonical embeddings and where $\theta$ is the equivalence relation on $\sum_{i \in I} Y_i$ generated by all pairs $(f_i(x), f_j(x))$ with $x \in X_i$ and $i, j \in I$.

**Lemma 3.7.5.** *Let $(\theta)_{i \in I}$ be a family of equivalence relations defined on a set $X$ and let $\theta$ be the relation $\theta := \bigvee_{i \in I} \theta_i$. Then the pushout of the family $(nat_{\theta_i} : X \to X/\theta_i)_{i \in I}$ is the set $X/\theta$ together with the family $\pi_i : X/\theta_i \to X/\theta$ defined by $\pi_i([x]_{\theta_i}) := [x]_\theta$ for all $i \in I$.*

**Proof:** We have $\pi_i(nat_{\theta_i}(x)) = \pi_i([x]_{\theta_i}) = [x]_\theta = \pi_j([x]_{\theta_j}) = \pi_j(nat_{\theta_j}(x))$ for all $i, j \in I$ and for all $x \in X$. Thus $\pi_i \circ nat\theta_i = \pi_j \circ nat\theta_j$ for all $i, j \in I$. Now if $Q$ is a set and $(q_i : X/\theta_i \to Q)_{i \in I}$ a family of mappings satisfying $g := q_i \circ nat\theta_i = q_j \circ nat\theta_j$, then $\theta_i \subseteq Kerg$ for all $i, j \in I$ and so $\theta \subseteq Kerg$. By the Diagram Lemma there is exactly one mapping $h : X/\theta \to Q$ with $h \circ \pi_\theta = g$. Therefore for every $i \in I$ we have $q_i \circ nat\theta_i = g = h \circ \pi_\theta = h \circ \pi_i \circ nat_{\theta_i}$, and we have $q_i = h \circ \pi_i$. ∎

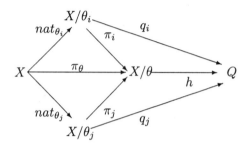

Fig. 3.11   Pushouts in $\mathbf{Set}$

## 3.8   Functors

So far we have considered morphisms between objects in one category. Now we turn to mappings between objects and morphisms of two different categories. A functor from a category $\mathbf{C}$ to a category $\mathbf{D}$ maps the objects of $\mathbf{C}$ to objects of $\mathbf{D}$ and the morphisms of $\mathbf{C}$ to morphisms of $\mathbf{D}$, in a way that is compatible with the composition of morphisms and preserves the identity morphism. The next definition makes this more precise.

**Definition 3.8.1.** Let $\mathbf{C}$ and $\mathbf{D}$ be categories. A *(covariant) functor $F$* : $\mathbf{C} \to \mathbf{D}$ defines a mapping $A \mapsto F(A)$ of objects $A$ from the object class $\mathcal{O}$ of the category $\mathbf{C}$ to objects $F(A)$ from the object class $\mathcal{O}'$ of the category $\mathbf{D}$; and also a mapping for morphisms which maps each morphism $f : A \to B$ from $\mathbf{C}$ to a morphism $F(f) : F(A) \to F(B)$ from the category $\mathbf{D}$, such that for all morphisms $f, g$ and all objects $A$ from $\mathbf{C}$ the following two conditions are satisfied:

(i) $F(g \circ f) = F(g) \circ F(f)$,

(ii) $F(id_A) = id_{F(A)}$.

The functor is called *contravariant* if instead of (i), (ii) the following conditions are satisfied:

(i') $F(g \circ f) = F(f) \circ F(g)$,

(ii') $F(id_A) = id_{F(A)}$.

We are interested here only in covariant functors, and shall refer to them merely as functors. We are particularly interested in *endofunctors* from **Set** to **Set**, which we illustrate with several examples.

**Example 3.8.2.**

1. $I$ with $I(X) = X$ for every set $X$ and $I(f) = f$ for every mapping $f$ is obviously a functor on the category **Set**. More generally, there is an identity functor $I_C$ for any category **C**.

2. Let $C$ be a given set. We denote by $C$ the functor mapping each set $X$ to the set $C$ and each morphism $f : X \to Y$ to the identity mapping $id_C$ of the set $C$. It is easy to check that $C$ is a functor.

3. The power set functor $\mathcal{P}$ maps each set $X$ to its power set $\mathcal{P}(X)$ and each mapping $f : X \to Y$ to the mapping $\mathcal{P}(f) : \mathcal{P}(X) \to \mathcal{P}(Y)$. It is easy to check that this is also a functor.

4. The power functor $(-)^C$ assigns to each set $X$ the set $X^C$ of all mappings from $C$ to $X$ and to each mapping $f : X \to Y$ the mapping $f^C$ defined by $f^C(u) := f \circ u$.

5. The $(3 - 2)$-functor maps each set $X$ to

$$(X)_2^3 := \{(x_1, x_2, x_3) \in X^3 \mid x_1 = x_2 \text{ or } x_1 = x_3 \text{ or } x_2 = x_3\}.$$

For an arbitrary mapping $f : X \to Y$ the mapping $(f)_2^3 : (X)_2^3 \to (Y)_2^3$ is defined component-wise by $(f)_2^3(x_1, x_2, x_3) := (f(x_1), f(x_2), f(x_3))$. Again, it can be checked that $(-)_2^3$ is a functor.

Having a stock of examples of functors, we can use the following constructions to combine them into new functors. Let $F_1$ and $F_2$ be functors from **Set** to **Set**.

(i) $F_1 \circ F_2$ defined by $A \mapsto F_1(F_2(A))$ for objects $A$ and $f \mapsto F_1(F_2(f))$ for morphisms $f$ is a functor. This construction is called *composition* of functors.

(ii) $F_1 \times F_2$ defined by $(F_1 \times F_2)(A) := F_1(A) \times F_2(A)$ for objects and $(F_1 \times F_2)(f)(u, v) := (F_1(f)(u), F_2(f)(v))$ for morphisms is a functor. This construction is called the *cartesian product* of functors. The cartesian product can also be defined for arbitrary families of endofunctors of **Set**.

(iii) $F_1 + F_2$ is defined on objects by $(F_1 + F_2)(A) := F_1(X) + F_2(X)$, where $F_1(X) + F_2(X)$ is the disjoint union of the sets $F_1(X), F_2(X)$. For each morphism $f$ we define

$$(F_1 + F_2)(f)(u) := \begin{cases} F_1(f)(u), & \text{if } u \in F_1(X) \\ F_2(f)(u), & \text{if } u \in F_2(X) \end{cases}.$$

Then $F_1 + F_2$ is a functor.

**Remark 3.8.3.**

1. If $F : \mathbf{Set} \to \mathbf{Set}$ is a functor and $f$ is injective (surjective) then $F(f)$ is also injective (respectively surjective).
2. Let **C** be a category and let $Hom(A, B)_{\mathbf{C}}$ be the set of all morphisms $f : A \to B$ in **C**. Every functor $F : \mathbf{C} \to \mathbf{D}$ defines a mapping $F_{A,B} : Hom(A, B)_{\mathbf{C}} \to Hom(F(A), F(B))_{\mathbf{D}}$.
3. A functor $F : \mathbf{C} \to \mathbf{D}$ is called *full* if for any two objects $A$ and $B$ of **C** the mapping $F_{A,B}$ is surjective. $F$ is called *faithful* if $F_{A,B}$ is injective for any two objects $A$ and $B$ of **C**.
4. The functor $F : \mathbf{C} \to \mathbf{D}$ is called an *isomorphism* if there is a functor $G : \mathbf{D} \to \mathbf{C}$ such that $F \circ G = I_{\mathbf{D}}$ and $G \circ F = I_{\mathbf{C}}$.

As an important application of these operations with functors, we show that every universal-algebraic type $\tau$ defines a functor. Let us first recall the necessary notation from Chapter 1. A type $\tau$ of a universal algebra $\mathcal{A} = (A; (f_j^A)_{j \in J})$ consists of a sequence $(f_j)_{j \in J}$ of operation symbols together with a sequence $(n_j)_{j \in J}$ of natural numbers, where $f_j$ is mapped to $n_j$ for $j \in J$. The non-negative integer $n_j$ is called the arity of the operation symbol $f_j$, and we usually write $\tau = (n_j)_{j \in J}$ for the type. To each operation symbol $f_j$ is associated an $n_j$-ary fundamental operation $f_j^A$ of the algebra $\mathcal{A}$.

Consider for instance the type $\tau = (2, 1, 0)$, with a binary operation symbol $f$, a unary operation symbol $g$ and a nullary operation symbol $e$. (Groups can be regarded as algebras of this type.) This type defines a set-valued functor $F$ with $F(X) := (X \times X) + X + \{e\}$ for any set $X$. For any mapping $f : X \to Y$ we define

$$F(f)(u) := \begin{cases} (f(x_0), f(x_1)), & \text{if } u = (x_0, x_1) \in X \times X \\ f(u), & \text{if } u \in X \\ e, & \text{if } u = e \end{cases}.$$

For an arbitrary type $\tau$ the mapping for the objects is defined by $F(X) := \sum_{j \in J} X^{n_j}$. This shows how we may describe an algebra of a type such as $(2, 1, 0)$ using a category theoretical notation and a functor $F$.

## 3.9 Natural Transformations

Natural transformations describe connections between two functors.

**Definition 3.9.1.** Let $F_1, F_2 : \mathbf{C} \to \mathbf{D}$ be functors from $\mathbf{C}$ to $\mathbf{D}$. A natural transformation $\eta$ from $F_1$ to $F_2$ maps each object $X$ from $\mathbf{C}$ to a morphism $\eta_X : F_1(X) \to F_2(X)$ such that for every morphism $f : X \to Y$ in $\mathbf{C}$ the equation

$$F_2(f) \circ \eta_X = \eta_Y \circ F_1(f)$$

is satisfied. If $\eta$ is a natural transformation from $F_1$ to $F_2$ we write $\eta : F_1 \to F_2$.

Natural transformations will be used in Chapter 6.

## 3.10 Exercises for Chapter 3

1. Let $f : A \to B$ be a morphism in a category $\mathbf{C}$. Prove that if $f$ has both a left and a right inverse, then these two inverses must be equal.

2. Prove that the initial object in a category is uniquely determined, up to isomorphism.

3. Prove Theorem 3.6.5, which states that if $\mathbf{C}$ is a category in which arbitrary products and equalizers exist, then arbitrary pullbacks also exist. Let $(f_i)_{i \in I}$ be a family of morphisms. For the canonical projections

$p_i : \prod_{i \in I} B_i \to B_i$, let the morphism $g : P \to \prod_{i \in I} B_i$ be the coequalizer of the family $(f_i \circ p_i)_{i \in I}$. Show that $\prod_{i \in I} B_i$ with the morphisms $(p_i \circ g)_{i \in I}$ is the pullback of the family $(f_i)_{i \in I}$.

4. Prove the analogue of Proposition 3.6.3 for pullbacks.

5. Show that the canonical injections of the sum are monomorphisms.

6. Prove that each right-invertible morphism is an epimorphism.

7. Let $\iota : \mathbb{Z} \to \mathbb{Q}$ be the embedding of the integers into the rational numbers. Show that in the category of rings (with ring homomorphisms as morphisms) the following facts hold:

(i) $\iota$ is an epimorphism.
(ii) $\iota$ is a monomorphism.
(iii) $\iota$ is not right-invertible.
(iv) $\iota$ is not left-invertible.

8. Prove that the following mappings are functors on **Set**:

(i) $F(X) = C$ for a fixed set $C$ and all sets $X$ and $F(f) = id_C$ for any mapping $f : X \to Y$.
(ii) $F(X) = X$ for all sets $X$ and $F(f) = f$ for all mappings $f : X \to Y$.
(iii) $F(X) = X^C$ for a fixed set $C$ and all sets $X$ and $F(f) := f_u : u \mapsto f \circ u$ for all mappings $f : X \to Y$ and a fixed mapping $u : C \to X$.
(iv) $F(X) = \mathcal{P}(X)$ for all sets $X$ and $F(f) = \mathcal{P}_f : X \to f(X) := \{f(x) \mid x \in X\}$ for all sets $X$ and all mappings $f : X \to Y$.

9. Let $F : \mathbf{Set} \to \mathbf{Set}$ be defined as follows. Let $F(X) = \mathcal{P}(X)$ for all sets $X$, and let $F(f) : U \mapsto \{x \in X \mid f(x) \in U\}$ for any $f : X \to Y$ and $U \in \mathcal{P}(Y)$. Why is $F$ not a functor?

10. Let $F_1$ and $F_2$ be functors from $\mathbf{Set} \to \mathbf{Set}$. Prove that the mappings $F_1 \circ F_2$, $F_1 + F_2$ and $F_1 \times F_2$ defined in Section 3.8 are functors on **Set**.

# Chapter 4

# $F$-Coalgebras

The concept of a *coalgebra* was first introduced in a 1971 paper by K. Drbohlav ([31]) in the following way. Let $A$ be a non-empty set. For each $n_j \geq 1$ we denote the $n_j$-th *copower* of $A$, that is the union of $n_j$ disjoint copies of $A$, by $A^{\sqcup n_j}$. (See example 3.4.2.) Specifically, $A^{\sqcup n_j} := \{1, \ldots, n_j\} \times A$, and an element $(i, a)$ corresponds to the element $a$ in the $i$-th copy of $A$. An $n_j$-ary *co-operation* on $A$ is then a mapping $f_j^A : A \to A^{\sqcup n_j}$. Each $n_j$-ary co-operation is uniquely determined by a pair $(f_1, f_2)$ of mappings, $f_1 : A \to \{1, \ldots, n_j\}$ and $f_2 : A \to A$. A *coalgebra of type* $\tau$ is then defined as a pair $(A; (f_j^A)_{j \in J})$ consisting of the non-empty set $A$ and a set of finitary co-operations $f_j^A : A \to A^{\sqcup n_j}$, where $n_j$ is the *coarity* of the co-operation $f_j^A$.

This concept of a coalgebra as the dual of an algebra turned out to be too narrow to model a wide range of structures occurring in computer science. As we saw in Chapter 2, for many state-based systems we need to define mappings from a set of states to various structures formed from that set, not just to the copowers. So a more general approach is needed, using categories and functors. In Chapter 2 we showed how state-based systems could be modelled by the concept of an $F$-coalgebra, where $F$ is a functor of the category $\mathbf{Set}$. Now we begin our introduction to the general theory of $F$-coalgebras. The traditional versions of universal algebras and coalgebras of type $\tau$ just described will appear as special cases of our more general definition.

## 4.1  Coalgebras and Homomorphisms

An $F$-coalgebra consists of a set and a mapping which is based on a set-valued functor. Our first definition makes this more precise.

**Definition 4.1.1.** Let $F : \mathbf{Set} \to \mathbf{Set}$ be a functor. An *F-coalgebra* or a coalgebra of type $F$ is a pair $\mathcal{A} := (A; \alpha_A)$, where $A$ is a set and $\alpha_A : A \to F(A)$ is a function. The set $A$ is called the *carrier* or *universe* of the coalgebra $\mathcal{A}$ or the set of *states*, while the mapping $\alpha_A$ is also called the *F-transition structure* or the *dynamics* of $\mathcal{A}$.

The essence of the definition of a coalgebra $\mathcal{A}$ is that the mapping $\alpha_A$ maps from the base set $A$ to $F(A)$ for some set-valued functor $F$. By contrast, an *F-algebra* (or algebra of type $F$) is a pair $(A; \beta_A)$ where $\beta_A$ is a mapping from $F(A)$ to $A$. We shall study $F$-algebras in more detail in the next chapter. This definition of an $F$-coalgebra can be generalized to any category $\mathbf{C}$, using any functor $F : \mathbf{C} \to \mathbf{C}$. The theory of coalgebras on arbitrary categories is considered in [53]. One can also study the category of all coalgebras over the category $Alg(\tau)$ of all algebras of type $\tau$ (see for instance [16]).

**Example 4.1.2.**

1. Coalgebras regarded as pairs $(A; (f_j^A)_{j \in J})$ consisting of a set $A$ and a set of co-operations, as described above, fit the more general definition of an $F$-coalgebra. A functor $F : \mathbf{Set} \to \mathbf{Set}$ is defined by setting $X \mapsto \prod_{j \in J} X^{\sqcup n_j}$ for every set $X$. If $f : X \to Y$ is a mapping, then $F(f)$ is canonically given by $F(f) : \prod_{j \in J} X^{\sqcup n_j} \to \prod_{j \in J} Y^{\sqcup n_j}$, where $F(f)$ maps $(k_j, a_j)_{j \in J}$ to $(k_j, f(a_j))_{j \in J}$, for $k_j \in \{1, \ldots, n_j\}$. Then the coalgebra $(A; (f_j^A)_{j \in J})$ is uniquely determined by $(A; \alpha_A)$ where $\alpha_A : A \to F(A)$ is given by $a \mapsto (f_j^A(a))_{j \in J}$ and vice versa.

2. A labeled *transition system* $(A; \to_A, S)$ consists of a set $A$ of states, a transition relation $\to_A \subseteq A \times S \times A$ and a set $S$ of labels, where $a \overset{s}{\to}_A a'$ means that $(a, s, a') \subseteq \to_A$. Such systems can also be regarded as $F$-coalgebras. The functor $F$ is defined by $F(X) = \mathcal{P}(S \times X) = \{V \mid V \subseteq S \times X\}$ for any set $X$, while each $f : X \to Y$ is mapped to $F(f) : \mathcal{P}(S \times X) \to \mathcal{P}(S \times Y)$. If we define $\alpha_A : A \to F(A)$ by $a \mapsto \{(s, a') \mid a \overset{s}{\to}_A a'\}$, then we obtain an $F$-coalgebra.

3. Every usual universal algebra $(A; (f_j^A)_{j \in J})$ is an example of an $F$-algebra. The functor $F$ here maps every set $X$ to $F(X) := \sum_{j \in J} X^{n_j}$ and every mapping $f : X \to Y$ to $F(f) : \sum_{j \in J} X^{n_j} \to \sum_{j \in J} Y^{n_j}$. The algebra $(A; (f_j^A)_{j \in J})$ is uniquely determined by the $F$-algebra $(A; \beta_A)$, where $\beta_A : F(A) \to A$ is given by $\beta_A(j, (a_1, \ldots, a_{n_j})) := f_j^A(a_1, \ldots, a_{n_j})$, and

conversely.

Homomorphisms of $F$-coalgebras are defined as follows.

**Definition 4.1.3.** Let $\mathcal{A} = (A; \alpha_A)$ and $\mathcal{B} = (B; \alpha_B)$ be $F$-coalgebras. A mapping $\varphi : A \to B$ is called a *homomorphism* from $\mathcal{A}$ to $\mathcal{B}$ if $\alpha_B \circ \varphi = F(\varphi) \circ \alpha_A$, that is if the diagram in Figure 4.1 is commutative.

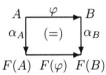

Fig. 4.1   Homomorphism of $F$-coalgebras

The following proposition then follows directly from the definition.

**Proposition 4.1.4.** Let $\mathcal{A} = (A; \alpha_A)$, $\mathcal{B} = (B; \alpha_B)$ and $\mathcal{C} = (C; \alpha_C)$ be any $F$-coalgebras. Then

(i) The mapping $id_A : A \to A$ is a homomorphism.
(ii) If $\varphi : A \to B$ and $\psi : B \to C$ are homomorphisms, then $\psi \circ \varphi : A \to C$ is a homomorphism.

**Proof:** (i) It is clear that $\alpha_A \circ id_A = \alpha_A = id_{F(A)} \circ \alpha_A = F(id_A) \circ \alpha_A$, by the functor property $id_{F(A)} = F(id_A)$.
(ii) Using the homomorphism properties for $\psi$ and $\varphi$ and the associativity of composition of mappings, we have $\alpha_C \circ (\psi \circ \varphi) = (\alpha_C \circ \psi) \circ \varphi = (F(\psi) \circ \alpha_B) \circ \varphi = F(\psi) \circ F(\varphi) \circ \alpha_A = F(\psi \circ \varphi) \circ \alpha_A$. (See Figure 4.2.)  ∎

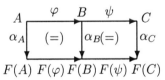

Fig. 4.2   Composition of Homomorphisms

**Corollary 4.1.5.** *The class of all $F$-coalgebras together with the coalgebra homomorphisms form a category, which we will call* $\mathbf{Set}_F$.

By choosing the category $\mathbf{Set}_F$ instead of $\mathbf{Set}$ as our base category, we could even consider coalgebras over coalgebras.

Isomorphisms between $F$-coalgebras are defined as isomorphisms in the category $\mathbf{Set}_F$, but we also have the following connection.

**Proposition 4.1.6.** *Any bijective homomorphism is an isomorphism in* $\mathbf{Set}_F$, *and conversely any isomorphism is a bijective homomorphism.*

**Proof:** Let $\varphi : \mathcal{A} \to \mathcal{B}$ be a homomorphism which is bijective. Then the mapping $\varphi^{-1} : \mathcal{B} \to \mathcal{A}$ exists, and we have to show that it is also a homomorphism. For this, we have $\alpha_A \circ \varphi^{-1} = id_{F(A)} \circ \alpha_A \circ \varphi^{-1} = F(id_A) \circ \alpha_A \circ \varphi^{-1}$
$= F(\varphi^{-1} \circ \varphi) \circ \alpha_A \circ \varphi^{-1} = F(\varphi^{-1}) \circ F(\varphi) \circ \alpha_A \circ \varphi^{-1} = F(\varphi^{-1}) \circ \alpha_B \circ \varphi \circ \varphi^{-1}$
$= F(\varphi^{-1}) \circ \alpha_B \circ id_B = F(\varphi^{-1}) \circ \alpha_B$. Since $\varphi^{-1} \circ \varphi = id_A$ and $\varphi \circ \varphi^{-1} = id_B$, the homomorphism $\varphi$ is an isomorphism. The converse is obviously true.

■

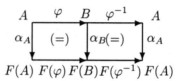

Fig. 4.3    Isomorphisms of $F$-coalgebras

**Proposition 4.1.7.** *Let* $\mathcal{A}$, $\mathcal{B}$ *and* $\mathcal{C}$ *be* $F$-coalgebras, and let $f : \mathcal{A} \to \mathcal{B}$ *and* $g : \mathcal{B} \to \mathcal{C}$ *be mappings such that the composition* $\varphi := g \circ f$ *from* $\mathcal{A}$ *to* $\mathcal{C}$ *is a homomorphism. Then*

(i) *if* $f$ *is a surjective homomorphism, then* $g$ *is also a homomorphism;*
(ii) *if* $g$ *is an injective homomorphism, then* $f$ *is a homomorphism.*

**Proof:** (i) Since $f$ and $\varphi$ are homomorphisms, the equations $\alpha_B \circ f = F(f) \circ \alpha_A$ and $\alpha_C \circ (g \circ f) = F(g) \circ F(f) \circ \alpha_A$ are satisfied. It follows from the second of these equations that $\alpha_C \circ (g \circ f) = F(g) \circ F(f) \circ \alpha_A = F(g) \circ \alpha_B \circ f$. Since the mapping $f$ is surjective, it is an epimorphism in the category $\mathbf{Set}$; therefore from $\alpha_C \circ g \circ f = F(g) \circ \alpha_B \circ f$ we obtain $\alpha_C \circ g = F(g) \circ \alpha_B$, which shows that $g$ is a homomorphism.
(ii) This proof is similar to that of (i).      ■

The Diagram Lemma from the category of sets also extends to the category $\mathbf{Set}_F$.

**Corollary 4.1.8.** *Let* $\mathcal{A}$, $\mathcal{B}$ *and* $\mathcal{C}$ *be* $F$-coalgebras, and let $\varphi : \mathcal{A} \to \mathcal{B}$ *and* $\psi : \mathcal{A} \to \mathcal{C}$ *be homomorphisms. Let* $\varphi$ *be surjective. Then there is a* *homomorphism* $\chi : \mathcal{B} \to \mathcal{C}$ *with* $\chi \circ \varphi = \psi$ *iff* $Ker\varphi \subseteq Ker\psi$.

Fig. 4.4   Diagram Lemma in $\mathbf{Set}_F$

**Proof**: By the Diagram Lemma for sets, we need now only show that when $Ker\varphi \subseteq Ker\psi$ the mapping $\chi$ is a homomorphism. Since $\varphi$ is a homomorphism, the equation $\alpha_B \circ \varphi = F(\varphi) \circ \alpha_A$ is satisfied, and the homomorphism property for $\psi$ gives us $\alpha_C \circ \psi = F(\psi) \circ \alpha_A$. Now using $\chi \circ \varphi = \psi$, the property of the functor $F$ and the first equation we obtain $\alpha_C \circ \chi \circ \varphi = F(\chi) \circ F(\varphi) \circ \alpha_A = F(\chi) \circ \alpha_B \circ \varphi$. Since $\varphi$ is surjective and therefore an epimorphism, this last equation implies that $\alpha_C \circ \chi = F(\chi) \circ \alpha_B$. This shows that $\chi$ is a homomorphism. $\blacksquare$

## 4.2   Subcoalgebras

Subcoalgebras are defined using coalgebra homomorphisms.

**Definition 4.2.1.** Let $\mathcal{A} = (A; \alpha_A)$ be a coalgebra of type $F$. Let $S \subseteq A$ be a subset. The set $S$ is said to be *open* in $\mathcal{A}$ if there is a mapping $\alpha_S : S \to F(S)$ such that the embedding (injection) $\subseteq_S^A : S \hookrightarrow A$ is a homomorphism. In this case $\mathcal{S} = (S; \alpha_S)$ is called a *subcoalgebra* of $\mathcal{A}$, and we write $\mathcal{S} \leq \mathcal{A}$.

As the diagram in Figure 4.5 indicates, if $\mathcal{S}$ is a subcoalgebra of $\mathcal{A}$ then the equation $\alpha_A \circ \subseteq_S^A = F(\subseteq_S^A) \circ \alpha_S$ is satisfied. Conversely if for a subset $S$ of $A$ and a mapping $\alpha_S : S \to F(S)$ this equation is satisfied, then $\mathcal{S}$ is a subcoalgebra of $\mathcal{A}$. In order to show that subcoalgebras are uniquely defined, we must show that the mapping $\alpha_S$ is uniquely determined. Thus instead of a subcoalgebra we can always speak of an open set.

**Lemma 4.2.2.** *Let $\mathcal{A}$ be an $F$-coalgebra. For every open set $S \subseteq A$, there is a uniquely determined mapping $\alpha_S : S \to F(S)$ such that $\mathcal{S} = (S; \alpha_S)$ is a subcoalgebra of $\mathcal{A}$.*

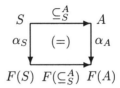

Fig. 4.5   Subcoalgebras

**Proof**: In the case that $S = \emptyset$ there is only one mapping available, the empty mapping. Assume now that $S \neq \emptyset$ and that there are two mappings $\sigma_1$ and $\sigma_2$ from $S$ to $F(S)$ for which the embedding $\subseteq_S^A \colon S \to A$ is a homomorphism. Then $\alpha_A \circ \subseteq_S^A = F(\subseteq_S^A) \circ \sigma_1$ and $\alpha_A \circ \subseteq_S^A = F(\subseteq_S^A) \circ \sigma_2$, and therefore $F(\subseteq_S^A) \circ \sigma_1 = F(\subseteq_S^A) \circ \sigma_2$. Since $\subseteq_S^A$ is injective, so is $F(\subseteq_S^A)$, forcing $\sigma_1 = \sigma_2$. ∎

It is an easy exercise to show that the empty set $\emptyset$ and $\mathcal{A}$ itself are always subcoalgebras of a coalgebra $\mathcal{A}$.

## 4.3   Homomorphic Images and Factorizations

In order to consider homomorphic images of coalgebras, we have to verify that the coalgebra structure on the image is uniquely determined.

**Lemma 4.3.1.** *Let $\mathcal{A} = (A; \alpha_A)$ and $\mathcal{B} = (B; \alpha_B)$ be $F$-coalgebras. If $\varphi : A \to B$ is a surjective homomorphism, then $\alpha_B$ is uniquely determined by $\varphi$ and $\alpha_A$, with $\alpha_B = \{(\varphi(a), F(\varphi)(\alpha_A(a))) \mid a \in A\}$.*

**Proof**: Assume that $\alpha_{B'} : B \to F(B)$ is another mapping with $\alpha_B \circ \varphi = F(\varphi) \circ \alpha_A$ and $\alpha_{B'} \circ \varphi = F(\varphi) \circ \alpha_A$. Then also $\alpha_B \circ \varphi = \alpha_{B'} \circ \varphi$. Since $\varphi$ is surjective, the map $F(\varphi)$ is also surjective and therefore an epimorphism. From this we obtain $\alpha_B = \alpha_{B'}$. The given mapping $\{(\varphi(a), F(\varphi)(\alpha_A(a))) \mid a \in A\}$ satisfies $\alpha_B(\varphi(a)) = F(\alpha)(\alpha_A(a))$ for all $a \in A$, and thus $\alpha_B$ has the form claimed. ∎

This result justifies the following definition.

**Definition 4.3.2.** If $\varphi : \mathcal{A} \to \mathcal{B}$ is a surjective coalgebra homomorphism, then the coalgebra $\mathcal{B}$ is called a homomorphic image of the coalgebra $\mathcal{A}$.

**Theorem 4.3.3. (The Factorization Theorem)** *Let $\mathcal{A}$ and $\mathcal{B}$ be F-coalgebras and let $\varphi : \mathcal{A} \to \mathcal{B}$ be a homomorphism. Let $\varphi = g \circ f$ be a factorization of $\varphi$ into the composition of a surjective mapping $f : A \to Q$ and an injective mapping $g : Q \to B$. Then there is a uniquely determined F-coalgebra structure on $Q$ for which $f$ and $g$ are homomorphisms.*

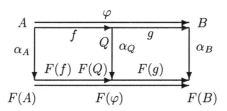

Fig. 4.6   Factorization Theorem

**Proof:** Since $\varphi = g \circ f$ is a homomorphism and $F$ is a functor, we have $\alpha_B \circ (g \circ f) = F(g \circ f) \circ \alpha_A = F(g) \circ F(f) \circ \alpha_A$. The injectivity of $g$ gives injectivity of $F(g)$ as well, and therefore the four mappings $f$, $\alpha_B \circ g$, $F(f) \circ \alpha_A$ and $F(g)$ form an $E - M$-square (see Lemma 3.7.3); note that $(\alpha_B \circ g) \circ f = F(g) \circ (F(f) \circ \alpha_A)$, where $f$ is surjective and therefore an epimorphism and $F(g)$ is injective and so a monomorphism. Then there exists a uniquely determined diagonal mapping $\alpha_Q : Q \to F(Q)$ for which the equations $\alpha_Q \circ f = F(f) \circ \alpha_A$ and $F(g) \circ \alpha_Q = \alpha_B \circ g$ are satisfied. The first equation means that $f$ is a homomorphism, while the last equation shows that $g : Q \to B$ is a homomorphism. ∎

If $f : A \to B$ is a mapping, then the set $f[A] := \{f(a) \mid a \in A\}$ is called the *image* of $A$ under $f$, and the mapping $f' : A \to f[A]$ is called the *restriction* of $f$ onto the image of $f$. Then

$$A \xrightarrow{f'} f[A] \; stackrel\subseteq_{f[A]}^{B} \to B$$

and $f$ can be written as a composition $f = \subseteq_{f[A]}^{B} \circ f'$, with $\subseteq_{f[A]}^{B}$ injective and $f'$ surjective. The proofs of the next two results are left as an exercise to the reader.

**Corollary 4.3.4.** *If $\varphi : \mathcal{A} \to \mathcal{B}$ is a homomorphism, the image $\varphi[A]$ is a homomorphic image of $\mathcal{A}$ and a subcoalgebra of $\mathcal{B}$.*

**Proposition 4.3.5.** *Let $\mathcal{A}$ and $\mathcal{B}$ be two F-coalgebras and let $\varphi : \mathcal{A} \to \mathcal{B}$ be a homomorphism.*

(i)  *If $S \leq A$ is a subcoalgebra of $A$, then $f(S)$ is a subcoalgebra of $B$.*
(ii) *If $T$ is a subcoalgebra of $B$, then $f^{-1}(T)$ is a subcoalgebra of $A$.*

## 4.4   Congruences and Factor Coalgebras

We saw in Chapter 1 that a congruence on an algebra is an equivalence relation on the base set of the algebra which is invariant under the operations of the algebra. Equivalently, congruences on algebras can be shown to coincide exactly with kernels of homomorphisms. In this section we show analogous results for congruences of $F$-coalgebras. We begin by defining a congruence as a kernel of a coalgebra homomorphism, and show that this definition is equivalent to an equivalence relation which is invariant under the coalgebra structure.

**Definition 4.4.1.** Let $A$ be an $F$-coalgebra. A *congruence* on $A$ is the kernel of a homomorphism $\varphi$ from $A$ into some $F$-coalgebra $B$.

That means that if $\theta$ is a congruence on an $F$-coalgebra $A$, then there is an $F$-coalgebra $B$ and a homomorphism $\varphi : A \to B$ such that $\theta = Ker\varphi$. If we assume that $\varphi$ is a surjective mapping (by restricting to the image if necessary) then together with the natural mapping $nat_\theta : A \to A/\theta$ taking $A$ to the quotient or factor set $A/\theta$ we have a bijection $f : A/\theta \to B$ for which $\varphi = f \circ nat_\theta$. This holds for arbitrary sets and mappings. Since $nat_\theta$ is surjective and $f$ is injective, we may apply the Factorization Theorem to obtain a uniquely determined coalgebra structure $\alpha_\theta$ on the set $A/\theta$. The coalgebra $(A/\theta; \alpha_\theta)$ is called a *factor coalgebra* of $A$. Since $f$ is bijective it is an isomorphism, and we have proved the following result.

**Theorem 4.4.2.** *If $\varphi : A \to B$ is a surjective homomorphism with kernel $\theta$, then the factor algebra $A/\theta$ is isomorphic to $B$.*

In fact the natural mapping $nat_\theta : A \to A/\theta$ has kernel equal to $\theta$, since

$$(a, b) \in \theta \Leftrightarrow [a]_\theta = [b]_\theta \Leftrightarrow nat_\theta(a) = nat_\theta(b) \Leftrightarrow (a, b) \in Kernat_\theta.$$

As a consequence we can show that every congruence is in fact an equivalence relation which is invariant under the coalgebraic structure.

**Corollary 4.4.3.** *Let $A$ be an $F$-coalgebra. An equivalence relation $\theta$ on $A$ is a congruence relation on $A$ iff for any two elements $a, b \in A$ we have*

$$(a, b) \in \theta \quad \Rightarrow \quad F(nat_\theta)(\alpha_A(a)) = F(nat_\theta)(\alpha_A(b)). \qquad (*)$$

**Proof:** First let $\theta$ be a congruence on $\mathcal{A}$. For any $(a, b) \in \theta$, we have $nat_\theta(a) = nat_\theta(b)$ and therefore also $\alpha_\theta(nat_\theta(a)) = \alpha_\theta(nat_\theta(b))$. By the homomorphism condition we have $\alpha_\theta \circ nat_\theta = F(nat_\theta) \circ \alpha_A$ and hence $F(nat_\theta)(\alpha_A(a)) = F(nat_\theta)(\alpha_A(b))$. Conversely, if the implication $(*)$ is satisfied, then $\theta \subseteq Ker(F(nat_\theta) \circ \alpha_A)$. Since $\theta = Kernat_\theta$ the Diagram Lemma ensures the existence of a mapping $\alpha_\theta : A/\theta \to F(A/\theta)$ such that $\alpha_\theta \circ nat_\theta = F(nat_\theta) \circ \alpha_\theta$. Therefore $nat_\theta$ is a homomorphism of the coalgebra $\mathcal{A}$ with $\theta$ as its kernel, making $\theta$ a congruence on $\mathcal{A}$. ∎

## 4.5 Colimits in Set$_F$

In this section we examine the various colimit constructions, the sum, co-equalizers and pushouts, in the category **Set**$_F$. For most but not all of these we shall see that the constructions from Chapter 3 for the category **Set** carry over to our new category.

We begin with the sum or coproduct of a family $(\mathcal{A}_i)_{i \in I}$ of $F$-coalgebras. We know from Section 3.4 that we can form the sum $\sum_{i \in I} A_i$ (in the category **Set**), and we now want to see if we can produce a structural mapping $\alpha : \sum_{i \in I} A_i \to F(\sum_{i \in I} A_i)$ to make this sum into an $F$-coalgebra. We have mappings $e_i : A_i \to \sum_{i \in I} A_i$ and $\alpha_i : A_i \to F(A_i)$. By definition of the sum the mappings $F(e_i) \circ \alpha_i$ then uniquely determine a morphism $\alpha : \sum_{i \in I} A_i \to F(\sum_{i \in I} A_i)$, and this map $\alpha$ is the uniquely determined structural mapping of $\sum_{i \in I} A_i$. From the definition of the sum in **Set**, we see also that $\alpha \circ e_i = F(e_i) \circ \alpha_i$; this means that the $e_i$'s are injective and are embeddings, and so each $\mathcal{A}_i$ is a subcoalgebra of $\sum_{i \in I} \mathcal{A}_i$. By definition the mapping $\alpha$ is given by $\alpha(i, a) = (F(e_i) \circ \alpha_i)(a)$, since $(\alpha \circ e_i)(a) = \alpha(i, a)$. These observations basically tell us that there is a way to make the disjoint sum of a family of $F$-coalgebras into an $F$-coalgebra.

**Theorem 4.5.1.** *Let $(\mathcal{A}_i)_{i \in I}$ be a family of $F$-coalgebras. The disjoint sum of the universes of the coalgebras $(\mathcal{A}_i)_{i \in I}$ in the category **Set** is the universe of the sum of the $(\mathcal{A}_i)_{i \in I}$ in the category **Set**$_F$.*

**Proof:** Let $\mathcal{Q} = (Q; \gamma)$ be an $F$-coalgebra and $(\varphi_i : A_i \to Q)_{i \in i}$ be morphisms in the category **Set**$_F$. The definition of the sum $\sum_{i \in I} A_i$ in the cat-

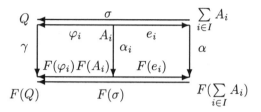

Fig. 4.7   Sums in $\mathbf{Set}_F$

egory $\mathbf{Set}$ guarantees the existence of a uniquely determined mapping $\sigma$ satisfying $\varphi_i = \sigma \circ e_i$ for all $i \in I$, and we need only show that this mapping $\sigma$ is a homomorphism. For every $i \in I$ we have

$$\sigma \circ e_i = \varphi_i \;\Rightarrow\; \gamma \circ \sigma \circ e_i = \gamma \circ \varphi_i = F(\varphi_i) \circ \alpha_i.$$

Thus the $\varphi_i$'s are homomorphisms, and $\gamma \circ \varphi_i = F(\varphi_i) \circ \alpha_i = F(\sigma \circ e_i) \circ \alpha_i = F(\sigma) \circ F(e_i) \circ \alpha_i = F(\sigma) \circ \alpha \circ e_i$, since $\sum_{i \in I} A_i$ is the disjoint union of the coalgebras $A_i$. Therefore $F(e_i) \circ \alpha_i = \alpha \circ e_i$, and $\gamma \circ \sigma \circ e_i = F(\sigma) \circ \alpha \circ e_i$. Since the $e_i$'s are epimorphisms this gives $\gamma \circ \sigma = F(\sigma) \circ \alpha$, making $\sigma$ a homomorphism. ∎

**Corollary 4.5.2.** *Let $A$ be an $F$-coalgebra. If $(\mathcal{U}_i)_{i \in I}$ is a family of sub-coalgebras of $A$, then $S := \bigcup_{i \in I} \mathcal{U}_i$ is a subcoalgebra of $A$.*

**Proof:** The mappings $e_i : \mathcal{U}_i \to \sum_{i \in I} \mathcal{U}_i$ for $i \in I$ and $\subseteq^A_{\mathcal{U}_i} : \mathcal{U}_i \to A$ determine a unique homomorphism $\varphi : \sum_{i \in I} \mathcal{U}_i \to A$ for which $\varphi \circ e_i = \subseteq^A_{\mathcal{U}_i}$ for all $i \in I$. Since $\varphi$ is a homomorphism, the image $\varphi(\sum_{i \in I} \mathcal{U}_i)$ is a homomorphic image of $A$ and a subcoalgebra of $A$ by Corollary 4.3.4. This image is

$$\varphi\left(\sum_{i \in I} U_i\right) = \{\varphi(i,u) \mid i \in I, u \in U_i\} = \bigcup_{i \in I}(\varphi \circ e_i)(U_i) = \bigcup_{i \in I} U_i.$$

∎

Let $A$ be an $F$-coalgebra, and let $S \subseteq A$ be a subset. We may consider the family of all subcoalgebras of $A$ contained in $S$. This family is non-empty, since $\emptyset$ is a subcoalgebra of $A$, and Corollary 4.5.2 ensures that there is a greatest such subcoalgebra contained in $S$. This coalgebra is called the *subcoalgebra cogenerated by $S$*, written as $[S]$.

**Corollary 4.5.3.** *The set of all subcoalgebras of a coalgebra $\mathcal{A}$ forms a lattice $Sub(\mathcal{A})$, with the empty set $\emptyset$ as its least element and $\mathcal{A}$ as its greatest element.*

To prove that $Sub(\mathcal{A})$ is a lattice, we need to show that the infimum and the supremum of any two subcoalgebras of $\mathcal{A}$ exist. Closure under arbitrary suprema is clear by the previous results. For the infimum we use the following fact.

**Lemma 4.5.4.** ([49]) *The intersection of finitely many subcoalgebras of a coalgebra $\mathcal{A}$ is a subcoalgebra of $\mathcal{A}$.*

**Proof:** It is clear enough to consider two subcoalgebras $U$ and $V$ of $\mathcal{A}$. When $U \cap V = \emptyset$ we obtain the empty subcoalgebra $(\emptyset; \emptyset)$. Now suppose that $U \cap V \neq \emptyset$, with $w \in U \cap V$. We define mappings $p_w : U \to U \cap V$ and $q_w : A \to V$ by

$$p_w(u) := \begin{cases} u & \text{if } u \in U \cap V \\ w & \text{otherwise} \end{cases}, \qquad q_w(a) := \begin{cases} a & \text{if } a \in V \\ w & \text{otherwise} \end{cases}.$$

Next we verify the two equations

$$\subseteq^V_{U \cap V} \circ p_w = q_w \circ \subseteq^A_U \qquad \text{and} \qquad q_w \circ \subseteq^A_V = id_V.$$

For the first equation we have

$$\subseteq^V_{U \cap V} (p_w(a)) = \begin{cases} \subseteq^V_{U \cap V}(a) & \text{if } a \in U \cap V \\ \subseteq^V_{U \cap V}(w) & \text{otherwise} \end{cases}$$

$$= \begin{cases} a & \text{if } a \in U \cap V \\ w & \text{otherwise} \end{cases}$$

and

$$(q_w \circ \subseteq^A_U)(a) = \begin{cases} \subseteq^A_U(a) & \text{if } \subseteq^A_U(a) \in V \\ w & \text{otherwise} \end{cases}$$

$$= \begin{cases} a & \text{if } a \in U \cap V \\ w & \text{otherwise.} \end{cases}$$

For the second equation we have $q_w(\subseteq^A_V(a)) = \subseteq^A_V(a) = a$ for all $a \in A$, so that $q_w \circ \subseteq^A_V = id_V$.

Finally we show that $\gamma := F(p_w) \circ \alpha_U \circ \subseteq^U_{U \cap V}$ defines a coalgebra structure on $U \cap V$ which makes $\mathcal{U} \cap \mathcal{V}$ a subcoalgebra of $\mathcal{A}$. We have

$$F(\subseteq^A_{U\cap V}) \circ \gamma$$
$$= \quad F(\subseteq^A_{U\cap V}) \circ F(p_w) \circ \alpha_U \circ \subseteq^U_{U\cap V}$$
$$= \quad F(\subseteq^A_V) \circ F(\subseteq^U_{U\cap V}) \circ F(p_w) \circ \alpha_U \circ \subseteq^U_{U\cap V}$$
$$\text{using } \subseteq^A_{U\cap V} = \subseteq^A_V \circ \subseteq^V_{U\cap V}$$
$$= \quad F(\subseteq^A_V) \circ F(q_w) \circ F(\subseteq^A_U) \circ \alpha_U \circ \subseteq^U_{U\cap V}$$
$$\text{using one of the previous equations}$$
$$= \quad F(\subseteq^A_V) \circ F(q_w) \circ \alpha_A \circ \subseteq^A_U \circ \subseteq^U_{U\cap V}$$
$$= \quad F(\subseteq^A_V) \circ F(q_w) \circ \alpha_A \circ \subseteq^A_V \circ \subseteq^V_{U\cap V}$$
$$= \quad F(\subseteq^A_V) \circ F(q_w) \circ F(\subseteq^A_V) \circ \alpha_V \circ \subseteq^V_{U\cap V}$$
$$= \quad F(\subseteq^A_V) \circ \alpha_V \circ \subseteq^V_{U\cap V} \quad \text{using one of the previous equations and}$$
$$F(q_w \circ \subseteq^A_V) = F(q_w) \circ F(\subseteq^A_V) = F(id_V) = id_{F(V)}$$
$$= \quad \alpha_A \circ \subseteq^A_{U\cap V}, \quad \text{using } \subseteq^A_V \circ \subseteq^V_{U\cap V} = \subseteq^A_{U\cap V}.$$

∎

It can be shown that families of subcoalgebras are not in general closed under arbitrary intersection. This means that the subcoalgebra lattice $\mathcal{S}ub(\mathcal{A})$ is not in general a complete lattice.

Next we turn to coequalizers in the category $\mathbf{Set}_F$, showing how they too can be extended from coequalizers in $\mathbf{Set}$.

**Lemma 4.5.5.** Let $(\varphi_i : \mathcal{A} \to \mathcal{B})_{i \in I}$ be a family of coalgebra homomorphisms. Let $nat_\theta : B \to B/\theta$ be the coequalizer of the mappings $\varphi_i$ in $\mathbf{Set}$. Then there is a uniquely determined structural mapping $\alpha_\theta$ on $B/\theta$ such that $nat_\theta : \mathcal{B} \to \mathcal{B}/\theta = (B/\theta; \alpha_\theta)$ is the coequalizer of the homomorphisms $\varphi_i$ in $\mathbf{Set}_F$.

**Proof:** For arbitrary $i, j \in I$,
$$F(nat_\theta) \circ \alpha_B \circ \varphi_i$$
$$= \quad F(nat_\theta) \circ F(\varphi_i) \circ \alpha_A$$
$$= \quad F(nat_\theta \circ \varphi_i) \circ \alpha_A$$
$$= \quad F(nat_\theta \circ \varphi_j) \circ \alpha_A$$
$$= \quad F(nat_\theta) \circ \alpha_B \circ \varphi_j.$$
Since $nat_\theta$ is a coequalizer in $\mathbf{Set}$ we get a structural mapping $\alpha_\theta$ on $B/\theta$, and $nat_\theta$ is a homomorphism. Now if $\mathcal{Q} = (Q; \alpha_Q)$ is any coalgebra and $\psi : \mathcal{B} \to \mathcal{Q}$ is any homomorphism with $\psi \circ \varphi_i = \psi \circ \varphi_j$ for all $i, j \in I$, then $Kernat_\theta = \theta \subseteq Ker\psi$. By the Diagram Lemma for coalgebras there is exactly one homomorphism $\chi : \mathcal{B}/\theta \to \mathcal{Q}$ with $\psi = \chi \circ nat_\theta$. Applying the Diagram Lemma for coalgebras again, we see that if $\varphi : \mathcal{A} \to \mathcal{B}$ and

$\psi : \mathcal{A} \to \mathcal{C}$ are homomorphisms and $\varphi$ is surjective, then there is exactly one homomorphism $\chi : \mathcal{B} \to \mathcal{C}$ with $\chi \circ \varphi = \psi$ if $Ker\varphi \subseteq Ker\psi$. This gives a uniquely determined coalgebra homomorphism $\chi : \mathcal{B}/\theta \to \mathcal{Q}$ with $\psi = \chi \circ nat_\theta$. It follows that $nat_\theta$ is the coequalizer of the $\varphi_i$'s.   ∎

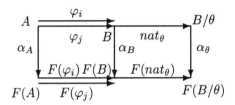

Fig. 4.8   Coequalizers in $\mathbf{Set}_F$

An analogous result holds for pushouts in $\mathbf{Set}_F$ as well.

**Proposition 4.5.6.** *Let $(\varphi_i : \mathcal{A} \to \mathcal{B}_i)_{i \in I}$ be a family of homomorphisms. Let $\psi_i : B_i \to P$ be the pushout of the mappings $\varphi_i$ in $\mathbf{Set}$. Then there is a uniquely determined coalgebra structure $\alpha_P$ on $P$ such that the $\psi_i$'s are homomorphisms to $\mathcal{P} = (P; \alpha_P)$, and $\mathcal{P}$ together with the $\psi_i$'s is the pushout of the $\varphi_i$'s in $\mathbf{Set}_F$.*

The proof of this Proposition is similar to the proof for coequalizers, and is left to the reader. We mention that these results can be generalized to arbitrary colimits as well: colimits in $\mathbf{Set}_F$ are formed as colimits in $\mathbf{Set}$ with the canonical mappings as homomorphisms. For limits, however, the situation is more complicated; whether for instance pullbacks in $\mathbf{Set}$ can be used to construct pullbacks in $\mathbf{Set}_F$ depends on the choice of the functor $F$. For more information see [75].

## 4.6   Bisimulations

The concept of a *bisimulation* (or *bisimilarity*) plays a central role in the theory of coalgebras, with [75] being an important reference. A bisimulation between two coalgebras is a binary relation between their base sets. As we saw in Chapter 2, bisimulations can be used to formalize the property of states of automata being indistinguishable.

**Definition 4.6.1.** Let $\mathcal{A}$ and $\mathcal{B}$ be coalgebras and let $R \subseteq A \times B$ be a binary relation from $A$ to $B$. The relation $R$ is called a *bisimulation* from

$\mathcal{A}$ to $\mathcal{B}$, or between $\mathcal{A}$ and $\mathcal{B}$, if a coalgebra structure $\varrho$ can be defined on the set $R$ in such a way that the projection mappings $\pi_A : R \to A$ and $\pi_B : R \to B$ are homomorphisms; that is, so that the equations

$$\alpha_A \circ \pi_A = F(\pi_A) \circ \varrho \quad \text{and} \quad \alpha_B \circ \pi_B = F(\pi_B) \circ \varrho$$

are satisfied. Figure 4.9 portrays this.

Note that technically what we have defined here is a bisimulation from $\mathcal{A}$ to $\mathcal{B}$. However, it is straightforward to show that when $R$ is a bisimulation from $\mathcal{A}$ to $\mathcal{B}$, the relation $R^{-1}$ is also a bisimulation from $\mathcal{B}$ to $\mathcal{A}$. This justifies the more usual wording of a bisimulation between $\mathcal{A}$ and $\mathcal{B}$. If $\mathcal{A} = \mathcal{B}$, we say that $R$ is a bisimulation on $\mathcal{A}$. In general $R$ is not an equivalence relation, but if $R$ happens to be an equivalence relation, we call it a *bisimulation equivalence* on $\mathcal{A}$.

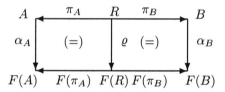

Fig. 4.9   Bisimulations in $\mathbf{Set}_F$

**Example 4.6.2.** Let $f : \mathcal{A} \to \mathcal{B}$ be a coalgebra homomorphism, so that $\alpha_B \circ f = F(f) \circ \alpha_A$. Let $G(f) := \{(a, f(a)) \mid a \in A\}$ be the graph of $f$. Then $G(f) \subseteq A \times B$ is a binary relation from $A$ to $B$, and we can define a coalgebra structure $\varrho$ on it by

$$\varrho((a, f(a))) = (F(\pi_A^{-1}) \circ \alpha_A)(a) = (F(\pi_A^{-1}) \circ \alpha_A \circ \pi_A)((a, f(a))).$$

It is easy to see that $f = \pi_B \circ \pi_A^{-1}$, since $(\pi_B \circ \pi_A^{-1})(a) = \pi_B((a, f(a))) = f(a)$ for all $a \in A$, and similarly that $f \circ \pi_A = \pi_B$. Using these two facts we can check that the projections $\pi_A$ and $\pi_B$ are homomorphisms:

$$
\begin{aligned}
& F(\pi_A) \circ \varrho \\
= \; & F(\pi_A) \circ F(\pi_A^{-1}) \circ \alpha_A \circ \pi_A \\
= \; & F(\pi_A \circ \pi_A^{-1}) \circ \alpha_A \circ \pi_A \\
= \; & \alpha_A \circ \pi_A
\end{aligned}
$$

and

$$F(\pi_B) \circ \varrho$$
$$= F(\pi_B) \circ F(\pi_A^{-1}) \circ \alpha_A \circ \pi_A$$
$$= F(\pi_B \circ \pi_A^{-1}) \circ \alpha_A \circ \pi_A$$
$$= F(f) \circ \alpha_A \circ \pi_A$$
$$= \alpha_B \circ f \circ \pi_A$$
$$= \alpha_B \circ \pi_B.$$

This example shows that the graph of any coalgebra homomorphism is a bisimulation. In fact the converse is also true.

**Theorem 4.6.3.** *Let $A$ and $B$ be F-coalgebras. A mapping $f : A \to B$ is a coalgebra homomorphism from $A$ to $B$ if and only if its graph $G(f):= \{(a, f(a)) \mid a \in A\}$ is a bisimulation from $A$ to $B$.*

**Proof:** One direction has already been proved in the previous example. For the other direction, we assume that $G(f)$ is a bisimulation from $A$ to $B$. In particular, the projections $\pi_A$ and $\pi_B$ are homomorphisms. The mapping $\pi_A : G(f) \to A$ is bijective, with the inverse mapping $\pi_A^{-1}$. In fact as a bijective homomorphism $\pi_A$ is an isomorphism, and the inverse mapping $\pi_A^{-1}$ is also an isomorphism. Also as we saw in the previous example $f = \pi_B \circ \pi_A^{-1}$. Now $f$ is the composition of two coalgebra homomorphisms, and hence is also a coalgebra homomorphism. ∎

Another characterization of bisimulations is given in the following theorem.

**Theorem 4.6.4.** *Let $A$ and $B$ be coalgebras. Let $P$ be a coalgebra and let $\varphi_A : P \to A$ and $\varphi_B : P \to B$ be homomorphisms. Then the set*

$$(\varphi_A, \varphi_B)(P) := \{(\varphi_A(p), \varphi_B(p)) \mid p \in P\}$$

*is a bisimulation from $A$ to $B$. Conversely, every bisimulation from $A$ to $B$ can be expressed in this form.*

**Proof:** First suppose that $R$ is a bisimulation from $A$ to $B$, meaning that there is a coalgebra structure $\mathcal{R} = (R; \varrho)$ for which the projections $\pi_A^R : \mathcal{R} \to A$ and $\pi_B^R : \mathcal{R} \to B$ are coalgebra homomorphisms. Then it is obvious that $(\pi_A^R, \pi_B^R)(R) = \{(\pi_A^R(p), \pi_B^R(p)) \mid p \in R\} = R$, and $R$ has the required form. Conversely, suppose that we are given coalgebra homomorphisms $\varphi_A : P \to A$ and $\varphi_B : P \to B$. We consider the set $R = (\varphi_A, \varphi_B)(P) := \{(\varphi_A(p), \varphi_B(p)) \mid p \in P\}$ and the surjective mapping $(\varphi_A, \varphi_B) : P \to R$. We can form a right-inverse $\mu$ of this mapping, by selecting (using the

Axiom of Choice) exactly one element from each class of elements from $P$ which have the same image in $R$. The set $R$ is a subset of $A \times B$, and we define on it a structural mapping

$$\varrho := F((\varphi_A, \varphi_B)) \circ \alpha_P \circ \mu.$$

Now it remains only to show that the projections $\pi_A$ and $\pi_B$ are homomorphisms. We have

$$
\begin{aligned}
& F(\pi_A) \circ \varrho \\
= \; & F(\pi_A) \circ F((\varphi_A, \varphi_B)) \circ \alpha_P \circ \mu \\
= \; & F(\pi_A \circ (\varphi_A, \varphi_B)) \circ \alpha_B \circ \mu \\
= \; & F(\varphi_A) \circ \alpha_P \circ \mu \\
= \; & \alpha_A \circ \varphi_A \circ \mu \qquad \text{since } \varphi_A \text{ is a coalgebra homomorphism} \\
= \; & \alpha_A \circ \pi_A \circ (\varphi_A, \varphi_B) \circ \mu \\
= \; & \alpha_A \circ \pi_A.
\end{aligned}
$$

It may be shown similarly that $F(\pi_B) \circ \varrho = \alpha_B \circ \pi_B$, establishing that $R$ is indeed a bisimulation. ∎

In universal algebra the compatible relations from $\mathcal{A}$ to $\mathcal{B}$ are precisely the subalgebras of the direct product $\mathcal{A} \times \mathcal{B}$. In the coalgebra setting, there is no structure on the cartesian product. Nevertheless bisimulation relations do behave somewhat like two-dimensional subcoalgebras, but bisimulation relations are closed under set-theoretical unions.

**Theorem 4.6.5.** *Let $(R_i)_{i \in I}$ be a family of bisimulations from $\mathcal{A}$ to $\mathcal{B}$. Then the union $\bigcup_{i \in I} R_i$ is also a bisimulation from $\mathcal{A}$ to $\mathcal{B}$.*

**Proof:** Let $\mathcal{R} := \sum_{i \in I} \mathcal{R}_i$ be the sum of the coalgebras $\mathcal{R}_i$. For each $i \in I$ the mappings $\pi_A^i : R_i \to A$ and $\pi_B^i : R_i \to B$ are homomorphisms. By the definition of the sum $\sum_{i \in I} \mathcal{R}_i$, there are uniquely determined homomorphisms $\pi_A : \mathcal{R} \to \mathcal{A}$ and $\pi_B : \mathcal{R} \to \mathcal{B}$ such that $\pi_A \circ e_i = \pi_A^i$ and $\pi_B \circ e_i = \pi_B^i$ for all $i \in I$. Then we have

$$
\begin{aligned}
& (\pi_A, \pi_B)(R) \\
= \; & \{(\pi_A((i, x)), \pi_B((i, x))) \mid i \in I, x \in R_i\} \\
= \; & \bigcup_{i \in I} \{((\pi_A \circ e_i)(x), (\pi_B \circ e_i)(x)) \mid x \in R_i\} \\
= \; & \bigcup_{i \in I} R_i.
\end{aligned}
$$

But by Theorem 4.6.3 the set $(\pi_A, \pi_B)(R)$ is a bisimulation from $\mathcal{A}$ to $\mathcal{B}$; therefore $\bigcup_{i \in I} R_i$ is also a bisimulation. ∎

**Corollary 4.6.6.** *Let $\mathcal{A}$ and $\mathcal{B}$ be coalgebras and let $R \subseteq A \times B$ be a binary relation. Then there is a greatest bisimulation contained in $R$.*

The greatest bisimulation from $\mathcal{A}$ to $\mathcal{B}$ to be contained in a relation $R$ is called the *bisimulation kernel* of $R$, and denoted by $[R]$. The special case of $R = A \times B$ gives the following important corollary.

**Corollary 4.6.7.** *There is a greatest bisimulation between any two coalgebras $\mathcal{A}$ and $\mathcal{B}$.*

**Definition 4.6.8.** Let $\mathcal{A}$ and $\mathcal{B}$ be coalgebras. The greatest bisimulation from $\mathcal{A}$ to $\mathcal{B}$ is denoted by $\sim_{A,B}$. Two elements $a \in A$ and $b \in B$ are called bisimilar if $(a, b) \in \sim_{A,B}$.

Theorem 4.6.4 gives the following result.

**Theorem 4.6.9.** *Let $\mathcal{A}$ and $\mathcal{B}$ be coalgebras. Two elements $a \in A$ and $b \in B$ are bisimilar iff there is a coalgebra $\mathcal{P}$ and there are homomorphisms $\varphi : \mathcal{P} \to \mathcal{A}$ and $\psi : \mathcal{P} \to \mathcal{B}$ and an element $p \in P$ such that $\varphi(p) = a$ and $\psi(p) = b$; that is, if $a$ and $b$ are the homomorphic images under $\varphi$ and $\psi$, respectively, of the same element $p \in P$.*

As a special case of a bisimulation between $\mathcal{A}$ and $\mathcal{B}$ we introduced for $\mathcal{A} = \mathcal{B}$ the concept of a bisimulation equivalence on $\mathcal{A}$. Again, there exists a greatest bisimulation equivalence in this case, usually denoted by $\sim_A$.

In Section 2.10 we defined a pair $(a, b) \in A \times B$ for two $F$-coalgebras $\mathcal{A} = (A; \alpha_A), \mathcal{B} = (B; \alpha_B)$ to be behaviourally equivalent ($a \sim_{beh} b$) if there is a $F$-coalgebra $(C; \alpha_C)$ and a pair of coalgebra homomorphisms $f : (A; \alpha_A) \to (C; \alpha_C)$ and $g : (B; \alpha_B) \to (C; \alpha_C)$ such that $f(a) = g(b)$. For Kripke structures the two relations of bisimulation and behavioural equivalence are logically equivalent (see Proposition 2.10.10 and Proposition 2.10.11). One direction of this equivalence is also clear in the general case: bisimulation is a sufficient condition for behavioural equivalence (see [85]). For the other direction we need the extra condition that the functor $F$ preserves weak pullbacks. Weak pullbacks are defined in much the same way as pullbacks, but without the uniqueness requirement.

**Theorem 4.6.10.** ([85]) *When the functor $F$ preserves weak pullbacks, the notions of bisimulation and behavioural equivalence for $F$-coalgebras coincide, and the collection of all bisimulations is closed under taking relational composition.*

Now as an example we consider bisimulations of Kripke structures, which were introduced in Chapter 2.

**Example 4.6.11.** Let $\Phi$ be a set. As we have seen in Chapter 2 a Kripke structure over $\Phi$ consists of a set $S$ of states, a binary relation $R \subseteq S \times S$ and a mapping $v : \Phi \to \mathcal{P}(S)$ (where as usual $\mathcal{P}$ denotes the formation of the power set). We think of the set $\Phi$ as consisting of some elementary or atomic propositions, and $v(s)$ as the set of such propositions which are true in state $s$. For states $s$ and $s'$ we write $sRs'$ or $s \overset{R}{\to} s'$ to mean that the system has a transition from state $s$ to state $s'$.

We have already described in Chapter 2 what it means for two states in such a system to be indistinguishable. Basically two states can be considered to be equivalent if the same set of atomic propositions is valid in them. For a "bisimulation" relation this means that we want $x \sim y$ to imply that $v(x) = v(y)$. Also, for each transition starting from the state $x$ there must exist a transition starting from $y$, otherwise $x$ and $y$ are distinguishable. This means that

$$\frac{x \sim y \wedge x \to x'}{\exists y'(y \to y' \wedge x' \sim y')} \quad \text{and} \quad \frac{x \sim y \wedge y \to y'}{\exists x'(x \to x' \wedge x' \sim y');}$$

i.e.

$$
\begin{array}{ccc}
x & \sim & y \\
\downarrow & & \downarrow \\
x' & \sim & \exists y'
\end{array}
\quad \text{and} \quad
\begin{array}{ccc}
x & \sim & y \\
\downarrow & & \downarrow \\
\exists x' & \sim & y'
\end{array}
$$

We have seen that a Kripke structure can be viewed as a coalgebra of the type $\mathcal{A} = (A; \alpha_A)$. The functorial mapping used here is constructed from both the power set functor which maps each set $\Phi$ to its power set $\mathcal{P}(\Phi)$ and the power functor $(-)^S$, which maps each set $X$ to the set $X^S$ of all mappings from $S$ to $X$ and each function $f : X \to Y$ to the function $f^S$ with $f^S(u) = f \circ u$. We combine these into the functor $F = \mathcal{P}(\Phi) \times \mathcal{P}((-)^S)$, so we can now define a structural mapping by $\alpha_A(a) := (\overset{A}{\to}(a), v_A(a))$. Then $(A; \alpha_A) = (A; \overset{A}{\to}, v_A)$ is a coalgebra. Let $\mathcal{B} = (B; \overset{B}{\to}, v_B)$ be another Kripke structure (coalgebra) and let $R$ be a bisimulation from $\mathcal{A}$ to $\mathcal{B}$. Then for all $(a, b) \in R$ there is a set $\Gamma \subseteq \Phi$ of propositions and a set $M \subseteq R$ such that

(i)  $v_A(a) = \Gamma = v_B(b)$,
(ii)  $\pi_A[M] = \overset{A}{\to}(a)$,
(iii)  $\pi_B[M] = \overset{B}{\to}(b)$.

Then $aRb$ means by (i) that $v_A(a) = v_B(b)$. The inclusions $\overset{A}{\to}(a) \subseteq \pi_A[M]$ and $\pi_B[M] \subseteq \overset{B}{\to}(b)$ mean that

$$a \overset{A}{\to} a' \;\Rightarrow\; \exists b' \; (b \overset{B}{\to} b' \wedge a'Rb').$$

Similarly, from $\overset{B}{\to}(b) \subseteq \pi_B[M]$ and $\pi_A[M] \subseteq \overset{A}{\to}(a)$ we get

$$b \overset{B}{\to} b' \;\Rightarrow\; \exists a' \; (a \overset{B}{\to} a' \wedge a'Rb').$$

## 4.7 Epi's and Mono's in $\mathbf{Set}_F$

In this section we use our knowledge about epimorphisms and monomorphisms in the category $\mathbf{Set}$ and the connections between categories $\mathbf{Set}$ and $\mathbf{Set}_F$ to characterize epi's and mono's in the latter category.

For epi's, we use the following close connection between epimorphisms and pushouts: a morphism $\varphi : A \to B$ is an epimorphism in a given category $\mathbf{C}$ iff the diagram from Figure 4.10 is a pushout.

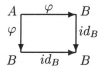

Fig. 4.10   Epimorphisms and Pushouts

In fact, in this case the two pushout conditions have the form:

(i) $id_B \circ \varphi = id_B \circ \varphi$,

(ii) if for all $Q$ and for all $q_i : B \to Q$ such that $q_i \circ \varphi = q_j \circ \varphi$, there is exactly one morphism $h : B \to Q$ such that $h \circ id_B = q_i$, then $q_i = q_j$ for all $i, j \in I$.

But we have seen that pushouts in $\mathbf{Set}$ are also the pushouts in $\mathbf{Set}_F$. Thus a mapping $f$ is an epimorphism in $\mathbf{Set}$ iff it is an epimorphism in $\mathbf{Set}_F$. Moreover, epimorphisms in $\mathbf{Set}$ are exactly the surjective mappings, giving us the following conclusion.

**Corollary 4.7.1.** *A coalgebra homomorphism is an epimorphism in $\mathbf{Set}_F$ iff it is surjective.*

For monomorphisms we have the following characterization involving the kernel of a mapping.

**Theorem 4.7.2.** *A coalgebra homomorphism $\varphi$ is a monomorphism in* $\mathbf{Set}_F$ *iff* $[Ker\varphi] = \Delta$; *that is, iff the kernel does not contain a nontrivial bisimulation.*

**Proof:** Let $\varphi : \mathcal{A} \to \mathcal{B}$ be a coalgebra homomorphism. We recall that $Ker\varphi = \{(a,b) \mid \varphi(a) = \varphi(b)\}$ is a binary relation on set $A$. By definition $[Ker\varphi]$ is the greatest bisimulation on the coalgebra $\mathcal{A}$ which is contained in $Ker\varphi$. Therefore the projections $\pi_1$ and $\pi_2$ from $[Ker\varphi]$ to $\mathcal{A}$ are homomorphisms. Suppose that $\varphi \circ \pi_1 = \varphi \circ \pi_2$. If $\varphi$ is a monomorphism, then $\pi_1 = \pi_2$ and so $[Ker\varphi] = \Delta$. Conversely let $[Ker\varphi] = \Delta$ and let $\psi_1$ and $\psi_2$ be homomorphisms from $\mathcal{Q}$ to $\mathcal{A}$ for which $\varphi \circ \psi_1 = \varphi \circ \psi_2$. Then for all $q \in Q$ we have $(\psi_1(q), \psi_2(q)) \in Ker\varphi$ and $(\psi_1, \psi_2)(Q) \subseteq Ker\varphi$. Since $(\psi_1, \psi_2)(Q)$ is a bisimulation by Theorem 4.6.4, and $[Ker\varphi]$ is the greatest bisimulation contained in $Ker\varphi$, we must have $(\psi_1, \psi_2)(Q) \subseteq [Ker\varphi]$. But the fact that $[Ker\varphi] = \Delta$ then forces $\psi_1 = \psi_2$.  ∎

## 4.8   Congruences

We have defined congruences on $F$-coalgebras as the kernels of homomorphisms. On any coalgebra $\mathcal{A}$ the diagonal relation $\Delta_A$ is a congruence relation, since it is the kernel of the identity homomorphism $id_A : \mathcal{A} \to \mathcal{A}$. But unlike the algebra case, the total relation $A \times A$ is not in general a congruence on $\mathcal{A}$.

**Theorem 4.8.1.** *Let $(\theta_i)_{i \in I}$ be a family of congruences on a coalgebra $\mathcal{A}$. Let $\underline{\theta} := (\bigcup_{i \in I} \theta_i)^*$ be the transitive closure of the union $\bigcup_{i \in I} \theta_i$. Then $\underline{\theta}$ is a congruence on $\mathcal{A}$ and is the supremum of the $\theta_i$'s.*

**Proof:** The collection of all equivalence relations on the base set $A$ forms a complete lattice. But the union of the family $(\theta_i)_{i \in I}$ of equivalence relations is in general not an equivalence relation. Instead, for the supremum in the equivalence relation lattice we must take the transitive closure of the union, the relation $\underline{\theta} = (\bigcup_{i \in I} \theta_i)^*$. We now verify that this transitive closure is also a congruence when the $\theta_i$'s are congruences on $\mathcal{A}$. We use the pushout $(\pi_i : A/\theta_i \to A/\underline{\theta})_{i \in I}$ of the family $(\pi_{\theta_i})_{i \in I}$ in $\mathbf{Set}$. By Proposition 4.5.6 this is also the pushout in $\mathbf{Set}_F$. Therefore $nat_{\underline{\theta}} = \pi_i \circ \pi_{\theta_i}$ is a homomorphism and its kernel $\theta$ is a congruence.  ∎

The infimum of a family of coalgebra congruences also exists, but unlike the case of algebra congruences it is not simply the set-theoretical intersection. Instead the infimum of the family $(\theta_i)_{i \in I}$ of congruences on $\mathcal{A}$ is defined as the transitive closure of the union of all congruences $\theta'$ for which $\theta' \subseteq \bigcap_{i \in I} \theta_i$. That is, $\bigwedge_{i \in I} \theta_i := (\bigcup \{\theta' \mid \theta' \subseteq \bigcap_{i \in I} \theta_i\})^*$.

**Corollary 4.8.2.** *The collection of all congruences on an F-coalgebra $\mathcal{A}$ forms a complete lattice, with $\nabla_A$ as its greatest element. In general, $\nabla_A \neq A \times A$.*

There is a connection between congruences and bisimulations on a coalgebra, as shown in the next Lemma.

**Lemma 4.8.3.** *If $R$ is a bisimulation on $\mathcal{A}$, then the smallest equivalence relation containing $R$ is a congruence on $\mathcal{A}$.*

**Proof:** When $R$ is a bisimulation on $\mathcal{A}$, there is a structural mapping $\alpha_R$ on $R$ such that $\mathcal{R} = (R; \alpha_R)$ is a coalgebra and $\pi_1, \pi_2 : \mathcal{R} \to \mathcal{A}$ are homomorphisms. The mappings $\pi_1$ and $\pi_2$ have the coequalizer $nat_\theta : A \to A/\theta$, where $\theta$ is the equivalence relation generated by the pairs $(\pi_1(r), \pi_2(r))$; that is, generated by $R$. The mapping $nat_\theta$ is then a homomorphism with kernel $\theta$, making $\theta$ a congruence. ∎

In particular, this means that any bisimulation equivalence is a congruence.

The greatest bisimulation $\sim_A$ on a coalgebra $\mathcal{A}$ is the bisimulation generated by $A \times A$. This relation is automatically reflexive and symmetric, but we need to take its transitive closure to get an equivalence relation.

**Corollary 4.8.4.** *Let $\mathcal{A}$ be any coalgebra, with $\sim_A$ the greatest bisimulation on $\mathcal{A}$. Then the transitive closure $(\sim_A)^*$ of $\sim_A$ is a congruence on $\mathcal{A}$.*

A natural question involves the converse direction: when is a congruence on $F$-coalgebra $\mathcal{A}$ a bisimulation equivalence? This depends on the functor $F$.

**Corollary 4.8.5.** *Let $F$ be a functor which preserves weak pullbacks. Then for any F-coalgebra $\mathcal{A}$, any congruence $\theta$ on $\mathcal{A}$ is a bisimulation equivalence.*

**Proof:** Congruences are defined as kernels of homomorphisms. Therefore for $\theta$ there is a homomorphism $\varphi : \mathcal{A} \to \mathcal{B}$ such that $\theta = Ker\varphi$. The kernel $Ker\varphi$ can be written as product of the inverse relation of $G(\varphi)$, the graph of $\varphi$, and $G(\varphi) : Ker\varphi = G(\varphi)^{-1} \circ G(\varphi)$. By Theorem 4.6.3 $G(\varphi)$ is a bisimulation. The inverse of a bisimulation is also a bisimulation (see Exercise 4.3). Since the functor $F$ preserves weak pullbacks, by Theorem 4.6.10 (see also Exercise 4.5), the relational product of two bisimulations is again a bisimulation. Therefore $Ker\varphi = G(\varphi)^{-1} \circ G(\varphi)$ is a bisimulation equivalence.

■

## 4.9   Covarieties

In this section we define covarieties of coalgebras, analogous to varieties of algebras. We recall from Section 1.4 that a variety of algebras is a class of algebras of the same type which is closed under the operations of formation of subalgebras, homomorphic images and products. In the coalgebra setting, we consider formation of isomorphic copies, homomorphic images, subcoalgebras and sums instead of products. Specifically, let $K$ be a class of $F$ coalgebras of the same type $F$. We define:
$I(K)$ to be the class of all isomorphic copies of coalgebras from $K$,
$H(K)$ to be the class of all homomorphic images of coalgebras from $K$,
$S(K)$ to be the class of all subcoalgebras of coalgebras from $K$,      and
$\Sigma(K)$ to be the class of all sums of coalgebras from $K$.

**Definition 4.9.1.** A *covariety* is a class of $F$-coalgebras which is closed under the operators $I$, $H$, $S$ and $\Sigma$.

It can be shown that the operators $HS$, $H\Sigma$ and $\Sigma S$ are closure operators and that

$$HS(K) \subseteq SH(K), \quad H\Sigma(K) \subseteq \Sigma H(K), \quad \text{and } \Sigma S(K) \subseteq S\Sigma(K).$$

Using these results, one can prove the following analogue for covarieties of Tarski's Theorem for varieties.

**Theorem 4.9.2.** *Let $K$ be a class of coalgebras of the same type $F$. Then the class $SH\Sigma(K)$ is the smallest covariety to contain $K$.*

As in the case of varieties, one may look for the smallest building blocks of a covariety. This question leads us to the notion of a *conjunct sum* which is dual to that of a subdirectly irreducible algebra.

**Definition 4.9.3.** Let $\mathcal{A}$ be an $F$-coalgebra. A *conjunct representation* of $\mathcal{A}$ by a family $\{\mathcal{A}_j \mid j \in J\}$ of $F$-coalgebras is a family of embeddings $\{e_j : \mathcal{A}_j \to \mathcal{A} \mid j \in J\}$ such that $A = \sum\limits_{j \in J} e_j(A_j)$. In this case the $F$-coalgebra $\mathcal{A}$ is called a *conjunct sum* of the $\mathcal{A}_j$'s. An $F$-coalgebra $\mathcal{A}$ is called *conjunctly irreducible* if each of its conjunct representations is trivial in the sense that one of the embeddings is an isomorphism.

Covarieties are closed under the formation of conjunct sums (see [85]). But in general, for arbitrary functors there is no analogue of Birkhoff's Theorem 1.2.10 (see [39]).

## 4.10   Exercises for Chapter 4

1. Prove part (ii) of Proposition 4.1.8:
Let $\mathcal{A}$, $\mathcal{B}$ and $\mathcal{C}$ be $F$-coalgebras, and let $f : \mathcal{A} \to \mathcal{B}$ and $g : \mathcal{B} \to \mathcal{C}$ be mappings such that the composition $\varphi := g \circ f$ from $\mathcal{A}$ to $\mathcal{C}$ is a homomorphism. Show that $f$ is a homomorphism if $g$ is an injective homomorphism.

2. Prove that for any coalgebra $\mathcal{A}$, the diagonal relation $\Delta_A$ is a bisimulation on $\mathcal{A}$.

3. Let $\mathcal{A}$ and $\mathcal{B}$ be coalgebras. Prove that if a relation $R$ is a bisimulation from $\mathcal{A}$ to $\mathcal{B}$, then the inverse relation $R^{-1}$ is also a bisimulation, from $\mathcal{B}$ to $\mathcal{A}$.

4. Let $\mathcal{A}$, $\mathcal{B}$ and $\mathcal{C}$ be coalgebras and let $f : \mathcal{A} \to \mathcal{B}$ and $g : \mathcal{A} \to \mathcal{C}$ be homomorphisms. Prove that the image

$$(f \otimes g)(A) \;\; = \;\; \{(f(a), g(a)) \mid a \in A\}$$

is a bisimulation from $\mathcal{B}$ to $\mathcal{C}$.

5. Let $\mathcal{A}$, $\mathcal{B}$ and $\mathcal{C}$ be $F$-coalgebras. Let $R \subseteq A \times B$ and $Q \subseteq B \times C$ be bisimulations. Prove that the relation $Q \circ R$ is a bisimulation from $\mathcal{A}$ to $\mathcal{C}$, if $F$ preserves weak pullbacks.

6. Let $\mathcal{A}$ and $\mathcal{B}$ be coalgebras, and let $f : \mathcal{A} \to \mathcal{B}$ be a homomorphism.
a) Prove that if $R$ is a bisimulation on $\mathcal{A}$ then the image $f(R)$ is a bisimulation on $\mathcal{B}$.

b) Prove that if $Q$ is a bisimulation on $\mathcal{B}$ then $f^{-1}(Q)$ is a bisimulation on $\mathcal{A}$.

7. Let $\mathcal{A}$ be an $F$-coalgebra. Prove that a subset $B$ of $A$ is the base of a subcoalgebra of $\mathcal{A}$ if and only if the diagonal relation $\Delta_B$ is a bisimulation on $\mathcal{A}$.

8. Prove that $\emptyset$ and $\mathcal{A}$ itself are always subcoalgebras of an $F$-coalgebra $\mathcal{A}$.

9. Prove Corollary 4.3.4:
If $\mathcal{A}$ and $\mathcal{B}$ are coalgebras and $\varphi : \mathcal{A} \to \mathcal{B}$ is a homomorphism, the image $\varphi[A]$ is a homomorphic image of $\mathcal{A}$ and a subcoalgebra of $\mathcal{B}$.

10. Prove Proposition 4.3.5:
Let $\mathcal{A}$ and $\mathcal{B}$ be two $F$-coalgebras and let $\varphi : \mathcal{A} \to \mathcal{B}$ be a homomorphism.
(i) If $\mathcal{S} \leq \mathcal{A}$ is a subcoalgebra of $\mathcal{A}$, then $f(\mathcal{S})$ is a subcoalgebra of $\mathcal{B}$.
(ii) If $\mathcal{T}$ is a subcoalgebra of $\mathcal{B}$, then $f^{-1}(\mathcal{T})$ is a subcoalgebra of $\mathcal{A}$.

11. Prove Proposition 4.5.6:
Let $\mathcal{A}$ and $(\mathcal{B})_{i \in I}$ be coalgebras, and let $(\varphi_i : \mathcal{A} \to \mathcal{B}_i)_{i \in I}$ be a family of homomorphisms. Let $\psi : B_i \to P$ be the pushout of the mappings $\varphi_i$ in **Set**. Then there is a uniquely determined coalgebra structure $\alpha_P$ on $P$ such that the $\psi_i$'s are homomorphisms to $\mathcal{P} = (P; \alpha_P)$, and $\mathcal{P}$ together with the $\psi_i$'s is the pushout of the $\varphi_i$'s in **Set**$_F$.

# Chapter 5

# $F$-Algebras

In Chapter 1 we considered algebras from the traditional universal-algebraic point of view, as sets of objects with indexed sets of fundamental operations. In this chapter we take a different approach: we show how algebras may be viewed as $F$-algebras, for some functor $F$ of the category of sets. Our definition of an $F$-algebra will be a dual to the definition of an $F$-coalgebra from Chapter 4. The advantage of this approach is that the "operations" of an algebra may now be expressed by a single mapping. The disadvantage is that we can not define terms, term operations and clones of term operations in this setting.

## 5.1 The Concept of an $F$-Algebra

We begin by defining an $F$-algebra, then show that this definition can be used to consider ordinary algebras.

**Definition 5.1.1.** Let $F : \mathbf{Set} \to \mathbf{Set}$ be a functor of the category $\mathbf{Set}$. Let $A$ be a set, and let $\beta_A : F(A) \to A$ be a mapping which maps the functorial image $F(A)$ to the set $A$. Then the pair $(A, \beta_A)$ is called an $F$-*algebra*.

We note that the empty $F$-algebra exists as a special case of this definition. When $A = \emptyset$, then $F(\emptyset) = \emptyset$ and $\beta_\emptyset = \emptyset : F(\emptyset) \to \emptyset$, giving us $(\emptyset, \emptyset)$ as an $F$-algebra.

To see how our definition of an $F$-algebra captures the concept of an algebra, let us first consider an algebra $\mathcal{A} = (A; g^A)$ of type $\tau = (n)$, having one $n$-ary operation $g^A$. We take $F : \mathbf{Set} \to \mathbf{Set}$ to be the functor $F(-) = (-)^n$ on $\mathbf{Set}$; that is, $F(X) = X^n$ for any set $X$, while for any mapping $f : X \to Y$ the morphism $F(f) : X^n \to Y^n$ is defined by

$F(f)(x_1, \ldots, x_n) = (f(x_1), \ldots, f(x_n))$ for all $x_1, \ldots, x_n \in X$. Then the algebra $(A; g^A)$ can be regarded as a uniquely determined $F$-algebra $(A; \beta_A)$, by taking $\beta_A : F(A) \to A$ to be defined by $\beta_A(a_1, \ldots, a_n) = g^A(a_1, \ldots, a_n)$ for all $a_1, \ldots, a_n \in A$.

Next we can extend this idea to view any universal algebra $(A; (g_i^A)_{i \in I})$ of arbitrary type $\tau$ as an example of an $F$-algebra. Our functor $F : \text{Set} \to \text{Set}$ in this case is defined as follows. For sets $X$, we use the disjoint sum of powers of $X$, letting $F(X) = \sum_{i \in I} X^{n_i}$. For any mapping $f : X \to Y$, we have the morphism $F(f) : \sum_{i \in I} X^{n_i} \to \sum_{i \in I} Y^{n_i}$ defined by $F(f)(i, (x_1, \ldots, x_{n_i})) = (i, (f(x_1), \ldots, f(x_{n_i})))$ for all $x_1, \ldots, x_{n_i} \in X$. Then we take $\beta_A : F(A) \to A$, defined by $\beta_A(i, (a_1, \ldots, a_{n_i})) = g_i^A(a_1, \ldots, a_{n_i})$ for all $a_1, \ldots, a_{n_i} \in A$. The algebra $(A; (g_i^A)_{i \in I})$ can now be regarded as the $F$-algebra $(A; \beta_A)$.

We remark in passing that if the single mapping is allowed to be partial, our new approach will also include partial algebras. Again, we consider first the simplest case, of a partial algebra $(A; g^A)$ having a single $n$-ary partial operation. This time the fundamental operation $g^A : A^n \to A$ is a partial operation on $A$, meaning that there are $n$-tuples $(a_1, \ldots, a_n) \in A^n$ for which $g^A(a_1, \ldots, a_n)$ does not exist. Again, our partial algebra $(A; g^A)$ is uniquely determined by the $F$-algebra $(A; \beta_A)$, where $\beta_A : F(A) \to A$ is defined by $\beta_A(x_1, \ldots, x_n) = g^A(x_1, \ldots, x_n)$, but now $\beta$ is also a partial mapping. This view of a type $(n)$ partial algebra as an $F$-algebra can also be extended to partial algebras of arbitrary type, just as we did above for ordinary algebras. Moreover, Definition 5.1.1 can be generalized to functors over arbitrary concrete categories.

## 5.2 Homomorphisms of $F$-Algebras

In this section we define homomorphisms of $F$-algebras, and study their properties.

**Definition 5.2.1.** Let $\mathcal{A} = (A; \beta_A)$ and $\mathcal{B} = (B; \beta_B)$ be $F$-algebras. A mapping $\varphi : A \to B$ is called a *homomorphism* from $\mathcal{A}$ to $\mathcal{B}$ if $\varphi \circ \beta_A = \beta_B \circ F(\varphi)$, meaning that the diagram in Figure 5.1 commutes.

The following proposition can be proved much as the analogous result for $F$-coalgebras was proved in Chapter 4, and we leave it as an exercise for the reader.

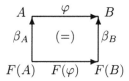

Fig. 5.1 Homomorphisms of $F$-algebras

**Proposition 5.2.2.** *Let* $\mathcal{A} = (A; \beta_A)$, $\mathcal{B} = (B; \beta_B)$ *and* $\mathcal{C} = (C; \beta_C)$ *be F-algebras. Then*

(i) *The mapping* $id_A : \mathcal{A} \to \mathcal{A}$ *is a homomorphism.*
(ii) *If* $\varphi : \mathcal{A} \to \mathcal{B}$ *and* $\psi : \mathcal{B} \to \mathcal{C}$ *are homomorphisms, then* $\psi \circ \varphi : \mathcal{A} \to \mathcal{C}$ *is a homomorphism.*

**Corollary 5.2.3.** *Let* $F$ *be a functor. The class of all F-algebras together with the F-algebra homomorphisms form a category, which we will denote by* $\mathrm{Set}^F$.

**Proposition 5.2.4.** *Any bijective homomorphism of F-algebras is an isomorphism in* $\mathrm{Set}^F$, *and conversely.*

**Proof:** Let $\varphi : \mathcal{A} \to \mathcal{B}$ be a bijective homomorphism of $F$-algebras. As a bijective mapping, $\varphi$ has an inverse mapping $\varphi^{-1} : B \to A$ which satisfies $\varphi \circ \varphi^{-1} = id_B$ and $\varphi^{-1} \circ \varphi = id_A$. This mapping $\varphi^{-1}$ is also a bijection, and we need to show that it is a homomorphism. From Proposition 5.2.2 we know that $id_A : \mathcal{A} \to \mathcal{A}$ is a homomorphism, and combining the corresponding homomorphism equation with the fact that $id_A = \varphi^{-1} \circ \varphi$ (and associativity) gives the equation $\varphi^{-1} \circ (\varphi \circ \beta_A) = (\beta_A \circ F(\varphi^{-1})) \circ F(\varphi)$. We can replace $\varphi \circ \beta_A$ on the left-hand side of this equation with $\beta_B \circ F(\varphi)$, using the homomorphism equation for $\varphi$, to get $(\varphi^{-1} \circ \beta_B) \circ F(\varphi) = (\beta_A \circ F(\varphi^{-1})) \circ F(\varphi)$. But $\varphi$ is a bijection and hence an epimorphism, so $F(\varphi)$ is also an epimorphism. This allows us to conclude that $\varphi^{-1} \circ \beta_B = \beta_A \circ F(\varphi^{-1})$, making $\varphi^{-1}$ a homomorphism.

Conversely, let $\chi : \mathcal{A} \to \mathcal{B}$ be an isomorphism. Of course $\chi$ is a homomorphism, and we need to show that $\chi$ is also both injective and surjective. Since $\chi$ is an isomorphism, there is a homomorphism $\psi : \mathcal{B} \to \mathcal{A}$ such that $\chi \circ \psi = id_B$ and $\psi \circ \chi = id_A$. These equations show that $\chi$ is surjective and injective respectively. ∎

**Proposition 5.2.5.** *Let* $\mathcal{A} = (A; \beta_A)$, $\mathcal{B} = (B; \beta_B)$ *and* $\mathcal{C} = (C; \beta_C)$ *be F-algebras. Let* $f : A \to B$ *and* $g : B \to C$ *be mappings such that* $\varphi := g \circ f : \mathcal{A} \to \mathcal{C}$ *is a homomorphism. Then*

(i) *If* $f$ *is a surjective homomorphism, then* $g$ *is also a homomorphism.*
(ii) *If* $g$ *is an injective homomorphism, then* $f$ *is a homomorphism.*

**Proof:** (i) Let $f : \mathcal{A} \to \mathcal{B}$ be a surjective homomorphism. The homomorphism equations for $f$ and $\varphi$ give us $f \circ \beta_A = \beta_B \circ F(f)$ and $\varphi \circ \beta_A = \beta_C \circ F(\varphi)$. We consider separately the two halves of the latter equation:

$$
\begin{aligned}
&\varphi \circ \beta_A \\
= \; &(g \circ f) \circ \beta_A \\
= \; &g \circ (f \circ \beta_A) \\
= \; &g \circ (\beta_B \circ F(f)) \\
= \; &(g \circ \beta_B) \circ F(f),
\end{aligned}
$$

while

$$
\begin{aligned}
&\beta_C \circ F(\varphi) \\
= \; &\beta_C \circ F(g \circ f) \\
= \; &\beta_C \circ (F(g) \circ F(f)) \\
= \; &(\beta_C \circ F(g)) \circ F(f).
\end{aligned}
$$

Since these two halves of the equation are equal, we get $(g \circ \beta_B) \circ F(f) = (\beta_C \circ F(g)) \circ F(f)$. But $f$ is surjective and $F$ is a functor, so that $F(f)$ is an epimorphism; this forces $g \circ \beta_B = \beta_C \circ F(g)$.

(ii) Since $g \circ f$ is a homomorphism, and composition of mappings is associative, we have $g \circ (f \circ \beta_A) = (\beta_C \circ F(g)) \circ F(f)$. When $g$ is a homomorphism, we also have $g \circ \beta_B = \beta_C \circ F(g)$. Substitution then gives $g \circ (f \circ \beta_A) = (g \circ \beta_B) \circ F(f) = g \circ (\beta_B \circ F(f))$. Since $g$ is injective, this forces $f \circ \beta_A = \beta_B \circ F(f)$.  ∎

**Corollary 5.2.6.** *Let* $\mathcal{A} = (A; \beta_A)$, $\mathcal{B} = (B; \beta_B)$ *and* $\mathcal{C} = (C; \beta_C)$ *be F-algebras and let* $\varphi : \mathcal{A} \to \mathcal{B}$ *and* $\psi : \mathcal{A} \to \mathcal{C}$ *be homomorphisms. If* $\varphi$ *is surjective, then there is a homomorphism* $\chi : \mathcal{B} \to \mathcal{C}$ *with* $\chi \circ \varphi = \psi$ *iff* $Ker\varphi \subseteq Ker\psi$.

**Proof:** By the Diagram Lemma for sets, we know that a mapping $\chi : B \to C$ with $\chi \circ \varphi = \psi$ exists iff $Ker\varphi \subseteq Ker\psi$. Thus we need only show here that $\chi$ is a homomorphism. Since $\varphi$ and $\psi$ are homomorphisms, we have $\varphi \circ \beta_A = \beta_B \circ F(\varphi)$ and $\psi \circ \beta_A = \beta_C \circ F(\psi)$. From the second equation, using $\chi \circ \varphi = \psi$, the property of the functor $F$ and the first equation, we get $\chi \circ \beta_B \circ F(\varphi) = \beta_C \circ F(\chi) \circ F(\varphi)$. But this implies that $\chi \circ \beta_B = \beta_C \circ F(\chi)$, since $\varphi$ is surjective and therefore an epimorphism and hence $F(\varphi)$ is an epimorphism. This shows that $\chi$ is a homomorphism. ∎

## 5.3 Subalgebras of *F*-Algebras

The concept of a subalgebra of an algebra can also be captured using $F$-algebras.

**Definition 5.3.1.** Let $\mathcal{A} = (A; \beta_A)$ be an $F$-algebra. Let $S \subseteq A$ be a subset. The set $S$ is said to be closed if there is a mapping $\beta_S : F(S) \to S$ such that the embedding (injection) mapping $\subseteq_S^A : S \hookrightarrow \mathcal{A}$ is a homomorphism. In this case $\mathcal{S} = (S; \beta_S)$ is called a *subalgebra* of $\mathcal{A}$ and we write $\mathcal{S} \preceq \mathcal{A}$.

If $\mathcal{S}$ is a subalgebra of $\mathcal{A}$ the homomorphism equation $\subseteq_S^A \circ \beta_S = \beta_A \circ F(\subseteq_S^A)$ is satisfied. Conversely, if for a subset $S$ of $A$ and a mapping $\beta_S : F(S) \to S$ this equation is satisfied, then $\mathcal{S}$ is a subalgebra of $\mathcal{A}$. We will prove that the mapping $\beta_S$ is uniquely determined.

**Lemma 5.3.2.** *For every closed set $S \subseteq A$ there is a uniquely determined mapping $\beta_S : F(S) \to S$ such that $\mathcal{S} = (S; \beta_S)$ is a subalgebra of $\mathcal{A}$.*

**Proof:** In the special case that $S = \emptyset$ there is certainly only one mapping possible, the empty mapping. Assume now that $S \neq \emptyset$, and suppose that there are two mappings $\delta_1$ and $\delta_2$ from $F(S)$ to $S$ for which the embedding $\subseteq_S^A : S \to A$ is a homomorphism. Then $\subseteq_S^A \circ \delta_1 = \beta_A \circ F(\subseteq_S^A)$ and $\subseteq_S^A \circ \delta_2 = \beta_A \circ F(\subseteq_S^A)$. Therefore $\subseteq_S^A \circ \delta_1 = \subseteq_S^A \circ \delta_2$, and now the injectivity of the mapping $\subseteq_S^A$ gives $\delta_1 = \delta_2$. ∎

## 5.4 Homomorphic Images and Factorizations

In this section we prove in the $F$-algebra setting some of the results on homomorphic images from Section 1.4.

**Lemma 5.4.1.** *Let $\mathcal{A} = (A; \beta_A)$ and $\mathcal{B} = (B; \beta_B)$ be F-algebras. If $\varphi : \mathcal{A} \to \mathcal{B}$ is a surjective homomorphism, then the structural mapping $\beta_B$ is uniquely determined by $\varphi$ and $\beta_A$, with $\beta_B = \{(F(\varphi)(a), \varphi \circ \beta_A(a)) \mid a \in F(A)\}$.*

**Proof:** Suppose that there is in addition to $\beta_B$ another structural mapping $\beta'_B : F(B) \to B$ making $\varphi$ a homomorphism. Thus we have both $\beta_B \circ F(\varphi) = \varphi \circ \beta_A$ and $\beta'_B \circ F(\varphi) = \varphi \circ \beta_A$. This gives $\beta_B \circ F(\varphi) = \beta'_B \circ F(\varphi)$. Since $\varphi$ is surjective so is $F(\varphi)$, and surjectivity implies that $F(\varphi)$ is an epimorphism. This shows that $\beta_B = \beta'_B$, and our mapping is unique. Moreover, it satisfies $\beta_B(F(\varphi)(a)) = \varphi \circ \beta_A(a)$ for all $a \in F(A)$, making $\beta_B = \{(F(\varphi)(a), \varphi \circ \beta_A(a)) \mid a \in F(A)\}$. ∎

This uniqueness of structure justifies the following definition.

**Definition 5.4.2.** Let $\mathcal{A} = (A; \beta_A)$ and $\mathcal{B} = (B; \beta_B)$ be $F$-algebras. If $\varphi : \mathcal{A} \to \mathcal{B}$ is a surjective homomorphism, then the algebra $\mathcal{B}$ is called a *homomorphic image* of the $F$-algebra $\mathcal{A}$.

**Theorem 5.4.3.** *(The Factorization Theorem) Let $\mathcal{A} = (A; \beta_A)$ and $\mathcal{B} = (B; \beta_B)$ be $F$-algebras, with a homomorphism $\varphi : \mathcal{A} \to \mathcal{B}$. Let $\varphi = g \circ f$ be a factorization of $\varphi$ into the composition of a surjective mapping $f : A \to Q$ and an injective mapping $g : Q \to B$, for some set $Q$. Then there is a uniquely determined $F$-algebra structure on $Q$ such that $f$ and $g$ are both homomorphisms.*

**Proof:** Since $\varphi = g \circ f$ is a homomorphism and $F$ is a functor, we have $(g \circ f) \circ \beta_A = \beta_B \circ F(g \circ f) = \beta_B \circ F(g) \circ F(f)$. This equation means that the four mappings $F(f)$, $f \circ \beta_A$, $\beta_B \circ F(g)$ and $g$ form an E-M-square (see Lemma 3.7.3), since $f$ and hence $F(f)$ are surjective and $g$ is injective and hence a monomorphism. Therefore by the E-M-Square Lemma there exists a uniquely determined diagonal mapping $\beta_Q : F(Q) \to Q$ for which the equations $f \circ \beta_A = \beta_Q \circ F(f)$ and $g \circ \beta_Q = \beta_B \circ F(g)$ are satisfied. These two equations tell us precisely that $f$ and $g$ are homomorphisms. ∎

We recall that for any mapping $f : A \to B$, the set $f[A] := \{f(a) \mid a \in A\}$ is called the *image* of $A$ under $f$ and the mapping $f' : A \to f[A]$ is called the restriction of $f$ onto the image of $f$. The mapping $f$ can then be factored into $f = \subseteq^B_{f[A]} \circ f'$, where the inclusion mapping $\subseteq^B_{f[A]}$ is injective and the restriction mapping $f'$ is surjective. Applying Theorem 5.4.3 to this factorization gives the following corollary.

**Corollary 5.4.4.** *Let $\mathcal{A} = (A; \beta_A)$ and $\mathcal{B} = (B; \beta_B)$ be $F$-algebras. If $\varphi : \mathcal{A} \to \mathcal{B}$ is a homomorphism, then the image set $\varphi[A]$ is a homomorphic image of $\mathcal{A}$ and a subalgebra of $\mathcal{B}$.*

We can also apply Theorem 5.4.3 to the case of a subalgebra $\mathcal{S} \preceq \mathcal{A}$, and the mapping $\varphi \circ \subseteq_{\mathcal{S}}^{\mathcal{A}}$. This gives a straightforward proof of part (i) of the following proposition. We shall prove part (ii) later.

**Proposition 5.4.5.** *Let $\mathcal{A}$ and $\mathcal{B}$ be two F-algebras and let $\varphi : \mathcal{A} \to \mathcal{B}$ be a homomorphism.*

(i) *If $\mathcal{S} \preceq \mathcal{A}$ is a subalgebra of $\mathcal{A}$, then $\varphi[S]$ is a subalgebra of $\mathcal{B}$.*
(ii) *If $\mathcal{T} \preceq \mathcal{B}$ is a subalgebra of $\mathcal{B}$, then $\varphi^{-1}[T]$ is a subalgebra of $\mathcal{A}$.*

## 5.5 Factor Algebras

As in the usual algebra case, congruence relations on $F$-algebras are equivalence relations which are invariant under the algebraic structure. This is equivalent to being the kernel of a homomorphism, which we use as our definition.

**Definition 5.5.1.** Let $\mathcal{A}$ be an $F$-algebra. A *congruence relation* on an $F$-algebra $\mathcal{A}$ is the kernel of any homomorphism $\varphi$ from $\mathcal{A}$ into some $F$-algebra $\mathcal{B}$.

This definition means that if $\theta$ is a congruence relation on the $F$-algebra $\mathcal{A}$, then there is an $F$-algebra $\mathcal{B}$ and a homomorphism $\varphi : \mathcal{A} \to \mathcal{B}$ such that $\theta = Ker\varphi$. In this case we also have the natural mapping $\pi_\theta : A \to A/\theta$, where $A/\theta$ is the quotient or factor set. Then we have an injection $f : A/\theta \to B$ and we can factor $\varphi$ as $f \circ \pi_\theta$. This factorization can be done for arbitrary sets and mappings. But since $\pi_\theta$ is surjective and $f$ is injective, we may apply the Factorization Theorem 5.4.3 to obtain a uniquely determined algebraic structure $\beta_\theta$ on the quotient set $A/\theta$. The resulting algebra $(A/\theta; \beta_\theta)$ is called the *factor* or *quotient algebra* of $\mathcal{A}$ modulo $\theta$.

**Theorem 5.5.2.** *Let $\mathcal{A}$ and $\mathcal{B}$ be F-algebras. If $\varphi : \mathcal{A} \to \mathcal{B}$ is a surjective homomorphism with kernel $\theta$, then $\mathcal{A}/\theta$ is isomorphic to $\mathcal{B}$.*

**Proof:** As discussed above, we can write $\varphi = f \circ \pi_\theta$, where $f$ is an injective homomorphism. Since $\varphi$ is surjective, $f$ is also surjective. Then by Proposition 5.2.4, $f$ is an isomorphism from $\mathcal{A}/\theta$ to $\mathcal{B}$. ∎

## 5.6    Limits in $\mathbf{Set}^F$

In this section we consider products, equalizers and pullbacks in the category $\mathbf{Set}^F$. Each of these has been defined in the category of sets, and we shall see how an appropriate $F$-algebra structure can be imposed to produce the analogous result for $\mathbf{Set}^F$.

We begin with the product construction. For any family $(\mathcal{A}_i)_{i \in I}$ of $F$-algebras, we can consider the $A_i$ as sets, and the product set $\Pi_{i \in I}\, A_i$ exists. To extend this to $\mathbf{Set}^F$, we need a structural mapping from $F(\Pi_{i \in I} A_i)$ to $\Pi_{i \in I} A_i$ to make $\Pi_{i \in I} A_i$ into an $F$-algebra. We use the mappings $\beta_i \circ F(p_i)$, where the $p_i : \Pi_{i \in I} A_i \to A_i$ are the projections and the $\beta_i : F(A_i) \to A_i$ are the structural mappings of the $\mathcal{A}_i$. From the definition of the product in $\mathbf{Set}$, there is a uniquely determined morphism $\beta : F(\Pi_{i \in I} A_i) \to \Pi_{i \in I} A_i$ such that $p_i \circ \beta = \beta_i \circ F(p_i)$. This is the uniquely determined structural mapping of $\Pi_{i \in I} A_i$. These observations allow us to show that the product of a family of $F$-algebras is an $F$-algebra.

**Theorem 5.6.1.** *Let $(\mathcal{A}_i)_{i \in I}$ be a family of $F$-algebras. The product of their universes in the category $\mathbf{Set}^F$ is the universe of their product in $\mathbf{Set}$.*

**Proof**: We show that the $F$-algebra structural mapping $\beta$ described above has the necessary properties. Let $\mathcal{Q}$ be any $F$-algebra and $(q_i : \mathcal{Q} \to \mathcal{A}_i)_{i \in I}$ be any family of $F$-algebra homomorphisms. Since $\Pi_{i \in I} A_i$ is a product, there is a uniquely determined mapping $\sigma : Q \to \Pi_{i \in I} A_i$ such that $q_i = p_i \circ \sigma$ for all $i \in I$; we need to show that $\sigma$ is an $F$-algebra homomorphism. Since $\Pi_{i \in I} A_i$ is a product and we have $\beta_i \circ F(q_i) : F(Q) \to A_i$, there is a uniquely determined mapping $\eta : F(Q) \to \Pi_{i \in I} A_i$ such that $\beta_i \circ F(q_i) = p_i \circ \eta$ for all $i \in I$. Using $q_i = p_i \circ \sigma$ and the homomorphism fact that $\beta_i \circ F(q_i) = q_i \circ \beta_Q$ gives us $p_i \circ \sigma \circ \beta_Q = \beta_i \circ F(q_i)$. The uniqueness of $\eta$ then shows that $\eta = \sigma \circ \beta_Q$. Similarly, we have $p_i \circ \beta \circ F(\sigma) = \beta_i \circ F(q_i)$, and again uniqueness of $\eta$ gives $\eta = \beta \circ F(\sigma)$. Hence $\sigma \circ \beta_Q = \beta \circ F(\sigma)$, making $\sigma$ a homomorphism.  ∎

For equalizers in $\mathbf{Set}^F$ we first discuss the equalizer of a family of mappings in $\mathbf{Set}$.

**Lemma 5.6.2.** *Let $A$ and $B$ be sets, with $(\varphi_i : A \to B)_{i \in I}$ a family of morphisms in $\mathbf{Set}$. Let $E = \{a \in A \mid \varphi_i(a) = \varphi_j(a)$ for all $i, j \in I\}$. Then the mapping $\subseteq_E^A : E \to A$ is the equalizer of the $\varphi_i$'s.*

**Proof**: By definition of $E$, we have $\varphi_i \circ \subseteq_E^A = \varphi_j \circ \subseteq_E^A$ for all $i, j \in I$. Let

$\psi : Q \to A$ be a morphism such that $\varphi_i \circ \psi = \varphi_j \circ \psi$ for all $i, j \in I$. Then $\psi(q) \in E$ for each $q \in Q$, so that $\psi[Q] \subseteq E$. Thus there is also a mapping $h : Q \to E$ such that $\subseteq_E^A \circ h = \psi$. By the injectivity of $\subseteq_E^A$ this mapping is uniquely determined. This means that $\subseteq_E^A : E \to A$ is the equalizer of the $\varphi_i$'s. ∎

**Lemma 5.6.3.** *Let $A$ and $B$ be $F$-algebras, with $(\varphi_i : A \to B)_{i \in I}$ a family of homomorphisms. Let $E = \{a \in A \mid \varphi_i(a) = \varphi_j(a) \text{ for all } i, j \in I\}$, and let $\subseteq_E^A : E \to A$ be the equalizer of the $\varphi_i$'s in **Set**. Then there is a uniquely determined structural mapping $\beta_E$ on $E$ such that $\subseteq_E^A : E \to A$ is the equalizer of the homomorphisms $\varphi_i$ in $\mathbf{Set}^F$.*

**Proof:** For any $i, j \in I$ we have $\varphi_i \circ \beta_A \circ F(\subseteq_E^A) = \beta_B \circ F(\varphi_i) \circ F(\subseteq_E^A) = \beta_B \circ F(\varphi_j) \circ F(\subseteq_E^A) = \varphi_j \circ \beta_A \circ F(\subseteq_E^A)$. Since $\subseteq_E^A$ is an equalizer in **Set** we have a structural mapping $\beta_E$ on $E$ and $\subseteq_E^A$ is a homomorphism. Now let $Q = (Q; \beta_Q)$ be any algebra, with a homomorphism $\psi : Q \to A$ such that $\varphi_i \circ \psi = \varphi_j \circ \psi$ for all $i, j \in I$. Then there is a mapping $h : Q \to E$ such that $\psi = \subseteq_E^A \circ h$, and we need to show that $h$ is a homomorphism. Since $\psi$ is a homomorphism, $\psi \circ \beta_Q = \beta_A \circ F(\psi)$, so $\subseteq_E^A \circ h \circ \beta_Q = \beta_A \circ F(\subseteq_E^A) \circ F(h) = \subseteq_E^A \circ \beta_E \circ F(h)$. The mapping $\subseteq_E^A$ is injective and so a monomorphism, which implies that $h \circ \beta_Q = \beta_E \circ F(h)$, as required. ∎

For pullbacks in $\mathbf{Set}^F$, we first examine the intersection of a family of subsets (or subalgebras) of a set (or algebra) $A$.

**Lemma 5.6.4.** *Let $A$ be a set, with $(U_i)_{i \in I}$ a family of subsets of $A$. Then the intersection $\cap_{i \in I} U_i$ along with the inclusion mappings $(\subseteq_{\cap U_i}^{U_i} : \cap U_i \to U_i)_{i \in I}$ is a pullback of $(\subseteq_{U_i}^A : U_i \to A)_{i \in I}$ in **Set**.*

**Proof:** We have two families of inclusion mappings, $(\subseteq_{\cap U_i}^{U_i} : \cap U_i \to U_i)_{i \in I}$ and $(\subseteq_{U_i}^A : U_i \to A)_{i \in I}$, which satisfy $\subseteq_{U_i}^A \circ \subseteq_{\cap U_i}^{U_i} = \subseteq_{U_j}^A \circ \subseteq_{\cap U_i}^{U_j}$ for all $i$ and $j$ in $I$. Now let $Q$ be a set and $(q_i : Q \to U_i)_{i \in I}$ be a family of mappings such that $\subseteq_{U_i}^A \circ q_i = \subseteq_{U_j}^A \circ q_j$ for all $i$ and $j$ in $I$. It follows that for each $y \in Q$ we have $\subseteq_{U_i}^A \circ q_i(y) = \subseteq_{U_j}^A \circ q_j(y)$ for all $i, j \in I$. Then $q_i(y) = q_j(y) \in \cap U_i$ for all $i, j \in I$, so $q_i[Q] \subseteq \cap U_i$ for all $i \in I$. There exists a mapping $h : Q \to \cap U_i$ such that $\subseteq_{\cap U_i}^{U_i} \circ h = q_i$ for all $i \in I$, and because $\subseteq_{\cap U_i}^{u_i}$ is an inclusion map, this mapping $h$ is uniquely determined. ∎

**Theorem 5.6.5.** *Let $A$ be an $F$-algebra and let $(U_i)_{i \in I}$ be a family of subalgebras of $A$. Then $\cap_{i \in I} U_i$ is a subalgebra of $A$.*

**Proof:** Since the $\mathcal{U}_i$'s are subalgebras of $\mathcal{A}$ the mappings $\subseteq_{U_i}^A : U_i \to A$ are homomorphisms for all $i \in I$. Let $(\subseteq_{\cap U_i}^{U_i} : \cap U_i \to U_i)_{i \in I}$ be the family of the inclusion mappings. Then $\subseteq_{U_i}^A \circ \subseteq_{\cap U_i}^{U_i} = \subseteq_{U_j}^A \circ \subseteq_{\cap U_i}^{U_j}$ for all $i, j \in I$. Now we want to show that these mappings $(\subseteq_{\cap U_i}^{U_i} : \cap U_i \to U_i)_{i \in I}$ are homomorphisms. Using the homomorphisms $\subseteq_{U_i}^A : U_i \to A$ for all $i \in I$ and the fact that $F$ is a functor we obtain $\subseteq_{U_j}^A \circ \beta_j \circ F(\subseteq_{\cap U_i}^{U_j}) = \beta_A \circ F(\subseteq_{U_j}^A) \circ F(\subseteq_{\cap U_i}^{U_j})$
$= \beta_A \circ F(\subseteq_{U_i}^A) \circ F(\subseteq_{\cap U_i}^{U_i}) = \subseteq_{U_i}^A \circ \beta_i \circ F(\subseteq_{\cap U_i}^{U_i})$. From Lemma 5.6.4 we know that $\cap_{i \in I} U_i$ with mappings $(\subseteq_{\cap U_i}^{U_i} : \cap U_i \to U_i)_{i \in I}$ is a pullback of $(\subseteq_{U_i}^A : U_i \to A)_{i \in I}$ in **Set**. Then there is a uniquely determined mapping $\beta_{\cap U_i} : F(\cap U_i) \to \cap U_i$ such that $\subseteq_{\cap U_i}^{U_j} \circ \beta_{\cap U_i} = \beta_j \circ F(\subseteq_{\cap U_i}^{U_j})$ and $\subseteq_{\cap U_i}^{U_i} \circ \beta_{\cap U_i} = \beta_i \circ F(\subseteq_{\cap U_i}^{U_i})$ for all $i, j \in I$. This shows that $\subseteq_{\cap U_i}^{U_i}$ and $\subseteq_{\cap U_i}^{U_j}$ are homomorphisms for all $i, j \in I$, so $(\subseteq_{\cap U_i}^{U_i} : \cap U_i \to U_i)_{i \in I}$ are also homomorphisms. This means that $\cap \mathcal{U}_i$ is a subalgebra of $\mathcal{U}_i$ for all $i \in I$, and $\cap \mathcal{U}_i$ is a subalgebra of $\mathcal{A}$. ∎

The fact that the intersection of subalgebras of a given $F$-algebra $\mathcal{A}$ is also a subalgebra has important consequences. Given a subset $S$ of $A$, we may consider the family of all subalgebras of $\mathcal{A}$ which contain $S$. This family is non-empty, since $\mathcal{A}$ itself is a subalgebra containing $S$. The intersection of this family is then a subalgebra of $\mathcal{A}$, and it is by definition the least subalgebra of $\mathcal{A}$ to contain $S$. This intersection algebra is called the subalgebra generated by $S$, and denoted by $< S >$.

**Proposition 5.6.6.** *Let $\mathcal{A}$ and $(\mathcal{B}_i)_{i \in I}$ be $F$-algebras, and let $(\varphi_i : B_i \to A)_{i \in I}$ be a family of homomorphisms. Let $(p_i : P \to B_i)_{i \in I}$ be the pullback of the mappings $\varphi_i$ in* **Set**. *Then there is a uniquely determined $F$-algebra structure $\beta_P$ on $P$ such that the $p_i$ are homomorphisms, and $P$ together with the $p_i$'s is the pullback of the $\varphi_i$'s in* **Set**$^F$.

**Proof:** For any $i, j \in I$ we have $\varphi_j \circ \beta_j \circ F(p_j) = \beta_A \circ F(\varphi_j) \circ F(p_j) = \beta_A \circ F(\varphi_i) \circ F(p_i) = \varphi_i \circ \beta_i \circ F(p_i)$. Since $P$ is a pullback in **Set**, there is a uniquely determined structure mapping $\beta_P$ on $P$ such that $p_i \circ \beta_P = \beta_j \circ F(p_j)$ and $p_i \circ \beta_P = \beta_i \circ F(p_i)$. This makes $(P; \beta_P)$ into an $F$-algebra, with the $p_i$'s being homomorphisms for all $i \in I$. Now we want to show that $P$ is a pullback in **Set**$^F$ as well. Let $\mathcal{Q}$ be an $F$-algebra and $(q_i : Q \to B_i)_{i \in I}$ a family of $F$-algebra homomorphisms such that $\varphi_i \circ q_i = \varphi_j \circ q_j$ for all $i, j \in I$. Since $P$ is a pullback in **Set** there is a uniquely determined mapping $\sigma : Q \to P$ such that $q_i = p_i \circ \sigma$ for all $i \in I$.

Since the $\varphi_i$'s are homomorphisms and $F$ is a functor, $\varphi_j \circ \beta_j \circ F(p_j) = \beta_A \circ F(\varphi_j) \circ F(p_j) = \beta_A \circ F(\varphi_i) \circ F(p_i) = \varphi_i \circ \beta_i \circ F(p_i)$ for all $i \in I$. Then there is a uniquely determined mapping $\eta : F(Q) \to P$ such that $p_i \circ \eta = \beta_i \circ F(q_i)$ for all $i \in I$. But we have $\beta_j \circ F(q_j) = q_j \circ \beta_Q = p_j \circ \sigma \circ \beta_Q$, from which the uniqueness of $\eta$ implies that $\eta = \sigma \circ \beta_Q$. Similarly, $\beta_j \circ F(q_j) = \beta_j \circ F(p_j) \circ F(\sigma) = p_j \circ \beta_p \circ F(\sigma)$, which implies that $\eta = \beta_p \circ F(\sigma)$. Hence $\sigma \circ \beta_Q = \beta_P \circ F(\sigma)$.

∎

## 5.7 Congruences

The congruences we introduced in Section 5.5 were a special kind of equivalence relations on the base set of an $F$-algebra. In this section we extend this idea, to allow for congruence relations from one $F$-algebra to another.

**Definition 5.7.1.** Let $\mathcal{A}$ and $\mathcal{B}$ be $F$-algebras and let $R \subseteq A \times B$ be a relation from $A$ to $B$. The relation $R$ is called a congruence relation between $\mathcal{A}$ and $\mathcal{B}$ if an $F$-algebra structure $\beta_R$ can be defined on the set $R$ in such a way that the projection mappings $\pi_A : R \to A$ and $\pi_B : R \to B$ are homomorphisms.

We remark that if $R \subseteq A \times B$ is a congruence, its inverse $R^{-1} \subseteq B \times A$ is also a congruence; this justifies referring to $R$ as a congruence between $\mathcal{A}$ and $\mathcal{B}$. A congruence relation between $\mathcal{A}$ and itself, or more naturally "on $\mathcal{A}$", is a congruence in the sense of Section 5.5 if it is also an equivalence relation on $A$. It is also straightforward to show that if $\mathcal{S} \preceq \mathcal{A}$, any congruence on $\mathcal{S}$ defines a congruence on $\mathcal{A}$ (see Exercises for Chapter 5).

A congruence from $\mathcal{A}$ to $\mathcal{B}$ in the $F$-algebra case is a dual of the concept of a bisimilarity relation in the $F$-coalgebra case. We shall see that as was the case for coalgebras, congruences coincide with the graphs of homomorphisms. The kernel of a homomorphism is also always a congruence. Let us recall that for $f : \mathcal{A} \to \mathcal{B}$ an $F$-algebra homomorphism, the graph of $f$ is $G(f) := \{(a, f(a)) \mid a \in A\}$, while the kernel of $f$ is $Ker f = \{(a, b) \in A \times B \mid f(a) = f(b)\}$.

**Lemma 5.7.2.** *Let $\mathcal{A}$ and $\mathcal{B}$ be F-algebras, and let $f : \mathcal{A} \to \mathcal{B}$ be a homomorphism. Then the graph and the kernel of $f$ are both congruences* between $\mathcal{A}$ and $\mathcal{B}$.

**Proof**: The graph $G(f)$ of $f$ is a subset of $A \times B$, and we can define an $F$-algebra structure $\beta$ on $G(f)$ by $\beta = \pi_A^{-1} \circ \beta_A \circ F(\pi_A)$. Then we have $\pi_A \circ \beta = \pi_A \circ \pi_A^{-1} \circ \beta_A \circ F(\pi_A) = id_A \circ \beta_A \circ F(\pi_A) = \beta_A \circ F(\pi_A)$. Similarly we have $\pi_\beta \circ \beta = \beta_B \circ F(\pi_B)$. This shows that $\pi_A$ and $\pi_B$ are homomorphisms, and so $G(f)$ is a congruence between $\mathcal{A}$ and $\mathcal{B}$.

For the kernel $Ker f$ of $f$, we note that $Ker f$ together with the projections $\pi_1$ and $\pi_2$ is a pullback of $f : \mathcal{A} \to \mathcal{B}$ in $\mathbf{Set}$. By Proposition 5.6.6 it is also a pullback in the category $\mathbf{Set}^F$. Thus there is an $F$-algebra structure $\beta_{Ker f}$ on $Ker f$ such that $\pi_1$ and $\pi_2$ are homomorphisms, making $Ker f$ a congruence on $\mathcal{A}$.                                        ∎

**Example 5.7.3.** Let $\Delta_A := \{(a, a) \mid a \in A\}$ be the diagonal relation on the universe of an $F$-algebra $\mathcal{A}$. Since $\Delta_A$ is the graph of the identity homomorphism $id_A$, it follows from Lemma 5.7.2 that $\Delta_A$ is a congruence on $\mathcal{A}$.

**Proposition 5.7.4.** *Let $\mathcal{A}$ be an $F$-algebra. A subset $S \subseteq A$ is a subalgebra of $\mathcal{A}$ if and only if the diagonal $\Delta_S$ of $S$ is a congruence on $\mathcal{A}$.*

**Proof**: Let $\mathcal{S} \preceq \mathcal{A}$. From Example 5.7.3 we know that $\Delta_S$ is a congruence on $\mathcal{S}$, and as remarked above this makes it also a congruence on $\mathcal{A}$. Conversely, let $\pi_1 : \Delta_S \to A$ be a projection homomorphism. We can write $\pi_1 = \subseteq_S^A \circ \pi_1'$ where $\pi_1' : \Delta_S \to S$ is also a projection. Since $\pi_1'$ is surjective and $\subseteq_S^A$ is injective, the Factorization Theorem tells us that $\mathcal{S} \preceq \mathcal{A}$.        ∎

**Theorem 5.7.5.** *Let $\mathcal{A}$ and $\mathcal{B}$ be $F$-algebras. A mapping $f : \mathcal{A} \to \mathcal{B}$ is an $F$-algebra homomorphism if and only if its graph $G(f)$ is a congruence between $\mathcal{A}$ and $\mathcal{B}$.*

**Proof**: We have already shown one direction of this theorem in Lemma 5.7.2; we leave the other direction as an exercise.                    ∎

We also leave as an exercise the proof of the following characterization of congruences.

**Theorem 5.7.6.** *Let $\mathcal{A}$, $\mathcal{B}$ and $\mathcal{P}$ be $F$-algebras, and let $\varphi_A : \mathcal{P} \to \mathcal{A}$ and $\varphi_B : \mathcal{P} \to \mathcal{B}$ be homomorphisms. Then the relation $(\varphi_A, \varphi_B)(\mathcal{P}) :=$*

$\{(\varphi_A(p), \varphi_B(p)) | p \in \mathcal{P}\}$ *is a congruence between $\mathcal{A}$ and $\mathcal{B}$; and moreover each congruence between F-algebras has this form.*

## 5.8 Mono's and Epi's in $\mathbf{Set}^F$

There is a close connection between monomorphisms and pullbacks. The morphism $\varphi : A \to B$ is a monomorphism in a given category $\mathbf{C}$ iff the diagram satisfying the following two conditions commutes:

(i) $\varphi \circ id_A = \varphi \circ id_A$;

(ii) If for all $Q$ and for all $q_i : Q \to A$ such that $\varphi \circ q_i = \varphi \circ q_j$ for all $i, j \in I$, there is exactly one morphism $h : Q \to A$ such that $id_A \circ h = q_i$ for all $i \in I$, then $q_i = q_j$ for all $i, j \in I$.

These conditions mean that the object $A$ together with the mapping $id_A : A \to A$ is a pullback of $\varphi : A \to B$. That is, in the category $\mathbf{Set}^F$, $\varphi : \mathcal{A} \to \mathcal{B}$ is a monomorphism if and only if $\mathcal{A}$ with the identity mapping is a pullback of $\varphi$. We have seen earlier that pullbacks in $\mathbf{Set}^F$ are precisely the pullbacks in $\mathbf{Set}$. As a result, we see that the same is true for monomorphisms in $\mathbf{Set}^F$: a mapping $\varphi$ is a monomorphism in $\mathbf{Set}^F$ iff it is a monomorphism in $\mathbf{Set}$. Moreover, monomorphisms in $\mathbf{Set}$ are exactly the injective mappings. We have thus proved the following result.

**Corollary 5.8.1.** *An F-algebra homomorphism is a monomorphism iff it is injective.*

## 5.9 Varieties of $F$-Algebras

We saw in Section 1.4 that a variety of algebras is a class closed under the formation of homomorphic images, subalgebras and products. Having analogues of these constructions for $F$-algebras, we can now define a variety of $F$-algebras.

Let $K$ be a class of $F$-algebras. We define:

$I(K)$ to be the class of all isomorphic copies of $F$-algebras from $K$,

$H(K)$ to be the class of all homomorphic images of $F$-algebras from $K$,

$S(K)$ to be the class of all subalgebras of arbitrary $F$-algebras from $K$, and

$P(K)$ to be the class of all products of $F$-algebras from $K$.

**Definition 5.9.1.** A *variety* (of $F$-algebras) is a class of $F$-algebras which is closed under the operators $H$, $S$ and $P$.

One can prove that the operators $H, S$ and $IP$ are closure operators and that $SH(K) \subseteq HS(K)$, $PS(K) \subseteq SP(K)$ and $PH(K) \subseteq HP(K)$. There is also an analogue for $F$-algebras of Tarski's Theorem for varieties:

**Theorem 5.9.2.** *Let $K$ be a class of $F$-algebras. Then $HSP(K)$ is the smallest variety of $F$-algebras which contains $K$.*

## 5.10    Exercises for Chapter 5

1. Prove Proposition 5.2.2:
Let $\mathcal{A} = (A; \beta_A)$, $\mathcal{B} = (B; \beta_B)$ and $\mathcal{C} = (C; \beta_C)$ be $F$-algebras. Then
a) The mapping $id_A : \mathcal{A} \to \mathcal{A}$ is a homomorphism.
b) If $\varphi : \mathcal{A} \to \mathcal{B}$ and $\psi : \mathcal{B} \to \mathcal{C}$ are homomorphisms, then $\psi \circ \varphi : \mathcal{A} \to \mathcal{C}$ is a homomorphism.

2. Prove Proposition 5.4.5(i):
If $\mathcal{S} \preceq \mathcal{A}$ is a subalgebra of $\mathcal{A}$, then $\varphi[S]$ is a subalgebra of $\mathcal{B}$.

3. Prove that if $\mathcal{A}$ is an $F$-algebra and $\mathcal{B}$ is a subalgebra of $\mathcal{A}$ and $\mathcal{C}$ is a subalgebra of $\mathcal{B}$, then $\mathcal{C}$ is also a subalgebra of $\mathcal{A}$.

4. Prove that if $\mathcal{S}$ is a subalgebra of $\mathcal{A}$ then any congruence on $\mathcal{S}$ defines a congruence on $\mathcal{A}$.

5. Prove the remaining direction of Theorem 5.7.5:
Let $\mathcal{A}$ and $\mathcal{B}$ be $F$-algebras. A mapping $f : \mathcal{A} \to \mathcal{B}$ is an $F$-algebra homomorphism if its graph $G(f)$ is a congruence between $\mathcal{A}$ and $\mathcal{B}$.

6. Prove Theorem 5.7.6:
Let $\mathcal{A}, \mathcal{B}$ and $\mathcal{P}$ be $F$-algebras, and let $\varphi_A : \mathcal{P} \to \mathcal{A}$ and $\varphi_B : \mathcal{P} \to \mathcal{B}$ be homomorphisms. Then the relation $(\varphi_A, \varphi_B)(\mathcal{P}) := \{(\varphi_A(p), \varphi_B(p)) \mid p \in \mathcal{P}\}$ is a congruence between $\mathcal{A}$ and $\mathcal{B}$. Moreover, any congruence between two $F$-algebras $\mathcal{A}$ and $\mathcal{B}$ has this form.

7. Prove that the operators $H$, $S$ and $IP$ from Section 5.9 are closure operators.

8. Prove that the operators $H$, $S$ and $P$ from Section 5.9 satisfy the inclusions

$$SH(K) \subseteq HS(K), \quad PS(K) \subseteq SP(K) \text{ and } PH(K) \subseteq HP(K).$$

# Chapter 6

# $(F_1, F_2)$-Coalgebras

In this chapter we introduce a new structure called an $(F_1, F_2)$-coalgebra, where $F_1$ and $F_2$ are set-valued endofunctors. This structure was defined in [25], in order to model tree automata; but as we shall see, it encompasses both $F$-algebras and $F$-coalgebras, along with other structures such as power algebras, power coalgebras and tree automata.

## 6.1   Tree Automata as $(F_1, F_2)$-Coalgebras

We begin with a definition and some results about tree automata, in order to motivate our definition of an $(F_1, F_2)$-coalgebra. As we saw in Section 2.4, automata can be used to accept sets of words, or expressions built up by concatenating letters from an input alphabet. Tree automata are a generalization of this, intended to accept sets of terms of any given type and alphabet. A $\Sigma - X_n$-tree automaton is a quintuple $\underline{A} = (\mathcal{A}, \Sigma, X_n, \alpha, A')$. The set $\Sigma$ is a set of operation symbols, and can be decomposed into $\Sigma = \bigcup_{m \geq 0} \Sigma_m$ so that every $f \in \Sigma_m$ is $m$-ary. $\mathcal{A}$ is a finite algebra with $\Sigma^{\mathcal{A}} := \{f^{\mathcal{A}} \mid f \in \Sigma\}$ as its set of fundamental operations, where $f^{\mathcal{A}}$ is the $m$-ary operation corresponding to the $m$-ary operation symbol $f$. The set $X_n := \{x_1, x_2, \ldots, x_n\}$ is a finite alphabet, $A' \subseteq A$ is a distinguished subset of $A$, and $\alpha : X_n \to A$ is a mapping. Using $W_\Sigma(X_n)$ as the set of all terms built up of operation symbols from $\Sigma$ and variables from $X_n$, we see that $\alpha$ can be uniquely extended to a homomorphism $\hat{\alpha} : \mathcal{F}_\Sigma(X_n) \to \mathcal{A}$ from the $n$-generated absolutely free algebra $\mathcal{F}_\Sigma(X_n)$ to the algebra $\mathcal{A}$. When the sets $\Sigma$ and $X_n$ are clear from the context, the $\Sigma - X_n$-tree automaton $\underline{A}$ is called simply a tree automaton. More details on this structure may be found in [34].

129

A tree automaton recognizes or accepts a set of terms in the following way: a term $t$ from $W_\Sigma(X_n)$ is recognizable iff $\hat{\alpha}(t) \in A'$. That is, the homomorphism $\hat{\alpha}$ which "processes" $t$ must result in an element of the distinguished set $A'$. This processing of term $t$ can be illustrated by a tree diagram for $t$, as in the example in Figure 6.1 below. Starting with the leaves, which are variables from the input alphabet $\Sigma$, the mapping $\alpha$ assigns values from $A$ to the variables, and then $\hat{\alpha}$ gives an inductive or top-down calculation, with the element $\hat{\alpha}(t)$ being calculated when the root of the tree is reached. If this element is in the set $A'$, the term is recognized, and otherwise it is not.

The usual algebraic constructions were considered for tree automata in [34]. We shall need the following definition of a tree automata homomorphism.

**Definition 6.1.1.** Let $\underline{A} = (\mathcal{A}, \Sigma, X_n, \alpha, A')$ and $\underline{B} = (\mathcal{B}, \Sigma, X_n, \beta, B')$ be tree automata. A mapping $f : A \to B$ is said to be a *homomorphism for tree automata* if the following conditions are satisfied:

(i) $f : \mathcal{A} \to \mathcal{B}$ is a homomorphism of the underlying algebras.
(ii) $f(\alpha(x)) = \beta(x)$ for all $x \in X_n$.
(iii) $f^{-1}(B') = A'$.

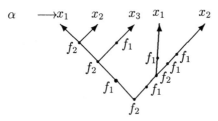

Fig. 6.1   Processing of a Term

We want to show next that tree automata can be viewed as automata, which can in turn be represented as $F$-coalgebras. In order to do this, we need the assumption that in the tree automaton $\underline{A} = (\mathcal{A}, \Sigma, X_n, \alpha, A')$ the underlying algebra $\mathcal{A}$ has only $n$-ary fundamental operations. For finite automata this is not a restriction: if the set $\Sigma^{\mathcal{A}}$ of fundamental operations is finite, we can take $n$ to be the maximal arity of any symbol from $\Sigma^{\mathcal{A}}$, and use the $n$-element alphabet $X_n$ for this $n$. With the addition of fictitious variables as necessary, we may then regard all the fundamental operations of the algebra as being $n$-ary. This produces a new algebra which is term equivalent to $\mathcal{A}$; this is sufficient for our purposes, since the two automata based on these algebras are equivalent to each other, in the sense that they accept the same language. With this argument, then, we consider only the case that $\Sigma = \Sigma_n$, and all operation symbols are $n$-ary. The generalization to arbitrary types can be found in [25].

Now we show how to use the components of the quintuple $\underline{A}$ to produce an automaton. We take $I := X_n \cup \Sigma$ as our set of inputs, $S := A^n$ as our set of states, and $O := \{0, 1\}$ as our set of outputs (binary, to correspond to recognizable or not). We define a state transition function $\delta : A^n \times (X_n \cup \Sigma) \to A^n$, which maps each pair consisting of a state and an input element to a new state, by the two conditions that

$$\delta((a_1, \ldots, a_n), x_i) = (a_i, \ldots, a_i) \text{ for } 1 \le i \le n, \quad \text{and}$$
$$\delta((a_1, \ldots, a_n), f) = (f^A(a_1, \ldots, a_n), \ldots, f^A(a_1, \ldots, a_n)) \quad \text{for } f \in \Sigma_n.$$

Then $\delta$ can be extended to a mapping $\hat{\delta} : A^n \times W_\Sigma(X_n) \to A^n$ by the following definition:

(i) $\hat{\delta}((a_1, \ldots, a_n), x_i) := \delta((a_1, \ldots, a_n), x_i)$,

(ii) $\hat{\delta}((a_1, \ldots, a_n), f_i(t_1, \ldots, t_{n_i}))$
$$:= (f_i(t_1, \ldots, t_{n_i})^A(a_1, \ldots, a_n), \ldots, f_i(t_1, \ldots, t_{n_i})^A(a_1, \ldots, a_n))$$
$$= (f_i^A(t_1^A(a_1, \ldots, a_n), \ldots, t_{n_i}^A(a_1, \ldots, a_n)), \ldots,$$
$$f_i^A(t_1^A(a_1, \ldots, a_n), \ldots, t_{n_i}^A(a_1, \ldots, a_n)))$$
$$= f_i^{A^n}(\hat{\delta}((a_1, \ldots, a_n), t_1), \ldots, \hat{\delta}((a_1, \ldots, a_n), t_{n_i})).$$

The initial state for the automaton is given by the $n$-tuple $(\alpha(x_1), \ldots, \alpha(x_n))$, and the set of final states is $(A')^n$. Finally, the output function $\gamma : A^n \times (X_n \cup \Sigma) \to \{0, 1\}$ is defined by

$$\gamma((a_1, \ldots, a_n), x_i) = \begin{cases} 0 & \text{if } a_i \notin A' \\ 1 & \text{if } a_i \in A' \end{cases}$$

and

$$\gamma((a_1, \ldots, a_n), f_i) = \begin{cases} 0 & \text{if } f_i(a_1, \ldots, a_n) \notin A' \\ 1 & \text{if } f_i(a_1, \ldots, a_n) \in A' \end{cases}.$$

Thus we have shown that tree automata correspond to certain automata, and since automata can be regarded as $F$-coalgebras, we see that every tree automaton can be transformed into an $F$-coalgebra. However, it does not follow from our construction of the automaton corresponding to a tree automaton that homomorphisms of tree automata are also $F$-coalgebra homomorphisms. We first give a lemma which relates algebraic homomorphisms with automata homomorphisms. Let

$$\underline{\mathcal{A}} = (\mathcal{A}, \Sigma, X_n, \alpha, A') \text{ and } \underline{\mathcal{B}} = (\mathcal{B}, \Sigma, X_n, \beta, B')$$

be two tree automata, and let $(A^n; \delta, \gamma, (\alpha(x_1), \ldots, \alpha(x_n)), (A')^n)$ and $(B^n; \delta', \gamma', (\beta(x_1), \ldots, \beta(x_n)), (B')^n)$ be the corresponding automata with initial states and sets of final states. We use the notation $\varphi^{\otimes n}$ for the tensor mapping $\varphi \otimes \ldots \otimes \varphi$, where $\varphi$ occurs $n$ times. We are interested in the properties of this mapping $\varphi^{\otimes n}$, in particular when it is an automata homomorphism. Translating the definition of an automata homomorphism from Definition 2.9.1 into the notation here for the two corresponding automata, we see that $\varphi^{\otimes n}$ is an automata homomorphism iff the following two conditions hold, for all $e \in \Sigma \cup X_n$ and all $a_1, \ldots, a_n \in A$:

(AH1)  $\gamma((a_1, \ldots, a_n), e) = \gamma'(\varphi^{\otimes n}((a_1, \ldots, a_n), e)),$
(AH2)  $\varphi^{\otimes n}(\delta((a_1, \ldots, a_n), e)) = \delta'(\varphi^{\otimes n}(a_1, \ldots, a_n), e).$

**Lemma 6.1.2.** *Let $\underline{\mathcal{A}}$ and $\underline{\mathcal{B}}$ be two tree automata, as above. A mapping $\varphi : A \to B$ is an algebra homomorphism $\varphi : A \to B$ iff*

$$\varphi^{\otimes n}(\delta((a_1, \ldots, a_n), f_i)) = \delta'(\varphi^{\otimes n}(a_1, \ldots, a_n), f_i)$$

*for all $i \in I$ and all $(a_1, \ldots, a_n) \in A^n$.*

**Proof:** The mapping $\varphi$ is an algebra homomorphism iff for all $i \in I$ and any $n$-tuple $a_1, \ldots, a_n$ from $A^n$,

$$
\begin{aligned}
&\varphi(f_i^A(a_1, \ldots, a_n)) = f_i^B(\varphi(a_1), \ldots, \varphi(a_n)) \\
\Leftrightarrow \quad &(\varphi(f_i^A(a_1, \ldots, a_n)), \ldots, \varphi(f_i^A(a_1, \ldots, a_n))) \\
= \quad &(f_i^B(\varphi(a_1), \ldots, \varphi(a_n)), \ldots, f_i^B(\varphi(a_1), \ldots, \varphi(a_n))) \\
\Leftrightarrow \quad &(\varphi^{\otimes n}(f_i^A(a_1, \ldots, a_n), \ldots, f_i^A(a_1, \ldots, a_n))) \\
= \quad &(f_i^B(\varphi^{\otimes n}(a_1, \ldots, a_n)), \ldots, \varphi^{\otimes n}(a_1, \ldots, a_n)) \\
\Leftrightarrow \quad &\varphi^{\otimes n}(\delta(a_1, \ldots, a_n), f_i) = \delta'(\varphi^{\otimes n}(a_1, \ldots, a_n), f_i).
\end{aligned}
$$
∎

This lemma now lets us prove the following result.

**Theorem 6.1.3.** *Let $\underline{\mathcal{A}}$ and $\underline{\mathcal{B}}$ be two tree automata, as above. A mapping $\varphi : A \to B$ is a tree automata homomorphism iff $\varphi^{\otimes n}$ is a homomorphism of the corresponding automata and preserves the initial element.*

**Proof:** First let $\varphi$ be a tree automata homomorphism, satisfying the three conditions of Definition 6.1.1; we show that $\varphi^{\otimes n}$ satisfies the conditions (AH1) and (AH2) above. By condition (i), $\varphi$ is at least an algebra homomorphism, and thus applying Lemma 6.1.2 we get that condition (AH2) is satisfied for any $e \in \Sigma$. This means we need only show that (AH2) is satisfied for any $x_i \in X_n$, and then that (AH1) is satisfied. For the first requirement, we have $\varphi^{\otimes n}(\delta(a_1, \ldots, a_n), x_i) = (\varphi(a_i), \ldots, \varphi(a_i)) = \delta'(\varphi^{\otimes n}(a_1, \ldots, a_n))$ for all $i \in I$ and all $a_1, \ldots, a_n \in A$.

To check (AH1), we note that by condition (iii) of Definition 6.1.1 we have $a \in A'$ iff $\varphi(a) \in B'$ for all $a \in A$. Therefore

$$\gamma((a_1, \ldots, a_n), x_i) = \begin{cases} 0 \text{ if } a_i \notin A' \\ 1 \text{ if } a_i \in A' \end{cases} = \gamma'(\varphi^{\otimes n}(a_1, \ldots, a_n), x_i)$$

$$= \begin{cases} 0 \text{ if } \varphi(a_i) \notin B' \\ 1 \text{ if } \varphi(a_i) \in B' \end{cases}$$

and

$$\gamma((a_1, \ldots, a_n), f_i) = \begin{cases} 0 \text{ if } f_i(a_1, \ldots, a_n) \notin A' \\ 1 \text{ if } f_i(a_1, \ldots, a_n) \in A' \end{cases}$$

$$= \gamma'(\varphi^{\otimes n}(a_1, \ldots, a_n), f_i)$$

$$= \begin{cases} 0 \text{ if } f_i^B(\varphi^{\otimes n}(a_1, \ldots, a_n)) \notin B' \\ 1 \text{ if } f_i^B(\varphi^{\otimes n}(a_1, \ldots, a_n)) \in B' \end{cases} .$$

The extra condition for initial states, that

$$\varphi^{\otimes n}(\alpha(x_1), \ldots, \alpha(x_n)) = (\beta(x_1), \ldots, \beta(x_n)),$$

follows from Definition 6.1.1 (ii).

Conversely, suppose that

$$\varphi^{\otimes n} : (A^n; \delta, \gamma) \to (B^n; \delta', \gamma')$$

is an automata homomorphism preserving the initial element, so that $\varphi(\alpha(x_1), \ldots, \alpha(x_n)) = (\beta(x_1), \ldots, \beta(x_n))$. Property (AH2) includes as one case the "if" condition of Lemma 6.1.2, allowing us to conclude that the mapping $\varphi$ is an algebra homomorphism. We also know that $\varphi(\alpha(x_i)) = \beta(x_i)$ for all $x_i \in X_n$, and it remains to show that condition (iii) from Definition 6.1.1 is satisfied. For any $a \in A'$, we have $\gamma((a, \ldots, a), x_i) = 1$, and since $\gamma((a, \ldots, a), x_i) = \gamma'(\varphi^{\otimes n}(a, \ldots, a), x_i)$ by assumption we get $\gamma'(\varphi^{\otimes n}(a, \ldots, a), x_i) = 1$. This means that $\varphi(a) \in B'$ and so $a \in \varphi^{-1}(B')$. Conversely, if $a \in \varphi^{-1}(B')$, then $\varphi(a) \in B'$ and $\gamma(\varphi^{\otimes n}(a, \ldots, a), x_i) = \gamma((a, \ldots, a), x_i) = 1$, forcing $a \in A'$. This shows that $\varphi^{-1}(B') = A'$, as required. ∎

A proof similar to that of Lemma 6.1.2 gives the following fact.

**Corollary 6.1.4.** *Let $\underline{A}$ and $\underline{B}$ be two tree automata, as above, and $\varphi : A \to B$ a mapping. Then $\varphi^{\otimes n}$ is an automata homomorphism iff the diagram in Figure 6.2 commutes.*

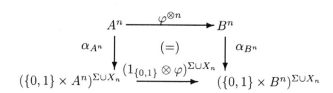

Fig. 6.2   Tree Automata Homomorphism

Now we put these pieces together to obtain the following theorem.

**Theorem 6.1.5.** *Let $\underline{A} = (A, \Sigma, X_n, \alpha, A')$ and $\underline{B} = (B, \Sigma, X_n, \beta, B')$ be two tree automata with $\Sigma = \Sigma_n$, and let $\varphi : A \to B$ be a mapping. Then $\varphi$ is a tree automata homomorphism iff the diagram in Figure 6.2 commutes and $\varphi^{\otimes n}(\alpha(x_1), \ldots, \alpha(x_n)) = (\beta(x_1), \ldots, \beta(x_n))$.*

**Proof:** If $\varphi$ is a tree automata homomorphism, then by Theorem 6.1.3 $\varphi^{\otimes n}(\alpha(x_1), \ldots, \alpha(x_n)) = (\beta(x_1), \ldots, \beta(x_n))$ holds. By the same theorem $\varphi^{\otimes n}$ is a homomorphism of the corresponding automata, and by Corollary 6.1.4 the diagram in Figure 6.2 commutes. The opposite direction also follows from Theorem 6.1.3 and Corollary 6.1.4. ∎

We emphasize here that the mapping $\alpha_{A^n}$ in the diagram in Figure 6.2 is not the structural mapping of a coalgebra, since it maps only $n$-th cartesian powers of sets to functorial images of these sets. Thus to think of tree automata and their homomorphisms in coalgebra terms, it is natural to consider using two functors. We need the functor $F_1$ mapping each set $X$ to the cartesian power $X^n$, and taking each mapping $f : X \to Y$ to the mapping $F_1(f) := f^{\otimes n} : X^n \to Y^n$. We also need a second functor $F_2$ which maps each set $X$ to $(\{0,1\} \times X)^{\Sigma \cup X_n}$ and each mapping $f : X \to Y$ to $F_2(f) := (1_{\{0,1\}} \otimes f)^{\Sigma \cup X_n} : (\{0,1\} \times X^n)^{\Sigma \cup X_n} \to (\{0,1\} \times Y^n)^{\Sigma \cup X_n}$. For each set $A$ we define a structural mapping $\alpha_A : A^n \to (\{0,1\} \times A^n)^{\Sigma \cup X_n}$. This structural mapping $\alpha_A$ then maps from $F_1(A)$ to $F_2(A)$, with $(\alpha_A(a_1, \ldots, a_n))(e) = (\gamma((a_1, \ldots, a_n), e), \delta((a_1, \ldots, a_n), e)) =$

$(\gamma \otimes \delta)((a_1, \ldots, a_n), e)$ for all $e \in \Sigma \cup X_n$ and all $(a_1, \ldots, a_n) \in A^n$. Now our original tree automaton $\underline{A}$ corresponds to the structure $(A; \alpha_A)$ with $(\alpha(x_1), \ldots, \alpha(x_n))$ as initial state and $(A')^n$ as set of final states. This construction leads us to the idea of an $(F_1, F_2)$-coalgebra, using these two different functors.

## 6.2 $(F_1, F_2)$-Coalgebras and Their Homomorphisms

This section offers a brief introduction to the topic of $(F_1, F_2)$-coalgebras. More details may be found in [25].

**Definition 6.2.1.** Let $F_1$, $F_2$ be endofunctors of the category $\mathbf{Set}$. An $(F_1, F_2)$-*coalgebra* is a pair $(A; \alpha_A)$ consisting of a set $A$ and a mapping $\alpha_A : F_1(A) \to F_2(A)$. If $a \in A$ is a distinguished element of $A$, then $(A; \alpha_A, a)$ is said to be a *pointed* $(F_1, F_2)$-coalgebra.

For any functor $F$, this definition encompasses both the $F$-coalgebras of Chapter 4 and the $F$-algebras of Chapter 5, in the following way. If we take $F_1$ to be the identity morphism and $F_2 = F$, then $(A; \alpha_A)$ is an $F$-coalgebra; and if $F_2$ is the identity morphism and $F_1 = F$, then $(A; \alpha_A)$ is an $F$-algebra. Tree automata with $\Sigma = \Sigma_n$, which were our motivation for this concept, form a third example of $(F_1, F_2)$-coalgebras: here we use the two special functors $F_1$ and $F_2$ given at the end of the previous section. If we consider a tree automaton with initial element we need the concept of a pointed $(F_1, F_2)$-coalgebra.

Several other structures can also be naturally represented as $(F_1, F_2)$-coalgebras. A *power algebra*, sometimes also called a *hyper-structure*, is a pair $\mathcal{A} = (A; (f_i^{\mathcal{A}})_{i \in i})$ consisting of a non-empty set $A$ and an indexed set of operations $f_i^{\mathcal{A}} : A^{n_i} \to \mathcal{P}(A)$. Power algebras can be used to prove that every non-deterministic tree automaton is equivalent to a deterministic one. We can regard power algebras as $(F_1, F_2)$-coalgebras, by taking $F_1$ to be the functor mapping each set $X$ to the coproduct $\sum_{i \in I} X^{n_i}$, and $F_2$ to be the functor which maps each set $X$ to its power set $\mathcal{P}(X)$. For the action of these functors on functions, for any morphism $f : X \to Y$, we let $F_1(f)$ map $\sum_{i \in I} X^{n_i}$ to $\sum_{i \in I} Y^{n_i}$ by $F_1(f)(i, (x_1, \ldots, x_{n_i})) = (i, f(x_1), \ldots, f(x_{n_i}))$ for all $i \in I$ and $(x_1 \ldots, x_{n_i}) \in A^{n_i}$; and let $F_2(f)$ map $\mathcal{P}(X)$ to $\mathcal{P}(Y)$.

Let $A^{\sqcup n_i}$ be the $n_i$-th copower of $A$. Mappings $f_i : \mathcal{P}(A) \to A^{\sqcup n_i}$ are called $n_i$-*ary power co-operations*. *Power coalgebras* are pairs $(A; (f_i^{\mathcal{A}})_{i \in I})$

where $(f_i^A)_{i \in I}$ is an indexed set of power co-operations. Such power coalgebras can also be viewed as $(F_1, F_2)$-coalgebras: we take $F_1$ to be the power set functor, and $F_2$ to be the functor mapping each set to the direct product $\prod_{i \in I} A^{\sqcup n_i}$ and each mapping $f : A \to B$ to $F(f) : \prod_{i \in I} A^{\sqcup n_i} \to \prod_{i \in I} B^{\sqcup n_i}$, which maps $(k_j, a_j)_{j \in J}$ to $(k_j, f(a_j))_{j \in J}$ for $k_j \in \{1, \dots, n_j\}$.

Similar approaches which combine algebras with coalgebras can be found in several papers, for instance in [7], [66] and [61].

Homomorphisms of $(F_1, F_2)$-coalgebras are defined as follows.

**Definition 6.2.2.** Let $\mathcal{A} = (A; \alpha_A)$ and $\mathcal{B} = (B; \alpha_B)$ be $(F_1, F_2)$-coalgebras. A mapping $\varphi : A \to B$ is called a *homomorphism* from $\mathcal{A}$ to $\mathcal{B}$ if $\alpha_B \circ F_1(\varphi) = F_2(\varphi) \circ \alpha_A$, that is, if the diagram in Figure 6.3 commutes.

The theory of $(F_1, F_2)$-coalgebras can be developed along the same lines as the theory of $F$-coalgebras (see for instance [75] or [44]). It is easy to check that for any set $A$ the identity mapping $1_A$ is a homomorphism and that the composition of two homomorphisms is again a homomorphism. These two facts ensure that the collection of all $(F_1, F_2)$-coalgebras together with all their homomorphisms forms a category, which we shall denote by $\mathbf{Set}_{(F_1, F_2)}$.

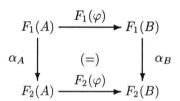

Fig. 6.3    Homomorphisms for $(F_1, F_2)$-coalgebras

$(F_1, F_2)$-coalgebras are related to natural transformations. We recall from Section 3.9 that for categories $\mathbf{C}$ and $\mathbf{D}$ with functors $F_1, F_2 : \mathbf{C} \to \mathbf{D}$, a natural transformation is a mapping $\eta$ from $F_1$ to $F_2$ which maps each object $X$ from $\mathbf{C}$ to a morphism $\eta_X : F_1(X) \to F_2(X)$, such that for every morphism $f : X \to Y$ in $\mathbf{C}$ the equation

$$F_2(f) \circ \eta_X = \eta_Y \circ F_1(f)$$

is satisfied. We write $\eta : F_1 \to F_2$ to denote this natural transformation. The following corollary shows that the category $\mathbf{Set}$ can be regarded as a category of $(F_1, F_2)$-coalgebras if there is a natural transformation $\eta : F_1 \to F_2$.

**Corollary 6.2.3.** *Let* $F_1, F_2 : \mathbf{Set} \to \mathbf{Set}$ *be two set-valued functors and let* $\eta : F_1 \to F_2$ *be a natural transformation. Then* $\eta$ *defines a category* $\mathbf{Set}_{(F_1, F_2)}$ *of* $(F_1, F_2)$*-coalgebras.*

**Proof:** For any set $X$, since $\eta_X : F_1(X) \to F_2(X)$ is a function, the pair $(X; \eta_X)$ is an $(F_1, F_2)$-coalgebra. Also for any mapping $f : X \to Y$, since $\eta$ is a natural transformation, the diagram in Figure 6.4 commutes.

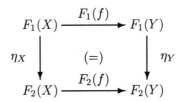

Fig. 6.4    Natural Transformations and $(F_1, F_2)$-coalgebras

But this means precisely that $f$ is a homomorphism from the $(F_1, F_2)$-coalgebra $(X; \eta_X)$ to the $(F_1, F_2)$-coalgebra $(Y; \eta_Y)$.                                    ∎

## 6.3    Limits and Colimits in $\mathbf{Set}_{(F_1, F_2)}$

As we saw in Chapter 3, an important question in any category is whether limits and colimits exist in the category. Here we examine the different limit and colimit constructions in the category $\mathbf{Set}_{(F_1, F_2)}$: the products, coproducts (sums), equalizers, coequalizers, pullbacks and pushouts.

Beginning with sums, we recall that $\sum_{i \in I} A_i$ together with the mappings $(e_i : A_i \to \sum_{i \in I} A_i)_{i \in I}$ is the sum of the family $(A_i)_{i \in I}$ in the category $\mathbf{Set}$. Now we want to see if we can produce a structural mapping $\alpha : \sum_{i \in I} A_i \to F(\sum_{i \in I} A_i)$ to make this sum into an $(F_1, F_2)$-coalgebra. We say that a functor $F : \mathbf{C} \to \mathbf{D}$ preserves sums if for any sum $\sum_{i \in I} A_i$ with

$(e_i : A_i \to \sum\limits_{i \in I} A_i)_{i \in I}$ in $\mathbf{C}$ the object $F(\sum\limits_{i \in I} A_i) = \sum\limits_{i \in I} F(A_i)$ with $(F(e_i) :$ $F(A_i) \to \sum\limits_{i \in I} F(A_i))_{i \in I}$ is the sum of the family $(F(A_i))_{i \in I}$ in the category $\mathbf{D}$.

**Proposition 6.3.1.** *If the functor $F_1 : \mathbf{Set} \to \mathbf{Set}$ preserves sums, then the sum $\sum\limits_{i \in I} A_i$ in the category $\mathbf{Set}$ of the universes of the $(F_1, F_2)$-coalgebras from the family $(\mathcal{A}_i)_{i \in I}$ is the sum of the family $(\mathcal{A}_i)_{i \in I}$ in $\mathbf{Set}_{(F_1, F_2)}$.*

**Proof**: Using the mappings $\alpha_{A_i} : F_1(\mathcal{A}_i) \to F_2(\mathcal{A}_i)$ and $F_2(e_i) : F_2(\mathcal{A}_i) \to F_2(\sum\limits_{i \in I} A_i)$ for all $i \in I$, we get a family $(F_2(e_i) \circ \alpha_{A_i} : F_1(A_i) \to F_2(\sum\limits_{i \in I} A_i))_{i \in I}$ of mappings. Since $F_1(\sum\limits_{i \in I} A_i)$ together with the family $F_1(e_i) : F_1(A_i) \to F_1(\sum\limits_{i \in I} A_i)$ is the sum of the family $(F_1(A_i))_{i \in I}$ in $\mathbf{Set}$, we have a uniquely determined morphism $\alpha : F_1(\sum\limits_{i \in I} A_i) \to F_2(\sum\limits_{i \in I} A_i)$ such that $\alpha \circ F_1(e_i) = F_2(e_i) \circ \alpha_{A_i}$ for all $i \in I$. This shows that $(\sum\limits_{i \in I} A_i; \alpha)$ is an $(F_1, F_2)$-coalgebra and the mappings $e_i : A_i \to \sum\limits_{i \in I} A_i$ are homomorphisms for all $i \in I$.

To show that $(\sum\limits_{i \in I} A_i; \alpha)$ together with this family $(e_i : A_i \to \sum\limits_{i \in I} A_i)_{i \in I}$ is the sum of the $A_i$'s in $\mathbf{Set}_{(F_1, F_2)}$, we consider any $(F_1, F_2)$-coalgebra $(Q; \alpha_Q)$ and any family $(q_i : A_i \to Q)_{i \in I}$ of homomorphisms. If we consider the underlying sets we get a uniquely determined mapping $h : \sum\limits_{i \in I} A_i \to Q$ such that $h \circ e_i = q_i$ for all $i \in I$. We need to show that $h$ is a homomorphism, that is, that $\alpha_Q \circ F_1(h) = F_2(h) \circ \alpha$. For all $i \in I$,

$$
\begin{aligned}
&\alpha_Q \circ F_1(h) \circ F_1(e_i) \\
=\ &\alpha_Q \circ F_1(h \circ e_i) \\
=\ &\alpha_Q \circ F_1(q_i) \\
=\ &F_2(q_i) \circ \alpha_{A_i} \\
=\ &F_2(h \circ e_i) \circ \alpha_{A_i} \\
=\ &F_2(h) \circ F_2(e_i) \circ \alpha_{A_i} \\
=\ &F_2(h) \circ \alpha \circ F_1(e_i).
\end{aligned}
$$

The injectivity of the mapping $e_i : A_i \to \sum\limits_{i \in I} A_i$, for $i \in I$, forces each $F_1(e_i)$ to be a monomorphism in $\mathbf{Set}$, and from the equation $\alpha_Q \circ F_1(h) \circ F_1(e_i) = F_2(h) \circ \alpha \circ F_1(e_i)$ we see that $\alpha_Q \circ F_1(h) = F_2(h) \circ \alpha$. ∎

As usual, the product construction can be seen as the dual of the sum construction, by reversing all the arrows. We leave the proof of the following result as an exercise.

**Proposition 6.3.2.** *If the functor $F_2 : \text{Set} \to \text{Set}$ preserves products, then for any family $(\mathcal{A}_i)_{i \in I}$ of $(F_1, F_2)$-coalgebras in $\text{Set}_{(F_1, F_2)}$, the product of the universes of the $(F_1, F_2)$-coalgebras in the category $\text{Set}$ is the product of the family $(\mathcal{A}_i)_{i \in I}$ in $\text{Set}_{(F_1, F_2)}$.*

As we have mentioned, $(F_1, F_2)$-coalgebras include as special cases both $F$-coalgebras and $F$-algebras, by using the identity functor for one of $F_1$ or $F_2$ and $F$ for the other. Since the identity functor preserves both sums and products, the previous two propositions encompass some of our theorems regarding sums and products from Chapter 3. Arbitrary sums exist in the category $\text{Set}_F$ of $F$-coalgebras, and are equal to sums in $\text{Set}$; and products exist in the category $\text{Set}^F$ of $F$-algebras, and are equal to products in $\text{Set}$. If $F$ preserves products, then arbitrary products in the category of $F$-coalgebras exist and if $F$ preserves sums, then arbitrary sums in the category of $F$-algebras exist.

Next we turn to coequalizers and equalizers in $\text{Set}_{(F_1, F_2)}$. It should be clear what it means for a functor $F : \text{Set} \to \text{Set}$ to preserve coequalizers.

**Proposition 6.3.3.** *Let $(f_i : \mathcal{A} \to \mathcal{B})_{i \in I}$ be a family of $(F_1, F_2)$-coalgebra homomorphisms. Let $\theta$ be the equivalence relation which is generated by the relation $R := \{(f_i(a), f_j(a)) \mid a \in A, i, j \in I\}$, and let $\pi_\theta : B \to B/\theta$ be the coequalizer of the mappings $f_i$ in $\text{Set}$. If the functor $F_1 : \text{Set} \to \text{Set}$ preserves coequalizers, then there is a uniquely determined structural mapping $\alpha_\theta : F_1(B/\theta) \to F_2(B/\theta)$ such that $\pi_\theta : B \to B/\theta$ is the coequalizer of the homomorphisms $f_i$ in the category $\text{Set}_{(F_1, F_2)}$ of $(F_1, F_2)$-coalgebras.*

**Proof:** Since $F_1 : \text{Set} \to \text{Set}$ preserves coequalizers and $\pi_\theta : B \to B/\theta$ is the coequalizer of the mappings $f_i$ in $\text{Set}$, the mapping $F_1(\pi_\theta) : F_1(B) \to F_1(B/\theta)$ is the coequalizer of the mappings $(F_1(f_i) : F_1(A) \to F_1(B))_{i \in I}$. There is a uniquely determined mapping $\alpha_\theta : F_1(B/\theta) \to F_2(B/\theta)$ such that $\alpha_\theta \circ F_1(\pi_\theta) = F_2(\pi_\theta) \circ \alpha_B$. This makes $(B/\theta; \alpha_\theta)$ an $(F_1, F_2)$-coalgebra, with $\pi_\theta : \mathcal{B} \to \mathcal{B}/\theta$ a homomorphism.

Since $\pi_\theta : \mathcal{B} \to \mathcal{B}/\theta$ is the coequalizer of the family $(f_i)_{i \in I}$ in the category $\text{Set}$, the first condition required for a coequalizer, that $\pi_\theta \circ f_i = \pi_\theta \circ f_j$ for all $i, j \in I$, is satisfied. It remains then to show the second condition for a coequalizer, that for any $(F_1, F_2)$-coalgebra $(Q; \gamma)$ and any homomorphism $q : \mathcal{B} \to \mathcal{Q}$ such that $q \circ f_i = q \circ f_j$ for all $i, j \in I$, there exists a unique homomorphism $h$ with $q = h \circ \pi_\theta$. Since $\pi_\theta : \mathcal{B} \to \mathcal{B}/\theta$ is the coequalizer of the family $(f_i)_{i\in I}$ in the category **Set**, there does exist a uniquely determined such mapping $h : B/\theta \to Q$, with $q = h \circ \pi_\theta$, and we need to verify that $h$ is a homomorphism. The diagram in Figure 6.5 illustrates this situation.

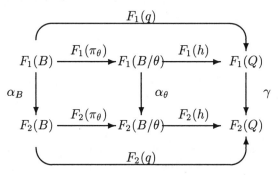

Fig. 6.5   Coequalizer Diagram

Now we have $\gamma \circ F_1(h) \circ F_1(\pi_\theta) = \gamma \circ F_1(h \circ \pi_\theta) = \gamma \circ F_1(q) = F_2(q) \circ \alpha_B$ $= F_2(h \circ \pi_\theta) \circ \alpha_B = F_2(h) \circ F_2(\pi_\theta) \circ \alpha_B = F_2(h) \circ \alpha_\theta \circ F_1(\pi_\theta)$. Since $\pi_\theta : B \to B/\theta$ is surjective, the mapping $\pi_\theta : B \to B/\theta$ is an epimorphism and therefore $F_1(\pi_\theta)$ is an epimorphism. It follows from this and the previous equalities that $\gamma \circ F_1(h) = F_2(h) \circ \alpha_\theta$. This shows that $h$ is a homomorphism, and hence that $\pi_\theta : \mathcal{B} \to \mathcal{B}/\theta$ is the coequalizer of the family $(f_i)_{i\in I}$ of homomorphisms in the category of $(F_1, F_2)$-coalgebras. ∎

Again this result can be dualized, to give a result about equalizers.

**Proposition 6.3.4.** *Let $(f_i : \mathcal{A} \to \mathcal{B})_{i\in I}$ be a family of $(F_1, F_2)$-coalgebra homomorphisms. Let $E = \{a \in A \mid f_i(a) = f_j(a), i, j \in I\}$, and let $g : E \to A$ be the equalizer of the mappings $f_i$ in the category* **Set***. If the functor $F_2 :$ **Set** $\to$ **Set** *preserves equalizers, then there is a uniquely determined structural mapping $\alpha_E$ on $E$ such that $g : \mathcal{E} \to \mathcal{A}$ is the equalizer of the homomorphisms $f_i$ in the category* **Set**$_{(F_1,F_2)}$.

Finally we turn to pullbacks and pushouts in **Set**$_{(F_1,F_2)}$.

**Theorem 6.3.5.** *Let $(\varphi_i : \mathcal{A} \to \mathcal{B}_i)_{i\in I}$ be a family of homomorphisms in* **Set**$_{(F_1,F_2)}$. *Let $P$ together with a family $(\psi_i : B_i \to P)_{i\in I}$ be the pushout of* the mappings $\varphi_i$ in the category **Set**. If the functor $F_1 : \mathbf{Set} \to \mathbf{Set}$ preserves pushouts, then there is a uniquely determined $(F_1, F_2)$-coalgebra structure $\alpha_P$ on $P$ such that the $(\psi_i)_{i \in I}$ are homomorphisms to $\mathcal{P} = (P; \alpha_P)$, and $\mathcal{P}$ together with the $(\psi_i)_{i \in I}$ is the pushout of the $\varphi_i$'s in $\mathbf{Set}_{(F_1, F_2)}$.

**Proof:** As always we start with the analogous construction in the category **Set**: since $F_1$ preserves pushouts, the object $F_1(P)$ together with the family $(F_1(\psi_i) : F_1(B_i) \to F_1(P))_{i \in I}$ is the pushout of the family $(F_1(\varphi_i))_{i \in I}$ in the category **Set**. For the object $F_2(P)$ and the morphisms $(F_2(\psi_i) \circ \alpha_{B_i})_{i \in I}$, we have for all $i, j \in I$

$$
\begin{aligned}
F_2(\psi_i) \circ \alpha_{B_i} \circ F_1(\varphi_i) &= F_2(\psi_i) \circ F_2(\varphi_i) \circ \alpha_A \\
&= F_2(\psi_i \circ \varphi_i) \circ \alpha_A \\
&= F_2(\psi_j \circ \varphi_j) \circ \alpha_A \\
&= F_2(\psi_j) \circ F_2(\varphi_j) \circ \alpha_A \\
&= F_2(\psi_j) \circ \alpha_{B_j} \circ F_1(\varphi_j).
\end{aligned}
$$

This can be seen in the diagram in Figure 6.6.

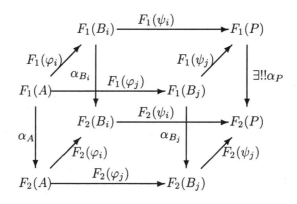

Fig. 6.6   Pushouts Diagram

We know therefore that there exists a uniquely determined structural mapping $\alpha_P : F_1(P) \to F_2(P)$ such that $\alpha_P \circ F_1(\psi_i) = F_2(\psi_i) \circ \alpha_{B_i}$ and $(P; \alpha_P)$ is an $(F_1, F_2)$-coalgebra and the maps $\psi_i : B_i \to \mathcal{P}$ for $i \in I$ are homomorphisms. To show that $\mathcal{P}$ together with $(\psi_i : B_i \to \mathcal{P})_{i \in I}$ is the pushout of the family $(\varphi_i)_{i \in I}$ of homomorphisms in $\mathbf{Set}_{(F_1, F_2)}$ we have to verify the following two conditions:

(i)  $\psi_i \circ \varphi_i = \psi_j \circ \varphi_j$  for all  $i, j \in I$ .
(ii) If  $Q$  is any  $(F_1, F_2)$ -coalgebra and  $(q_i : \mathcal{B}_i \to Q)_{i \in I}$  is any family of homomorphisms satisfying  $q_i \circ \varphi_i = q_j \circ \varphi_j$  for all  $i, j \in I$ , then there exists a uniquely determined homomorphism  $h : \mathcal{P} \to Q$  such that  $h \circ \psi_i = q_i$  for all  $i \in I$ .

Condition (i) is immediate from the category **Set**. For condition (ii), let  $(Q; \alpha_Q)$  be any  $(F_1, F_2)$ -coalgebra and let  $(q_i : \mathcal{B}_i \to Q)_{i \in I}$  be any family of homomorphisms in **Set**$_{(F_1, F_2)}$  such that  $q_i \circ \varphi_i = q_j \circ \varphi_j$  for all  $i, j \in I$ . Again since  $\mathcal{P}$  together with the family  $(\psi_i : \mathcal{B}_i \to P)_{i \in I}$  is the pushout of the family  $(\varphi_i)_{i \in I}$  of mappings in **Set**, there is a uniquely determined mapping  $h : P \to Q$  such that  $h \circ \psi_i = q_i$  for all  $i \in I$ , and we have to show that  $h$  is a homomorphism. For this we need  $\alpha_Q \circ F_1(h) = F_2(h) \circ \alpha_P$ . Since the object  $F_1(P)$  together with the family  $(F_1(\psi_i) : F_1(B_i) \to F_1(P))_{i \in I}$  is the pushout of the family  $(F_1(\varphi_i) : F_1(A) \to F_1(B_i))_{i \in I}$  in **Set** we have the situation shown in the diagram in Figure 6.7.

Since for all  $i, j \in I$ 

$$
\begin{aligned}
F_2(q_i) \circ \alpha_{B_i} \circ F_1(\varphi_i) &= F_2(q_i) \circ F_2(\varphi_i) \circ \alpha_A \\
&= F_2(q_i \circ \varphi_i) \circ \alpha_A \\
&= F_2(q_j \circ \varphi_j) \circ \alpha_A \\
&= F_2(q_j) \circ F_2(\varphi_j) \circ \alpha_A \\
&= F_2(q_j) \circ \alpha_{B_j} \circ F_1(\varphi_j),
\end{aligned}
$$

there is precisely one morphism  $\sigma : F_1(P) \to F_2(Q)$  such that  $\sigma \circ F_1(\psi_i) = F_2(q_i) \circ \alpha_{B_i}$  for all  $i \in I$ . Since  $F_2(h) \circ \alpha_P \circ F_1(\psi_i) = F_2(h) \circ F_2(\psi_i) \circ \alpha_{B_i} = F_2(q_i) \circ \alpha_{B_i}$ , the uniqueness of  $\sigma$  gives us  $\sigma = F_2(h) \circ \alpha_P$ . Moreover,  $\alpha_Q \circ F_2(h) \circ F_1(\psi_i) = \alpha_Q \circ F_1(q_i) = F_2(q_i) \circ \alpha_{B_i}$ , so that  $\sigma = \alpha_Q \circ F_1(h)$  also. Altogether,  $F_1(h) \circ \alpha_P = \alpha_Q \circ F_1(h)$  as required, and  $h$  is a homomorphism. ∎

Again there is a dual result.

**Theorem 6.3.6.** *Let*  $(\varphi_i : \mathcal{A}_i \to \mathcal{B})_{i \in I}$  *be a family of homomorphisms in* **Set**$_{(F_1, F_2)}$ . *Let  $P$  together with  $(\psi_i : P \to A_i)_{i \in I}$  be the pullback of the family  $(\varphi_i)_{i \in I}$  of mappings in* **Set**. *If the functor  $F_2 :$* **Set** $\to$ **Set** *preserves pullbacks, then there is a uniquely determined  $(F_1, F_2)$ -coalgebra structure  $\alpha_P$  on  $P$  such that the  $\psi_i$ 's are homomorphisms and  $\mathcal{P} = (P; \alpha_P)$  together* with the family $(\psi_i)_{i \in I}$ is the pullback of the family $(\varphi_i : \mathcal{A}_i \to \mathcal{B})_{i \in I}$ in $\mathbf{Set}_{(F_1, F_2)}$.

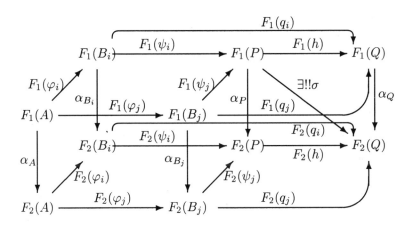

Fig. 6.7   Second Pushout Diagram

As with sums and products, our theorems include as special cases results for $F$-coalgebras and $F$-algebras. In the category $\mathbf{Set}_F$ of $F$-coalgebras arbitrary coequalizers and arbitrary pushouts exist, and are equal to coequalizers and pushouts respectively in $\mathbf{Set}$; and in the category $\mathbf{Set}^F$ of $F$-algebras arbitrary equalizers and pullbacks exist, and are equal to equalizers and pullbacks in $\mathbf{Set}$. We also know that in any category arbitrary limits exist if products and equalizers exist, and colimits exist if coproducts and coequalizers exist, which gives the following theorem.

**Theorem 6.3.7.** *If $F_1$ preserves colimits, then arbitrary colimits exist in* $\mathbf{Set}_{(F_1, F_2)}$. *If $F_2$ preserves limits, then arbitrary limits exist in* $\mathbf{Set}_{(F_1, F_2)}$.

We mention that the "preservation properties" used for one of the functors $F_1, F_2$ in results 6.3.2 to 6.3.7 are very strong, and in many cases are not satisfied. For example, it is easy to see that the functor $F_1$ which was used to define tree automata as $(F_1, F_2)$-coalgebras at the end of Section 6.1 does not preserve coproducts.

## 6.4 Exercises for Chapter 6

1. Prove the assertion from Section 6.2 that for an automaton $(S; \alpha_S)$ the automata homomorphism conditions

(i) $\gamma(\varphi(a)) = \varphi(\gamma(a))$,
(ii) $\delta(\varphi(a), e) = \varphi(\delta(a, e))$.

correspond to the coalgebra homomorphism condition $\alpha_{S'} \circ \varphi = F(\varphi) \circ \alpha_S$.

2. Prove that for any set $A$ the identity mapping $1_A$ is a homomorphism of $(F_1, F_2)$-coalgebras.

3. Prove that the composition of two $(F_1, F_2)$-coalgebras is also an $(F_1, F_2)$-coalgebra.

4. Give examples of functors $F_1$ and $F_2$ for which both limits and colimits exist in the category $\mathbf{Set}_{(F_1, F_2)}$.

5. Define homomorphisms for power algebras and power coalgebras. Then prove that the collections of all power algebras and of all power coalgebras of a given type $\tau$, together with homomorphisms between them, form categories. behaviour

# Chapter 7

# Terminal Coalgebras

In universal algebra the absolutely free algebras play an important role. Every algebra of type $\tau$ is a homomorphic image, under exactly one homomorphism, of an absolutely free algebra of type $\tau$ with an appropriate generating set. As a result, the free algebra of type $\tau$ with a countably infinite generating system is an initial object in the category $Alg(\tau)$ of all algebras of type $\tau$. In the theory of coalgebras, the dual concept of terminal coalgebras is important: elements of a terminal coalgebra can be viewed as interpreting the behaviour of state-based systems.

## 7.1 Terminal Coalgebras and Bisimulations

A terminal object of a category is an object $T$ with the property that there is exactly one morphism from each object in the category to $T$. For the category $\mathbf{Set}_F$ of $F$-coalgebras, then, an $F$-coalgebra $\mathcal{T}$ is terminal if for every $F$-coalgebra $\mathcal{A}$ there is exactly one homomorphism $\rho$ from $\mathcal{A}$ to $\mathcal{T}$. We proved in Section 3.3 that terminal objects are uniquely determined up to isomorphism, so we can speak of "the" terminal object of $\mathbf{Set}_F$.

As discussed in Chapter 2, bisimulations are used to describe the input-output behaviour of a state-based system. The next theorem gives a key connection between bisimulations and terminal coalgebras.

**Theorem 7.1.1.** *Let $\mathcal{T}$ be a terminal $F$-coalgebra. For every $F$-coalgebra $\mathcal{A} \in \mathbf{Set}_F$ and for every $a \in A$ there is exactly one element $t \in T$ such that $t \sim_{A,T} a$, that is such that $a$ and $t$ are bisimilar.*

**Proof**: We know from Theorem 4.6.9 that bisimulations are precisely the graphs of homomorphisms. Since $\mathcal{T}$ is terminal, for any coalgebra $\mathcal{A}$ there is exactly one homomorphism $\rho : \mathcal{A} \to \mathcal{T}$. Thus we have $a \sim_{A,T} \rho(a) = t$. For the uniqueness, suppose that there is another element $t' \in T$ with $a \sim_{A,T} t'$. By Theorem 4.6.9, bisimulation of the elements $a \in A$ and $t' \in T$ is equivalent to the existence of an $F$-coalgebra $\mathcal{P}$, homomorphisms $\varphi : \mathcal{P} \to \mathcal{A}$ and $\psi : \mathcal{P} \to \mathcal{T}$, and an element $p \in P$ such that $\varphi(p) = a$ and $\psi(p) = t'$. But now we have two homomorphisms from $\mathcal{P}$ to $\mathcal{T}$: the mapping $\psi$ and the composition $\rho \circ \varphi$. The terminal property of $\mathcal{T}$ then forces $\psi = \rho \circ \varphi$. From this we conclude that

$$t' = \psi(p) = (\rho \circ \varphi)(p) = \rho(\varphi(p)) = \rho(a) = t$$

and thus $t = t'$. ∎

Let $R$ be a bisimulation on a state-based system. We have seen that bisimilar states are those with the same behaviour. If the coalgebra representing the system is a terminal coalgebra $\mathcal{T}$, then each state represents a different behaviour. In fact the corresponding projection homomorphisms $\pi_1, \pi_2 : \mathcal{R} \to \mathcal{T}$ must satisfy $\pi_1 = \pi_2$, which means that $R = \Delta_T$ is the diagonal relation on $T$. As a rule of inference this becomes

$$\frac{s \sim s'}{s = s'}.$$

A bisimulation relation $R$ on a coalgebra $\mathcal{A}$ is said to satisfy the *principle of coinduction*, or the *coinduction proof principle* if it satisfies $R \subseteq \Delta_A$; that is, if it satisfies the rule that $s \sim s'$ implies $s = s'$. Satisfaction of this principle is equivalent to the coalgebra being *simple*, meaning that it has no proper homomorphic images. More information about these properties may be found in [75].

What we have shown above then is that any terminal coalgebra is simple, and satisfies the principle of coinduction.

## 7.2 Terminal Automata

In this section we describe an automaton which acts as the terminal coalgebra for the family of finite deterministic automata. We begin by recapping from Section 2.4 how an automaton may be viewed as an $F$-coalgebra for a particular choice of functor $F$.

An automaton is a system $\mathcal{A} = (I, S, O; \delta, \gamma)$, consisting of a set $I$ of inputs, a set $S$ of states, and a set $O$ of outputs, along with a state transition function $\delta : S \times I \to S$ and an output function $\gamma : S \to O$. A recognizer is a similar system $\mathcal{H} = (I, S; \delta)$, but with no output set and no output function; we showed in Section 2.4 that a recognizer can also be considered as an automaton.

The functor used to describe automata is built up of two simpler functors. First, for any set $X$ let $X^I$ be the set of all mappings from $I$ to $X$. The power functor $(-)^I$ maps each set $X$ to the set $X^I$ and each morphism $f : X \to Y$ to the mapping $f^I$ defined by $f^I(u) := f \circ u$. The constant functor $F_O$ maps each set $X$ to the constant set $O$ and each morphism to the identity morphism of $O$. Now we use the construction of the product functor, from Section 3.8: for functors $F$ and $G$, the product functor $F \times G$ maps each set $X$ to the cartesian product $F(X) \times G(X)$ and each morphism $f$ to $(F \times G)(f)(u, v) = (F(f)(u), G(f)(v))$.

This product applied to $F_O$ and $(-)^I$ gives us our functor $F(-) := F_O \times (-)^I$. On our automaton $\mathcal{A} = (I, S, O; \delta, \gamma)$, we impose a coalgebra structure on the set $S$ of states as follows. We define a mapping $\alpha_S : S \to O \times S^I$ with the property that the projection mappings $\pi_1 : O \times X^I \to O$ and $\pi_2 : O \times X^I \to X^I$ satisfy the following two conditions:

(i) $\pi_1(\alpha_S(s)) = \gamma(s)$,
(ii) $(\pi_2(\alpha_S))(e) = \delta(s, e)$.

Then the coalgebra $(S; \alpha_S)$ corresponds to our automaton. A homomorphism $\varphi$ of an automaton is usually given by the conditions

(i) $\gamma(\varphi(a)) = \varphi(\gamma(a))$,
(ii) $\delta(\varphi(a), e) = \varphi(\delta(a, e))$.

It can be shown that these conditions correspond to the coalgebra homomorphism condition $\alpha_{S'} \circ \varphi = F(\varphi) \circ \alpha_S$.

Now we want to describe the terminal automaton $\mathcal{T}$ for the class of automata, within this framework. For the base set (the set of states) we use the set of all infinite trees $t$ of the following form. Each node of a tree has $|I|$ neighbours; and nodes of trees are labeled by elements of the output set $O$. The root of tree $t$ is labeled by $\gamma(t)$. We use $\delta(t, e)$ for the subtree $t_e$ of $t$ whose root is the branch $t_e$ of $t$. Each node is uniquely described by the path going from the node to the root; for instance $d'' = \gamma(\hat{\delta}(t, e_3 e_2 \varepsilon))$ where $\varepsilon$ is the empty word. This path corresponds to a word $w \in I^*$. Therefore, the $O$-labeled trees with $|I|$ branches correspond to the mappings $\rho : I^* \to O$. The universe for our terminal algebra $\mathcal{T}$ is thus the set $O^{I^*}$ of all such mappings.

**Theorem 7.2.1.** *Let $I$ and $O$ be fixed sets of inputs and outputs respectively. Let $\mathcal{T}$ be the coalgebra $(O^{I^*}; \delta, \gamma)$ where $\delta(\rho, e)(w) = \rho(ew)$ and $\gamma(\tau) = \rho(\varepsilon)$. Then $\mathcal{T}$ is the terminal automaton for the automata on $I$ and $O$.*

**Proof**: Let $\mathcal{A} = (A; \delta', \gamma')$ be an arbitrary automaton with input alphabet $I$ and output alphabet $O$. A homomorphism $\varphi : \mathcal{A} \to \mathcal{T}$ has to satisfy

(i) $\varphi(a)(\varepsilon) = \gamma(\varphi(a)) = \gamma'(a)$ and
(ii) $\varphi(a)(ew) = \delta(\varphi(a), e)(w) = \varphi(\delta'(a, e))(w)$.

Conversely, these two conditions define a unique mapping $\varphi : A \to T$. This mapping is defined by induction on the complexity of $w \in I^*$. ∎

## 7.3 Existence of Terminal Coalgebras

Having seen an example of a terminal $F$-coalgebra, for the functor $F$ used to describe finite automata as coalgebras, we now show that there are some functors $F$ for which no terminal $F$-coalgebra exists. We do this by means of the following lemma, which gives an important necessary condition on the structural mapping of a terminal coalgebra.

**Lemma 7.3.1.** (*Lambek's Lemma*) *If $\mathcal{T}$ is a terminal $F$-coalgebra, then its structural mapping $\alpha_T : T \to F(T)$ must be bijective.*

**Proof**: Let $\mathcal{T} = (T; \alpha_T)$ be a terminal $F$-coalgebra. There is also a natural $F$-coalgebra structure on the set $F(T)$, given by $F(\alpha_T) : F(T) \to F(F(T))$. This makes the system $(F(T); F(\alpha_T))$ into an $F$-coalgebra. The structural mapping $\alpha_T$ is trivially a coalgebra homomorphism from the coalgebra $(T; \alpha_T)$ to the coalgebra $(F(T); F(\alpha_T))$. Since $\mathcal{T}$ is terminal, there is exactly one homomorphism $\beta : \mathcal{F}(T) \to \mathcal{T}$. But now we have two coalgebra homomorphisms from $\mathcal{T}$ to $\mathcal{T}$, the identity homomorphism $id_T$ and $\beta \circ \alpha_T$. By the property of terminality then $\beta \circ \alpha_T = id_T$. Since $\beta$ is a homomorphism, we have $\alpha_T \circ \beta = F(\beta) \circ F(\alpha_T) = F(\beta \circ \alpha_T) = F(id_T) = id_{F(T)}$, and so $\alpha_T$ is bijective. ∎

The requirement that the structural mapping $\alpha_T$ of a terminal coalgebra $\mathcal{T}$ be a bijection is very strong. For example, it means that we cannot form terminal coalgebras when the functor $F$ is the power set functor, since there cannot be a bijection between a set and its power set. We saw in Section 2.7 that Kripke structures of the form $(S; R, \ prop)$ over a set $\Phi$ of propositions consist of a set $S$ of states, a binary relation $R \subseteq S \times S$ and a mapping $prop : S \to \mathcal{P}(\Phi)$. The power set functor is needed here to model Kripke structures, and we conclude that there is no terminal Kripke structure.

## 7.4 Weakly Terminal Coalgebras

A terminal coalgebra $\mathcal{T}$ has the property that for every coalgebra $\mathcal{A}$ there is a unique homomorphism from $\mathcal{A}$ to $\mathcal{T}$. In the following definition we generalize the concept of a terminal coalgebra to a *weakly terminal coalgebra*.

**Definition 7.4.1.** An $F$-coalgebra $\mathcal{W}$ is said to be *weakly terminal* if for every $F$-coalgebra $\mathcal{A}$ there is at least one homomorphism $\varphi : \mathcal{A} \to \mathcal{W}$.

There is a close connection between weakly terminal and terminal coalgebras.

**Lemma 7.4.2.** *Let $\mathcal{W}$ be a weakly terminal $F$-coalgebra, and let $\nabla$ be the greatest congruence relation on $\mathcal{W}$. Then the factor coalgebra $\mathcal{W}/\nabla$ is terminal.*

**Proof**: Let $\mathcal{A}$ be any $F$-coalgebra. Since $\mathcal{W}$ is weakly terminal there is a homomorphism $\varphi : \mathcal{A} \to \mathcal{W}$. Then the composition of $\varphi$ with the natural homomorphism $\pi_\nabla$ gives a homomorphism $\pi_\nabla \circ \varphi : \mathcal{A} \to \mathcal{W}/\nabla$ from $\mathcal{A}$ to $\mathcal{W}/\nabla$. To show that $\mathcal{W}/\nabla$ is terminal then we need only show that this homomorphism is unique. Suppose that $\varphi_1$ and $\varphi_2$ are both homomorphisms from $\mathcal{A}$ to $\mathcal{W}/\nabla$. We can form their coequalizer $\psi : \mathcal{W}/\nabla \to \mathcal{B}$, and the first part of the definition of a coequalizer tells us that $\psi \circ \varphi_1 = \psi \circ \varphi_2$. But $\nabla$ is the greatest congruence on $\mathcal{W}$ so there can be no non-trivial congruences on $\mathcal{W}/\nabla$; every homomorphism $\psi : \mathcal{W}/\nabla \to \mathcal{B}$ must be injective. It follows that $\varphi_1 = \varphi_2$. ∎

We recall from Section 3.9 the definition of a natural transformation between functors. Let $\mathbf{C}$ and $\mathbf{D}$ be two categories and let $F$ and $G$ be two functors from $\mathbf{C}$ to $\mathbf{D}$. Then a natural transformation $\eta$ from $F$ to $G$ maps each object $X$ from $\mathbf{C}$ to a $\mathbf{D}$-morphism $\eta_X : F(X) \to G(X)$ such that for every morphism $f$ in $\mathbf{C}$ we have $G(f) \circ \eta_X = \eta_Y \circ F(f)$.

**Lemma 7.4.3.** *Let $F$ and $G$ be two endofunctors of the category* **Set** *and let $\eta : F \to G$ be a surjective natural transformation. If $\mathcal{T} = (T; \pi)$ is a weakly terminal $G$-coalgebra, then $\mathcal{T}_\eta = (T; \eta_T \circ \pi)$ is a weakly terminal $F$-coalgebra.*

**Proof:** Let $\mathcal{A} = (A; \alpha_A)$ be a non-empty $F$-coalgebra. By the Axiom of Choice we may select from the homomorphism set $Hom(F(\mathcal{A}), G(\mathcal{A}))$ a homomorphism $h : F(\mathcal{A}) \to G(\mathcal{A})$ with $\eta_A \circ h = id_{F(\mathcal{A})}$. Now we consider the $G$-coalgebra $\mathcal{A}_G = (A; h \circ \alpha_A)$. There exists a $G$-homomorphism $\varphi : \mathcal{A}_G \to \mathcal{T}$, and

$$
\begin{aligned}
&F(\varphi \circ \alpha) \\
= \ &F(\varphi) \circ \eta_A \circ (h \circ \alpha) \\
= \ &\eta_T \circ G(\varphi) \circ (h \circ \alpha) \\
= \ &(\eta_T \circ \pi) \circ \varphi.
\end{aligned}
$$

Therefore $\varphi : \mathcal{A} \to \mathcal{T}_\eta$ is an $F$-homomorphism, and $T_\eta$ is weakly terminal. ∎

## 7.5  Restricted Functors

We saw earlier that there are functors $F$ for which no terminal $F$-coalgebra can exist. Now we determine a condition on $F$ under which terminal $F$-coalgebras can be shown to exist. We start by restricting the power set functor $P(-)$ to apply only to those sets whose cardinality is smaller than a given cardinality $\aleph$:

$$
\mathcal{P}_\aleph(X) := \{U \mid U \subseteq X \text{ and } |U| < \aleph\}.
$$

An important special case of this is $\mathcal{P}_\omega$, where $\omega$ is the cardinality of the set of natural numbers.

Let $\Phi$ and $S$ be sets, with $R \subseteq S \times S$ and $prop : S \to \mathcal{P}(\Phi)$. A Kripke structure $(S; R, prop)$ is called *image-finite* if any point $s$ has only finitely many successors $s'$, with a transition from $s$ to $s'$. An image-finite Kripke structure can be regarded as a coalgebra over the functor $F(-) = \mathcal{P}(\Phi) \times \mathcal{P}_\omega(-)$. We shall see in this section that although there are no terminal Kripke structures, there do exist terminal image-finite Kripke structures.

**Definition 7.5.1.** Let $\aleph$ be a cardinal. An endofunctor $F$ of **Set** is called $\aleph$-*restricted* if for any element $a$ from an arbitrary $F$-coalgebra $\mathcal{A}$ there exists a subcoalgebra $\mathcal{U} \leq \mathcal{A}$ such that $a \in U$ and $|U| < \aleph$. The functor

$F$ is called *weakly $\aleph$-restricted* if every non-empty $F$-coalgebra has a non-empty subcoalgebra $\mathcal{U}$ such that $|U| < \aleph$. A functor $F$ is called *restricted* or *weakly restricted*, if there is a cardinal $\aleph$ such that $F$ is $\aleph$-restricted or weakly $\aleph$-restricted respectively.

For any set $I$, we shall denote by $|I|^*$ the maximum of the cardinalities $\omega$ and $|I|$. It can be shown that the functor $(-)^I$ is $|I|^*$-restricted.

**Lemma 7.5.2.** ([44]) *Let $F$ be an $\aleph$-restricted endofunctor of* **Set**. *Then*

(i) *For every fixed set $O$ the functor $F_O \times F(-)$ is also $\aleph$-restricted.*
(ii) *If $\eta : F \to G$ is a surjective natural transformation, then $G$ is also $\aleph$-restricted.*

**Theorem 7.5.3.** *For a functor $F :$ **Set** $\to$ **Set** *the following are equivalent:*

(i) *$F$ is restricted.*
(ii) *$F$ is weakly restricted.*
(iii) *There is a surjective natural transformation $\eta : F_O \times (-)^I \to F$ for some sets $O$ and $I$.*

**Proof:** (i) $\Rightarrow$ (ii): If the functor $F$ is restricted by the cardinality $\kappa$, then for every $F$-coalgebra $\mathcal{A}$ and every $a \in A$ there is a subcoalgebra $\mathcal{U} \subseteq \mathcal{A}$ with $|U| < \kappa$ and $a \in U$. Then if $A$ is non-empty there is a subcoalgebra $\mathcal{U} \subseteq \mathcal{A}$ which is non-empty and has $|U| < \kappa$. Therefore, $F$ is weakly restricted.

(ii) $\Rightarrow$ (iii): Let $F$ be weakly $\kappa$-restricted. We take $I$ to be any set of cardinality $\kappa$, and define $O := F(I)$. Then the functor $G := F_O \times (-)^I$ transforms any mapping $f : X \to Y$ into a mapping $G(f) : O \times X^I \to O \times Y^I$, with $G(f)(d, \tau) := (d, f \circ \tau)$ where $\tau : I \to X$. We define a mapping $\eta : F_O \times (-)^I \to F$ by $\eta_X(d, \tau) := F(\tau)(d)$ for every set $X$. Showing that $\eta$ is a natural transformation means showing that the diagram in Figure 7.1 commutes.

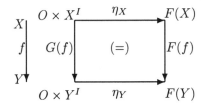

Fig. 7.1 Natural Transformation Diagram

For this commutativity we have $(F(f) \circ \eta_X)(d, \tau) = F(f)(F(\tau)(d) = F(f \circ \tau))(d) = \eta_Y(d, f \circ \tau) = (\eta_Y \circ (G(f)))(d, f)$. Therefore $\eta$ is a natural transformation.

Now we show that $\eta_X$ is surjective for $X \neq \emptyset$. Let $w \in F(X)$ be an arbitrary element from $F(X)$. We have to find an element $u \in F(I)$ and a mapping $\tau \in X^I$ with $\eta_X(u, \tau) = w$. The constant mapping $\alpha_w : X \to F(X)$ with $\alpha_w(x) := w$ for some $w \in F(X)$ defines an $F$-coalgebra $\mathcal{X} = (X; \alpha_w)$ on $X$. Since $F$ is weakly $\kappa$-restricted, there is a non-empty subcoalgebra $\mathcal{S} \subseteq \mathcal{X}$ with $|S| < \kappa$, and hence a surjective mapping $\gamma : I \to S$. The embedding $\subseteq_S^X$ is a homomorphism, satisfying $\alpha_w \circ \subseteq_S^X = F(\subseteq_S^X) \circ \alpha_S$. We define $\tau := \subseteq_S^X \circ \gamma$. For an arbitrary element $s \in S$ define $v := \alpha_S(s) \in F(S)$. Since $\gamma$ is surjective, it is right-invertible, and (by the Axiom of Choice) there is a mapping $\gamma'$ such that $\gamma \circ \gamma' = 1_S$. Then $F(\gamma) \circ F(\gamma') = 1_{F(S)}$, making $F(\gamma)$ also right-invertible and thus surjective. Then for each $v = \alpha_S(s)$ there exists an element $u \in F(\kappa)$ such that $F(\gamma)(u) = v_x$, and we have $\eta_X(u, \tau) = F(\tau)(u) = F(\subseteq_S^X \circ \gamma)(u) = (F(\subseteq_S^U) \circ F(\gamma))(u) = F(\subseteq_S^X)(v) = (F(\subseteq_S^X) \circ \alpha_S)(s) = (\alpha_w \circ \subseteq_S^X)(s) = \alpha_w(\subseteq_S^X(s)) = w$. This can be seen in Figure 7.2. ∎

**Corollary 7.5.4.** *Let $F$ be a weakly $\kappa$-restricted functor, and let $I$ be a set of cardinality $\kappa$. Let $T_F^\kappa$ be the set of all infinite trees with $\kappa$ branches at each node, where the nodes are labeled by elements of $F(I)$. Then $T_F^\kappa$ is the base set of a weakly terminal $F$-coalgebra. The structural mapping for this coalgebra is given by $\alpha(t) = (F(\delta(t)))(\gamma(t))$, where $\gamma$ is the labeling of the root of $t$ and $\delta(t)$ is the mapping which assigns to each $e \in I$ the subtree $t_e$ of $t$.*

**Proof:** Theorem 7.2.1 showed that $\mathcal{T} = (O^{\Gamma^*}; \delta, \gamma)$ is the terminal Moore automaton on the input alphabet $\Gamma$ and output alphabet $O$, with mappings $\delta(\tau, e)(w) = \tau(ew)$ and $\gamma(\tau) = \tau(\varepsilon)$ for $\tau \in O^{\Gamma^*}$ and $\varepsilon$ the empty word. The set $O^{\Gamma^*}$ can be regarded as the set of all infinite trees with root $\gamma(t)$ and with $|\Gamma|$ branches at each node. The nodes are labeled by elements from $O$. The element $\delta(t, e_i)$ corresponds to the subtree $t_{e_i}$ whose root starts with the node at the end of $e_i$. Then the corresponding $F'$-coalgebra $\mathcal{T} = (O^{\Gamma^*}; \pi)$ with $\pi : O^{\Gamma^*} \to F'(O^{\Gamma^*})$, and with the functor $F'(-) = F_O \times (-)^\Gamma$ is also terminal and thus weakly terminal. By assumption, the functor $F$ is weakly $\kappa$-restricted. Then by Theorem 7.5.3 (ii) $\Rightarrow$ (iii) there are sets $E$ and $I$ and there is a surjective natural transformation $\eta : F_E \times (-)^I \to F$. By the proof of Theorem 7.5.3 we have $E = F(I)$. By Lemma 7.4.3 with this natural transformation $\eta$ the coalgebra $T_\eta = (E^{I^*}; \eta_T \circ \pi)$ is a weakly terminal $F$-coalgebra. Here $T_F^\kappa := F(I)^{\Gamma^*}$ is the set of all infinite trees with $\kappa$ branches at each node, where the nodes are labeled by elements of $F(I)$. The form of the structural mapping can be derived from the coalgebra representation of the Moore automaton. ∎

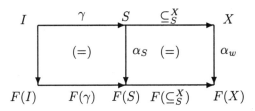

Fig. 7.2   Restricted Functors

All these observations give the following conclusion.

**Theorem 7.5.5.** *If $F$ is a weakly $\kappa$-restricted functor, then the terminal $F$-coalgebra exists.*

**Proof:** By Corollary 7.5.4 the weakly terminal $F$-coalgebra exists for a weakly $\kappa$-restricted functor $F$. Now by Lemma 7.4.2 the factor coalgebra $T_F^\kappa/\nabla$, where $\nabla$ is the greatest congruence on $T_F^\kappa$, is terminal. ∎

An important example of a weakly $\kappa$-restricted functor is the functor $\mathcal{P}_\omega$ which maps each set to the set of all its finite subsets.

## 7.6   Exercises for Chapter 7

1. Prove Lemma 7.5.2:

(i) For an $\aleph$-restricted endofunctor of **Set** and every fixed set $O$ the functor $F_O \times F(-)$ is also $\aleph$-restricted.

(ii) If $\eta : F \to G$ is a surjective natural transformation, then $G$ is also $\aleph$-restricted.

2. Let $F : \mathbf{Set} \to \mathbf{Set}$ be a functor for which the terminal $F$-coalgebra $\mathcal{T}$ exists. Show that an $F$-coalgebra $\mathcal{A}$ satisfies the principle of coinduction if and only if the morphism from $\mathcal{A}$ into $\mathcal{T}$ is a monomorphism.

## Chapter 8

# Cofree $F$-coalgebras and Coequations

In this chapter we investigate the coalgebra version of Birkhoff's Theorem, the Main Theorem of Equational Theory from Section 1.4. This theorem tells us that the equationally definable classes of algebras are precisely the varieties, which are classes of algebras closed under the formation of homomorphic images, subalgebras and products. One direction of this theorem is easy to prove, that equational classes are varieties. The harder direction is to show that every variety is indeed definable by some set of equations or identities. The concept of free algebras is used here, and as we have seen, free algebras are free objects in the category $Alg(\tau)$ of all algebras of type $\tau$. The relation of satisfaction of an identity by an algebra defines a Galois connection $(Mod, Id)$ between classes of algebras and sets of identities.

In Section 8.1 we define cofree $F$-coalgebras. These coalgebras are then used to define the concept of a coequation satisfied by a coalgebra. The relation of satisfaction of a coequation by a coalgebra is then a relation between classes of coalgebras and sets of coequations. The satisfaction relation thus defines a Galois connection $(Mod, Ceq_X)$ between the class of all $F$-coalgebras and the set of all coequations. This connection produces two closure operators $ModCeq_X$ and $Ceq_X Mod$, whose sets of fixed points form two dually isomorphic complete lattices. With this machinery we prove a co-Birkhoff result, that the coequationally definable classes of $F$-coalgebras are precisely the covarieties.

## 8.1 Cofree $F$-coalgebras

To develop a coequational theory for $F$-coalgebras we first need to define cofree $F$-coalgebras as cofree objects in the category $\mathbf{Set}_F$.

**Definition 8.1.1.** Let $F : \mathbf{Set} \to \mathbf{Set}$ be a functor and let $X$ and $T_X$ be sets with $F(T_X) = X$. Let $\varepsilon_X : T_X \to F(T_X) = X$ be a structural mapping. The $F$-coalgebra $\mathcal{T}_X = (T_X; \varepsilon_X)$ is called *cofree on* $X$ if for every $F$-coalgebra $\mathcal{A}$ and every mapping $g : A \to X$ there is exactly one homomorphism $\bar{g} : \mathcal{A} \to \mathcal{T}_X$ with $g = \varepsilon_X \circ \bar{g}$. The set $X$ is called a set of *colours* and the mappings $g$ and $\varepsilon_X$ are called *colourings* of $A$ and $T_X$ respectively (see Figure 8.1).

This definition means that for every colouring $g$ of $A$ with colours from $X$, there is exactly one homomorphism $\bar{g} : \mathcal{A} \to \mathcal{T}_X$ which extends $g$. Conversely, every homomorphism $\varphi : \mathcal{A} \to \mathcal{T}_X$ is the extension of a colouring of $A$ with colours from $X$.

**Corollary 8.1.2.** *Let $\mathcal{T}_X$ be cofree on $X$, and let $\mathcal{A}$ be an $F$-coalgebra. Then*

(i) *Every homomorphism $\varphi : \mathcal{A} \to \mathcal{T}_X$ has the form $\varphi = \bar{g}$ with $g = \varepsilon_X \circ \varphi$.*
(ii) *If $\varphi_1$ and $\varphi_2$ are homomorphisms from $\mathcal{A}$ to $\mathcal{T}_X$ with $\varepsilon_X \circ \varphi_1 = \varepsilon_X \circ \varphi_2$, then $\varphi_1 = \varphi_2$.*

**Proof:** (i) By the definition of cofreeness, the colouring $\varepsilon_X \circ \varphi$ satisfies $\varepsilon_X \circ (\overline{\varepsilon_X \circ \varphi}) = \varepsilon_X \circ \varphi$. But for the given colouring $g := \varepsilon \circ \varphi$ there is only one homomorphism $\psi : \mathcal{A} \to \mathcal{T}_X$, so $\varphi = \bar{g}$. See Figure 8.2.
(ii) Since there is exactly one homomorphism which satisfies $g = \varepsilon_X \circ \varphi_1$, the condition $\varepsilon_X \circ \varphi_1 = \varepsilon_X \circ \varphi_2$ must imply $\varphi_1 = \varphi_2$. ∎

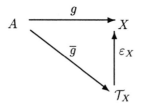

Fig. 8.1    Cofree $F$-coalgebras

Cofree $F$-coalgebras need not exist for arbitrary functors $F$, so we must verify the existence of such coalgebras. First we note that when cofree $F$-coalgebras do exist, they are uniquely determined, up to isomorphism.

**Corollary 8.1.3.** *Cofree $F$-coalgebras over a set $X$, if they exist, are uniquely determined up to isomorphism.*

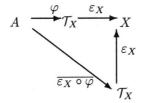

Fig. 8.2   Extension Homomorphism

**Proof:** Let $\mathcal{T}_X$ and $\mathcal{T}'_X$ be cofree $F$-coalgebras over $X$. Then for $\varepsilon'_X :$ $T'_X \to X$ there is exactly one homomorphism $\overline{\varepsilon'_X} : \mathcal{T}'_X \to \mathcal{T}_X$ which makes the diagram of Figure 8.3 commute. Commutativity of this diagram means that $\varepsilon_X \circ \overline{\varepsilon'_X} = \varepsilon'_X$. Similarly, for $\varepsilon_X : \mathcal{T}_X \to X$ there is also exactly one homomorphism $\overline{\varepsilon_X} : \mathcal{T}_X \to \mathcal{T}'_X$ which makes the diagram of Figure 8.4 commute, by satisfying the condition $\varepsilon'_X \circ \overline{\varepsilon_X} = \varepsilon_X$.

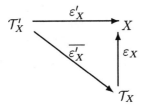

Fig. 8.3   Uniqueness for $\overline{\varepsilon'_X}$

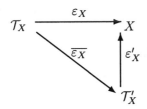

Fig. 8.4   Uniqueness for $\overline{\varepsilon_X}$

Combining these commutativity equations gives $\varepsilon_X \circ \overline{\varepsilon'_X} \circ (\overline{\varepsilon_X}) = (\varepsilon_X \circ \overline{\varepsilon'_X}) \circ \overline{\varepsilon_X} = \varepsilon'_X \circ \overline{\varepsilon_X} = \varepsilon_X$, and dually $\varepsilon'_X \circ (\overline{\varepsilon_X} \circ \overline{\varepsilon'_X}) = (\varepsilon'_X \circ \overline{\varepsilon_X}) \circ \overline{\varepsilon'_X}$ $= \varepsilon_X \circ \overline{\varepsilon'_X} = \varepsilon'_X$. It is also true that $\varepsilon_X \circ 1_{T_X} = \varepsilon_X$ and $\varepsilon'_X \circ 1_{T'_X} = \varepsilon'_X$. It follows that $\varepsilon_X \circ 1_{T_X} = \varepsilon_X \circ (\overline{\varepsilon'_X} \circ \overline{\varepsilon_X})$ and $\varepsilon'_X \circ 1_{T'_X} = \varepsilon'_X \circ (\overline{\varepsilon_X} \circ \overline{\varepsilon'_X})$. From

Corollary 8.1.2 we get $1_{T_X} = \overline{\varepsilon'_X} \circ \overline{\varepsilon_X}$ and $1_{T'_X} = \overline{\varepsilon_X} \circ \overline{\varepsilon'_X}$. This means that $\overline{\varepsilon_X}$ is a bijection; and since bijective homomorphisms are isomorphisms, we have $T_X \cong T'_X$.                                                                                    ∎

**Remark:** Let $T_X$ be a cofree $F$-coalgebra over a one-element set $X = \{x\}$ of colours. In this case there can only be one mapping $g : A \to \{x\}$ and one homomorphism $g : A \to T_X$, for every $F$-coalgebra $A$. This shows that if the cofree coalgebra $T_X$ exists in this special case, then it is a terminal coalgebra.

Now we consider pairs consisting of an $F$-coalgebra and a colouring with colours from $X$. Such pairs can be regarded as coalgebras using the functor $F_X \times F(-)$. From this observation we get the following fact (see for instance [44]).

**Lemma 8.1.4.** *The cofree $F$-coalgebra $T_X$ over the set $X$ of colours is the terminal coalgebra for the functor $F_X \times F(-)$.*

Lemma 7.5.2 tells us that if the functor $F$ is restricted, so is the functor $F_X \times F(-)$ for any set $X$. Theorems 7.5.3 and 7.5.5 then guarantee the existence of terminal $F$-coalgebras for the functor $F_X \times F(-)$. This proves the following theorem.

**Theorem 8.1.5.** *If $F$ is a restricted functor, then cofree $F$-coalgebras exist over arbitrary sets $X$ of colours.*

Suppose that $F$ is a functor for which cofree $F$-coalgebras exist. Then each set $X$ can be mapped to the cofree $F$-coalgebra $T_X$ on the set $X$ of colours. If $f : X \to Y$ is any mapping, then we can show that $T(f) := \overline{f \circ \varepsilon_X} : T_X \to T_Y$ is a homomorphism. Since $T_Y$ is cofree on the set $Y$ of colours, the mapping $f \circ \varepsilon_X : T_X \to X \to Y$ can be uniquely extended to a homomorphism $\overline{f \circ \varepsilon_X}$, with the property that $f \circ \varepsilon_X = \varepsilon_Y \circ \overline{f \circ \varepsilon_X}$.

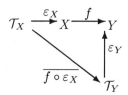

Fig. 8.5   Extension of $f \circ \varepsilon_X$

Then we have $\varepsilon_Y \circ T(f) = \varepsilon_Y \circ (\overline{f \circ \varepsilon_X}) = f \circ \varepsilon_X$. This makes the diagram in Figure 8.6 commute and hence shows that $T(f)$ is a homomorphism.

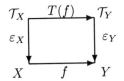

Fig. 8.6   The Functor $T$

**Theorem 8.1.6.** *If for every set $X$ of colours the cofree $F$-coalgebra on $X$ exists, then $T : \text{Set} \to \text{Set}$ where $T$ maps each set $X$ to $\mathcal{T}_X$ and each mapping $f : X \to Y$ to $T(f) := \overline{f \circ \varepsilon_X}$, is a functor.*

**Proof**:  By definition $T$ maps sets to cofree $F$-coalgebras and mappings to homomorphisms. Since $T(1_{T_X})$ is a homomorphism, it satisfies $\varepsilon_X \circ T(1_X) = 1_X \circ \varepsilon_X = \varepsilon_X$. It is also true that $\varepsilon_X \circ 1_{T_X} = \varepsilon_X$, and it follows that $\varepsilon_X \circ 1_{T_X} = \varepsilon_X \circ T(1_X)$. Applying Corollary 8.1.2(ii) then gives $1_{T_X} = T(1_X)$. Similarly, for $f : X \to Y$, $g : Y \to Z$ and $g \circ f : X \to Z$ we obtain the homomorphisms $T(f) : \mathcal{T}_X \to \mathcal{T}_Y$, $T(g) : \mathcal{T}_Y \to \mathcal{T}_Z$, $T(g \circ f) : \mathcal{T}_X \to \mathcal{T}_Z$ and $T(g) \circ T(f) : \mathcal{T}_X \to \mathcal{T}_Z$, and the equation $\varepsilon_Z \circ T(g \circ f) = \varepsilon_Z \circ T(g) \circ T(f)$. Again by Corollary 8.1.2 we obtain $T(g \circ f) = T(g) \circ T(f)$. ∎

## 8.2   Coequations

Let $\mathcal{T}$ be a terminal $F$-coalgebra. One of the important consequences of the terminal definition is the following property: for every $F$-coalgebra $\mathcal{A}$ and every $a \in A$, there is exactly one element $s \in T$ such that $s$ is related to $a$ by the largest bisimulation relation $\sim_{A,T}$. This bisimulation in some sense expresses patterns of behaviour. Since as we saw in the previous section cofree $F$-coalgebras over $X$ are terminal $F_X \times F(-)$-coalgebras, the elements of a cofree $F$-coalgebra can thus be considered as patterns of behaviour.

   Such patterns of behaviour can be used to define classes of coalgebras, much the same way as equations defined classes in universal algebra in Chapter 1. But, in contrast to universal algebra, classes of coalgebras are defined by avoidance of particular behaviours. This idea is made more precise by the concept of a coequation.

**Definition 8.2.1.** A *coequation* is an element $p \in T_X$, for some set $X$. Let $p \in T_X$ be a coequation and let $g : A \to X$ be a colouring. An element $a \in A$ is said to *satisfy* the coequation $p$ under the colouring $g$ if $\bar{g}(a) \neq p$. In this case we write $(\mathcal{A}, a \models_g p)$. We say that an element $a \in A$ satisfies $p$, and write $(\mathcal{A}, a \models p)$, if $(\mathcal{A}, a \models_g p)$ holds for every colouring $g : A \to X$. Finally, we say that the $F$-coalgebra $\mathcal{A}$ satisfies $p$ if $(\mathcal{A}, a \models p)$ for all $a \in A$. For a class $K$ of $F$-coalgebras and a set $P$ of elements of $T_X$ we say that $K$ satisfies $P$ if every coalgebra in $K$ satisfies every coequation in $P$; in this case we write $K \models P$.

We recall from Theorem 8.1.5 that when cofree $F$-coalgebras exist for all sets $X$, there is an induced functor $T$ on the category $\mathbf{Set}$.

**Lemma 8.2.2.** *Let $p \in T_X$ be a coequation and let $X \subseteq Y$ be sets. Then for any $F$-coalgebra $\mathcal{A}$ we have*

$$\mathcal{A} \models p \quad \Leftrightarrow \quad \mathcal{A} \models T(\subseteq_X^Y)(p).$$

**Proof:** For the injection $\subseteq_X^Y$ there is a partial mapping $\pi : Y \to X$ such that $\pi \circ \subseteq_X^Y = 1_X$. Applying the functor $T$ to this equation gives $T(\pi \circ \subseteq_X^Y) = T(\pi) \circ T(\subseteq_X^Y) = T(1_X) = 1_{T_X}$. Then for any coequation $p$ we have $T(\pi)(T(\subseteq_X^Y))(p) = 1_{T_X}(p) = p$. For any element $a \in A$ there is a homomorphism $\varphi$ with $\varphi(a) = p$ if and only if there is a homomorphism $\psi$ with $\psi(a) = T(\subseteq_X^Y)(p)$; therefore $\mathcal{A} \not\models T(\subseteq_X^Y)(p)$ if and only if $\mathcal{A} \not\models p$. ∎

When we consider terms of a fixed type, in the algebra setting, we can always view the terms as elements of one absolutely free algebra, taken over a countably infinite alphabet $X$. The following lemma describes a similar property for coalgebras and coequations.

**Lemma 8.2.3.** *Let $P$ be a set of coequations. Then there is a fixed set $X$ and a set $P' \subseteq T_X$, where $T_X$ is a cofree $F$-coalgebra on $X$, such that for every $F$-coalgebra $\mathcal{A}$ we have*

$$\mathcal{A} \models P \quad \Leftrightarrow \quad \mathcal{A} \models P'.$$

**Proof:** For each coequation $p \in P$ there is a set $X_p$ such that $p \in T_{X_p}$. We take for $X$ the disjoint union $X := \sum_{p \in P} X_p$, with the canonical embeddings $e_p : X_p \to X$ for $p \in P$. Then by Lemma 8.2.2 for each $p \in P$ we have

$$\mathcal{A} \models p \quad \Leftrightarrow \quad \mathcal{A} \models T(e_p)(p),$$

where $T(e_p) : \mathcal{T}_{X_p} \to \mathcal{T}_X$. So taking $P' = \{T(e_p)(p) \mid p \in P\}$ gives the property that $\mathcal{A} \models P$ if and only if $\mathcal{A} \models P'$. ∎

In Section 7.5 we described how terminal coalgebras, for restricted functors, can be viewed as sets of infinite trees, with edges labeled by elements of $F(\Sigma)$ for some set $\Sigma$ of cardinality $\kappa$ and with colours from $X$. By Lemma 8.1.4, cofree coalgebras are terminal coalgebras, and coequations are elements of such coalgebras. This means that it is also possible to interpret coequations as infinite trees.

## 8.3   The Co-Birkhoff Theorem

Any binary relation between two sets determines a Galois connection between the sets (see [29] for details). We saw an important example of this in Section 1.4, where the relation of satisfaction of an identity by an algebra determined the Galois connection $(Id, Mod)$ between sets of identities and classes of algebras. We can now use the relation of satisfaction of a coequation by a coalgebra to produce another such Galois connection.

The relation of satisfaction, that $\mathcal{A} \models p$ for a coalgebra $\mathcal{A}$ and a coequation $p \in T_X$, is a binary relation $\models \subseteq \mathbf{Set}_F \times T_X$. For the induced Galois connection we consider classes $K \subseteq \mathbf{Set}_F$ and sets $P \subseteq T_X$. We define

$$Mod(P) := \{\mathcal{A} \mid \mathcal{A} \in \mathbf{Set}_F \text{ and } \forall p \in P(\mathcal{A} \models p)\}$$

and

$$Ceq_X(K) := \{p \mid p \in T_X \text{ and } \forall \mathcal{A} \in K(\mathcal{A} \models p)\}.$$

We leave it as an exercise for the reader to verify that the pair $(Ceq_X, Mod)$ is indeed a Galois connection. Specifically, the following properties are satisfied, for any $P, P_1, P_2 \subseteq T_X$ and any $K, K_1, K_2 \subseteq \mathbf{Set}_F$:

$$P_1 \subseteq P_2 \quad \Rightarrow \quad Mod(P_2) \subseteq Mod(P_1),$$
$$K_1 \subseteq K_2 \quad \Rightarrow \quad Ceq_X(K_2) \subseteq Ceq_X(K_1),$$
$$P \subseteq Ceq_X Mod(P), \quad K \subseteq Mod Ceq_X(K).$$

These properties of a Galois connection imply that the two products $Mod Ceq_X$ and $Ceq_X Mod$ are closure operators. The resulting fixed points, or Galois-closed sets, are called coequational classes and coequational theories respectively. We set

$$\mathcal{L}_F := \{K \mid K \subseteq \mathbf{Set}_F \text{ and } Mod Ceq_X(K) = K\}$$

and

$$\mathcal{L}_{Ceq} := \{P \mid P \subseteq T_X \text{ and } Ceq_X Mod(P) = P\}.$$

These two sets are complete lattices which are dually isomorphic to each other. Any element of $\mathcal{L}_F$ has the form $Mod(P)$ for some $P \subseteq T_X$, and dually any element of $\mathcal{L}_{Ceq}$ has the form $Ceq_X(K)$ for some set $K$ of coalgebras.

Our main interest is to give a coalgebraic characterization of the Galois-closed sets $Mod(P)$ and $Ceq_X(K)$, respectively. This will lead to our co-Birkhoff Theorem, comparing coequational classes of the form $Mod(P)$ to covarieties. We recall that covarieties are classes of $F$-coalgebras which are closed under arbitrary applications of the operators $H$, $S$ and $\Sigma$ of formation of homomorphic images, subcoalgebras and direct sums respectively.

**Theorem 8.3.1.** *Classes of the form $Mod(P)$ for some set $P$ of coequations are covarieties.*

**Proof:** Let $p \in T_X$ be a coequation. We show first that if $\mathcal{A} \models p$ and $\varphi : \mathcal{A} \to \mathcal{B}$ is an epimorphism, then $\mathcal{B} \models p$. Suppose that $\mathcal{B} \not\models p$. Then there is an element $b \in B$ and a homomorphism $\psi : \mathcal{B} \to \mathcal{T}_X$ such that $\psi(b) = p$. Since $\varphi$ is surjective, there is also an element $a \in A$ with $\varphi(a) = b$. But then $(\psi \circ \varphi)(a) = p$ and thus $\mathcal{A} \not\models p$, which is a contradiction.

For direct sums, let $(\mathcal{A}_i)_{i \in I}$ be an indexed family of $F$-coalgebras and assume that $\mathcal{A}_i \models p$ for all $i \in I$. For every $a \in \sum_{i \in I} \mathcal{A}_i$ there is an element $a_i \in A_i$ such that $e_i(a) = a_i$, where $e_i : \mathcal{A}_i \to \sum_{i \in I} \mathcal{A}_i$ is the canonical injection. Suppose that $\sum_{i \in I} \mathcal{A}_i \not\models p$. Then there is an element $a \in \sum_{i \in I} \mathcal{A}_i$ and a homomorphism $\psi : \sum_{i \in I} \mathcal{A}_i \to \mathcal{T}_X$ with $\psi(a) = p$. The composition of $\psi$ with $e_i$ gives a homomorphism $\psi \circ e_i : \mathcal{A}_i \to \mathcal{T}_X$ with $(\psi \circ e_i)(a) = p$. This shows that $\mathcal{A}_i \not\models p$, again a contradiction.

Finally, let $\mathcal{U} \leq \mathcal{A}$ and let $\mathcal{A} \models p$. If $\mathcal{U} \not\models p$, there is an element $u \in U$ and a homomorphism $\psi : \mathcal{U} \to \mathcal{A}$ with $\psi(u) = p$. The homomorphism $\psi$ extends a mapping $g : U \to X$, where $g = \varepsilon_X \circ \psi$. But also any mapping $g : U \to X$ can be extended to a mapping $g' : A \to X$ with $g' \circ \subseteq_U^A = g$. By the cofreeness property of $\mathcal{T}_X$ we must have $g' = \varepsilon_X \circ \overline{g'}$. We set $\psi' := \overline{g'}$. Combining the previous information, we get $\varepsilon_X \circ \psi' \circ \subseteq_U^A = g' \circ \subseteq_U^A = g = \varepsilon_X \circ \psi$. Corollary 8.1.2(ii) then forces $\psi' \circ \subseteq_U^A = \psi$. From this and the fact that $\psi(u) = p$ we see that $\psi'(u) = p$, which means that $\mathcal{A} \not\models p$ and gives us a contradiction. ∎

**Theorem 8.3.2.** *Let $K$ be a class of $F$-coalgebras, let $X$ be a fixed set of colours and let $F$ be a $\kappa$-restricted functor. Then $Mod(Ceq_X(K)) = SH\Sigma(K)$.*

**Proof:** We have just proved in Theorem 8.3.1 that $Mod(Ceq_X(K))$ is a covariety, that is, closed under applications of the operators $S$, $H$ and $\Sigma$. This means that $SH\Sigma(K) \subseteq Mod(Ceq_X(K))$. For the opposite inclusion, let $\mathcal{A}$ be an element of $Mod(Ceq_X(K))$. Since the functor $F$ is $\kappa$-restricted, for each $a \in A$ there is a subcoalgebra $\mathcal{U}_a \leq \mathcal{A}$ such that $a \in U_a$ and $|U_a| < \kappa$. Then $A = \bigcup_{a \in A} U_a$ and there is a surjective homomorphism $\sum_{a \in A} \mathcal{U}_a \to \mathcal{A}$. It is enough to show that $\mathcal{U}_a \in SH\Sigma(K)$ for every $\mathcal{U}_a$, and so we now consider only $F$-coalgebras $\mathcal{A}$ with $|A| < \kappa$. For $\Gamma$ a set of cardinality $\kappa$, there is an injective mapping $g : A \to \Gamma$. Then the extension $\overline{g} : \mathcal{A} \to \mathcal{T}_\Gamma$ is also injective, since

$$\overline{g}(a) = \overline{g}(b) \quad \Rightarrow \quad \varepsilon_\Gamma(\overline{g}(a)) = \varepsilon_\Gamma(\overline{g}(b)) \quad \Rightarrow \quad g(a) = g(b) \quad \Rightarrow \quad a = b$$

because $g = \varepsilon_\Gamma \circ \overline{g}$. Since $\overline{g} : \mathcal{A} \to \mathcal{T}_\Gamma$ is injective, the coalgebra $\mathcal{A}$ is isomorphic to the image subcoalgebra $\mathcal{C} := \overline{g}[\mathcal{A}]$ of $\mathcal{T}_\Gamma$. Every element $c \in C$ is a coequation which is not satisfied by $\mathcal{A}$. Therefore, there is a coalgebra $\mathcal{B}_c \in K$ with $\mathcal{B}_c \not\models c$; and there is a homomorphism $\psi_c : \mathcal{B}_c \to \mathcal{T}_\Gamma$ with $c \in \psi_c[\mathcal{B}_c]$. Any $\psi_c[\mathcal{B}_c]$ is a subcoalgebra of $\mathcal{T}_\Gamma$, so the union $\mathcal{B} := \bigcup_{c \in C} \psi_c[\mathcal{B}_c]$ is a subcoalgebra of $\mathcal{T}_\Gamma$. Since all the $\mathcal{B}_c$ are in $K$, we get $\mathcal{B} \in H(\Sigma(K)) \subseteq SH\Sigma(K)$. Since $\mathcal{A}$ is isomorphic to a subcoalgebra $\mathcal{C}$ of $\mathcal{B}$, we have $\mathcal{A} \in SH\Sigma(K)$. ∎

Let $\mathcal{U}$ be a subcoalgebra of the cofree coalgebra $\mathcal{T}_X$ on $X$. Then it can be shown that the class

$$Q(\mathcal{U}) := \{\mathcal{A} \in \mathbf{Set}_F \mid \forall \varphi : A \to X \ (\overline{\varphi}[\mathcal{A}] \leq \mathcal{U})\}$$

is a covariety (see exercise 8.2).

Let $K$ be a class of $F$-coalgebras, and let $\mathcal{T}_X(K)$ be a subcoalgebra of $\mathcal{T}_X$ with embedding $\varphi : \mathcal{T}_X(K) \to \mathcal{T}_X$. Then $\mathcal{T}_X(K)$ is said to be cofree for class $K$ if for every $\mathcal{B} \in K$ and every $f : \mathcal{B} \to \mathcal{T}_X$ there is a homomorphism $h : \mathcal{B} \to \mathcal{T}_X(K)$ such that $f = \varphi \circ h$. If $(s_i : \mathcal{B} \to \mathcal{T}_X)$ is the class of all morphisms with domain in $K$, then $\mathcal{T}_X(K)$ is the union of all images of the $s_i$'s.

It is also possible to consider the class $\mathcal{V}_X(K) := Q(\mathcal{T}_X(K))$. The subcoalgebra $\mathcal{T}_X(K)$ is closed under all endomorphisms of $\mathcal{T}_X$. Such subcoalgebras of $\mathcal{T}_X$ are called *invariant*. Clearly, if $K$ is closed under sums and homomorphic images then we have $T_X(K) \in K$. Let $V_X(K)_{\leq \kappa}$ be the class of all coalgebras in $V_X(K)$ which have cardinalities less than or equal to $\kappa$.

**Lemma 8.3.3.** *For any* $\kappa \leq |X|$, *if* $T_X$ *exists, then* $V_X(K)_{\leq \kappa} \supseteq SH\Sigma(K)_{\leq \kappa}$.

**Proof:** The class $V_X(K)$ is a covariety containing $K$, so it contains $SH\Sigma(K)$. Thus $SH\Sigma(K) \subseteq V_X(K)$ and also $SH\Sigma(K)_{\leq \kappa} \subseteq V_X(K)_{\leq \kappa}$. ∎

The converse inclusion $V_X(K)_{\leq \kappa} \subseteq SH\Sigma(K)_{\leq \kappa}$ is also satisfied, but the proof of this fact needs the concept of a conjunct sum.

If the functor $F$ is bounded by the cardinality of $X$, then the cofree $F$-coalgebra $T_X$ exists, and the following properties hold:

(i) for every covariety $K$ there exists an invariant subcoalgebra $U$ of $T_X$ with $Q(U) = K$;

(ii) for every invariant subcoalgebra $U$ of $T_X$, there exists a covariety $K$ with $U = T_X(K)$.

Let $p \in T_X$ be a coequation and let $E \subseteq T_X$ be a set of coequations. We write $E \vdash p$ to mean that every coalgebra $A$ satisfying all the coequations in $E$ also satisfies the coequation $p$. We know that for the functor $F = F_D \times (-)^\Sigma$, coequations can be regarded as infinite trees. Then it can be proven that

$$E \vdash p \quad \Leftrightarrow \quad \exists \varphi : T_X \to T_X$$

where $\varphi$ is an endomorphism mapping a subtree of $p$ into $E$.

## 8.4 Exercises for Chapter 8

1. Verify that the pair $(Ceq_X, Mod)$ is indeed a Galois connection. Specifically, show that the following properties are satisfied, for any $P$, $P_1$, $P_2 \subseteq T_X$ and any $K$, $K_1$, $K_2 \subseteq Set_F$:

$$P_1 \subseteq P_2 \quad \Rightarrow \quad Mod(P_2) \subseteq Mod(P_1),$$
$$K_1 \subseteq K_2 \quad \Rightarrow \quad Ceq_X(K_2) \subseteq Ceq_X(K_1).$$

2. Let $\mathcal{U}$ be a subcoalgebra of the cofree coalgebra $\mathcal{T}_X$ on $X$. Prove that the class

$$Q(\mathcal{U}) := \{\mathcal{A} \in \mathbf{Set}_F \mid \forall \varphi : A \to X \ (\overline{\varphi}[\mathcal{A}] \leq \mathcal{U})\}$$

is a covariety.

3. Prove that the "mapping" $T$ defined by $T(X) = \mathcal{T}_X, T(f) : \mathcal{T}_X \to \mathcal{T}_Y$ for $f : X \to Y$ is a set-valued functor if for every set $X$ the cofree $F$-coalgebra $\mathcal{T}_X$ exists.

# Chapter 9

# Coalgebras of Type $\tau$

In this chapter we consider coalgebras of type $\tau$, as an analogue of algebras of type $\tau$ from Chapter 1. We mentioned such algebras briefly in Chapter 4, where we showed that they were too narrow to fully model all state-based systems. Instead we enlarged our definition there to $F$-coalgebras, which included coalgebras of type $\tau$ as a special case. Thus many of the results in this chapter will be instances of the $F$-coalgebra results from Chapter 4. Nevertheless it is interesting and instructive to consider directly the theory of coalgebras of type $\tau$. Moreover, for coalgebras of type $\tau$ we can form terms, identities and varieties as we did earlier for algebras of type $\tau$.

Let $\tau = (n_i)_{i \in I}$ be a type, indexed by some set $I$. A coalgebra of type $\tau$ is a system $(A; (f_i^A)_{i \in I})$ consisting of a non-empty set $A$ and a set of finitary co-operations $f_i^A : A \to A^{\sqcup n_i}$, where $n_i$ is the *arity* of the co-operation $f_i^A$ and $A^{\sqcup n_i}$ is the $n_i$-th copower of $A$. We recall that such copowers are defined by $A^{\sqcup n_i} := \{1, \ldots, n_i\} \times A$. This means that each $n_i$-ary co-operation $f_i^A$ is uniquely determined by a pair $((f_i^A)_1, (f_i^A)_2)$ of mappings, $(f_i^A)_1$ from $A$ to $\{1, \ldots, n_i\}$ and $(f_i^A)_2$ from $A$ to $A$.

As noted above, any coalgebra of type $\tau$ can be regarded as an $F$-coalgebra for a suitable functor $F : \mathbf{Set} \to \mathbf{Set}$. The functor maps sets $X$ to $\prod_{j \in J} X^{\sqcup n_j}$, and takes mappings $f : X \to Y$ to mappings $F(f) : \prod_{i \in I} X^{\sqcup n_i} \to \prod_{i \in I} Y^{\sqcup n_i}$ defined by $(k_i, a)_{i \in I} \mapsto (k_i, f_i^A(a))_{i \in I}$, where $k_i \in \{1, \ldots, n_i\}$. Then the type $\tau$ coalgebra $(A; (f_i^A)_{i \in I})$ is uniquely determined by $(A; \alpha_A)$ where $\alpha_A : A \to F(A)$ is given by $a \mapsto (f_i^A(a))_{i \in I}$, and vice versa.

Any coalgebra $(A; (f_i^A)_{i \in I})$ can be regarded as a recognizer, by taking $A$ as its set of states, a distinguished subset $A' \subseteq A$ as its set of accepting states, and $\{f_i^A \mid i \in I\}$ as its input alphabet. Recognizers can be regarded as automata with output alphabet $O = \{0, 1\}$. In Section 7.5 it was shown that the functor $F_O \times (-)^E$, which expresses automata with output alphabet $O$ and input alphabet $E$ as $F_O \times (-)^E$-coalgebras, is $\kappa$-restricted. If there is a surjective natural transformation from $F_O \times (-)^E$ onto a functor $F$, then $F$ is also $\kappa$-restricted. Thus coalgebras of type $\tau$ are important not only as another kind of state-based systems, but because they play a special role in the theory of $F$-coalgebras.

## 9.1 Homomorphisms, Subcoalgebras and Sums

We begin by considering the formation of homomorphic images, subcoalgebras and sums of coalgebras of type $\tau$, with the corresponding operators $H$, $S$ and $\Sigma$. In each case our definition is based on the type $\tau$ coalgebra structure, and we show that our definition is in fact equivalent to the $F$-coalgebra version for the appropriate functor $F$.

**Definition 9.1.1.** Let $\mathcal{A} = (A; (f_i^A)_{i \in I})$ and $\mathcal{B} = (B; (f_i^B)_{i \in I})$ be coalgebras of type $\tau$. A mapping $\varphi : A \to B$ is called a *homomorphism* from $\mathcal{A}$ to $\mathcal{B}$ if the following equations are satisfied for all $i \in I$ and $a \in A$:

(i) $(f_i^A)_1(a) = (f_i^B)_1(\varphi(a))$,  and
(ii) $\varphi((f_i^A)_2(a)) = (f_i^B)_2(\varphi(a))$.

Let us set $\varphi^{\sqcup n_i}(f_i^A(a)) = ((f_i^A)_1(a), \varphi((f_i^A)_2(a)))$. Then we see that our definition of homomorphism means that the diagram in Figure 9.1 commutes, since $f_i^B(\varphi(a)) = ((f_i^B)_1(\varphi(a)), (f_i^B)_2(\varphi(a))) = ((f_i^A)_1(a), \varphi(f_i^A)_2(a)) = \varphi^{\sqcup n_i}(f_i^A(a))$.

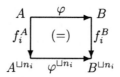

Fig. 9.1  Homomorphisms of Coalgebras of Type $\tau$

As a result, our definition of homomorphism for type $\tau$ coalgebras is equivalent to the definition of an $F$-coalgebra homomorphism. This tells us that all the results of Section 4.3 are valid in the case of type $\tau$ coalgebras as well, and it is not necessary to reprove them in our new context. In particular, let us note that the identity mapping $id_A$ is a homomorphism on any type $\tau$ coalgebra $\mathcal{A}$; the composition of two homomorphisms is a homomorphism (where defined); and any bijective homomorphism is an isomorphism. The class of all coalgebras of type $\tau$ therefore forms a concrete category, which we shall call $\mathbf{Coalg}(\tau)$.

**Example 9.1.2.** Consider the set $A = \{a, b, c, d\}$ and its copower $A^{\sqcup 2}$ $= \{(1, a), (2, a), (1, b), (2, b), (1, c), (2, c), (1, d), (2, d)\}$. We define a binary co-operation $f^A : A \to A^{\sqcup 2}$ by $a \mapsto (2, a)$, $b \mapsto (2, c)$, $c \mapsto (1, d)$ and $d \mapsto (2, b)$. Now let the set $B = \{x, y, u, v\}$ and let $f^B : B \to B^{\sqcup 2}$ be given by $x \mapsto (2, u)$, $y \mapsto (1, x)$, $u \mapsto (2, y)$ and $v \mapsto (2, v)$. Consider the mapping $\varphi : A \to B$ given by $a \mapsto v$, $b \mapsto u$, $c \mapsto y$ and $d \mapsto x$. Then we have

$$
\begin{aligned}
\varphi(f^A(a)) &= \varphi((2, a)) &= (2, \varphi(a)) &= (2, v) &= f^B(v) \\
&= f^B(\varphi(a)) \\
\varphi(f^A(b)) &= \varphi((2, c)) &= (2, \varphi(c)) &= (2, y) &= f^B(u) \\
&= f^B(\varphi(b)) \\
\varphi(f^A(c)) &= \varphi((1, d)) &= (1, \varphi(d)) &= (1, x) &= f^B(y) \\
&= f^B(\varphi(c)) \\
\varphi(f^A(d)) &= \varphi((2, b)) &= (2, \varphi(b)) &= (2, u) &= f^B(x) \\
&= f^B(\varphi(d)).
\end{aligned}
$$

This shows that $\varphi$ is a homomorphism, and as a bijection is thus an isomorphism.

To define subcoalgebras of coalgebras of type $\tau$, we use the restriction $f_i^A|B := \{((f_i^A)_1(b), (f_i^A)_2(b)) \mid b \in B\}$ of a co-operation on a set $A$ to a subset $B$ of $A$.

**Definition 9.1.3.** Let $\mathcal{A} = (A; (f_i^A)_{i \in I})$ and $\mathcal{B} = (B; (f_i^B)_{i \in I})$ be coalgebras of type $\tau$, with $B \subseteq A$. Then $\mathcal{B}$ is called a *subcoalgebra* of $\mathcal{A}$ if $f_i^B := f_i^A|B$ for all $i \in I$. We use the notation $\mathcal{B} \leq \mathcal{A}$ to indicate that $\mathcal{B}$ is a subcoalgebra of $\mathcal{A}$.

To show that this definition is equivalent to the definition of a subcoalgebra for $F$-coalgebras, we must verify that the embedding $\varphi : B \to A$ is a homomorphism. But for any $b \in B$ and any $i \in I$, we have $\varphi^{\sqcup n_i}(f_i^B(b)) = \varphi^{\sqcup n_i}((f_i^A)(b)) = ((f_i^A)_1(b), (f_i^A)_2((\varphi(b))) = f_i^A(\varphi(b))$, since $(f_i^A)_1((\varphi(b))) = (f_i^A)_1(b)$.

**Example 9.1.4.** Let $A = \{a, b, c\}$, with copower

$$A^{\sqcup 2} = \{(1, a), (2, a), (1, b), (2, b), (1, c), (2, c)\}.$$

We define the binary co-operation $f^A : A \to A^{\sqcup 2}$ by $a \mapsto (1, a)$, $b \mapsto (2, b)$ and $c \mapsto (1, c)$. For $B = \{a, b\}$, the restriction $f^A|B$ maps $a \mapsto (1, a)$ and

$b \mapsto (2, b)$. Then taking $f^B := f^A|B$ makes the coalgebra $(B; f^B)$ a sub-coalgebra of $\mathcal{A}$ of type $\tau = (2)$.

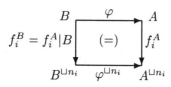

Fig. 9.2  Subcoalgebras of Coalgebras of Type $\tau$

There is a "subcoalgebra criterion" for subcoalgebras of type $\tau$, similar to the one for algebras of type $\tau$.

**Lemma 9.1.5.** *Let* $\mathcal{A} = (A; (f_i^A)_{i \in I})$ *be a coalgebra of type* $\tau$ *and let* $B \subseteq A$ *be a subset of* $A$. *Then the coalgebra* $(B; (f_i^B)_{i \in I})$ *of type* $\tau$ *is a subcoalgebra of* $(A; (f_i^A)_{i \in I})$ *if and only if* $B$ *is closed under all the co-operations* $f_i^A$ *for* $i \in I$; *that is if and only if* $f_i^A(a) \in B^{\sqcup n_i}$ *for all* $a \in B$ *and all* $i \in I$.

**Proof:** When $(B; (f_i^B)_{i \in I})$ is a subcoalgebra of $(A; (f_i^A)_{i \in I})$, the mapping $f_i^B = f_i^A|B$ is an $n_i$-ary co-operation on $B$ for all $i \in I$. Therefore $f_i^B(b) = (f_i^A|B)(b) \in B^{\sqcup n_i}$ for all $b \in B$ and all $i \in I$.

Conversely, suppose that $B$ is closed with respect to $f_i^A$ for all $i \in I$. Then $(f_i^A|B)(B) \subseteq B^{\sqcup n_i}$, so $f_i^A|B$ is an $n_i$-ary co-operation on $B$ and $(B; (f_i^B)_{i \in I})$ with $f_i^B = f_i^A|B$ is a subcoalgebra of $(A; (f_i^A)_{i \in I})$. ∎

The following usual properties of subalgebras also hold for subcoalgebras, and we leave them for the reader to verify.

**Corollary 9.1.6.** *Let* $\mathcal{A}$, $\mathcal{B}$ *and* $\mathcal{C}$ *be coalgebras of type* $\tau$. *Then*

(i) *If* $\mathcal{A} \leq \mathcal{B}$ *and* $\mathcal{B} \leq \mathcal{C}$, *then* $\mathcal{A} \leq \mathcal{C}$.
(ii) *If* $\mathcal{A} \subseteq \mathcal{B} \subseteq \mathcal{C}$ *and* $\mathcal{A} \leq \mathcal{C}$ *and* $\mathcal{B} \leq \mathcal{C}$, *then* $\mathcal{A} \leq \mathcal{B}$.

One of the main results of Section 4.5 was that for $F$-coalgebras, the union of subcoalgebras of a given coalgebra $\mathcal{A}$ is a subcoalgebra, and so is the intersection of finitely many subcoalgebras of $\mathcal{A}$. By the equivalence of the definition of subcoalgebras for $F$-coalgebras and coalgebras of type $\tau$, these results carry over. We can also use the subcoalgebra criterion to prove directly that the union of subcoalgebras is a subcoalgebra.

**Theorem 9.1.7.** *If* $(\mathcal{B}_j)_{j \in J}$ *is a family of subcoalgebras of a coalgebra* $(A; (f_i^A)_{i \in I})$ *of type* $\tau$, *then* $\bigcup\limits_{j \in J} \mathcal{B}_j$ *is a subcoalgebra of* $\mathcal{A}$.

**Proof:** Since each $B_j$ is a subset of $A$, we clearly have $\bigcup_{j \in J} B_j \subseteq A$.
For any $b \in \bigcup_{j \in J} B_j$, there exists an index $j_0 \in J$ such that $b \in B_{j_0}$.
Since $f_i^{B_{j_0}}(b) \in B_{j_0}^{\sqcup n_i} \subseteq (\bigcup_{j \in J} B_j)^{\sqcup n_i}$ and $f_i^A(b) = f_i^{B_{j_0}}(b)$, we have
$f_i^A(b) \in (\bigcup_{j \in J} B_j)^{\sqcup n_i}$. Therefore $(\bigcup_{j \in J} B_j; (f_i^A|(\bigcup_{j \in J} B_j))_{i \in I})$ is a subcoalgebra
of $(A; (f_i^A)_{i \in I})$. ∎

This result allows us to define the subcoalgebra $[S]$ cogenerated by a subset $S$ of a coalgebra $\mathcal{A}$.

**Definition 9.1.8.** Let $\mathcal{A} = (A; (f_i^A)_{i \in I})$ be a coalgebra of type $\tau$ and let $S \subseteq A$ be a subset. Then the union of all the subcoalgebras of $\mathcal{A}$ contained in $S$ forms the greatest subcoalgebra of $\mathcal{A}$ contained in $S$. This subcoalgebra is called the *coalgebra cogenerated by* $S$, and denoted by $[S]$.

**Example 9.1.9.** Let $A = \{a, b, c, d, e\}$ with co-operation $f^A : A \to A^{\sqcup 2}$ given by $a \mapsto (1, a)$, $b \mapsto (1, c)$, $c \mapsto (2, b)$, $d \mapsto (1, d)$ and $e \mapsto (2, e)$. What is the subcoalgebra of $\mathcal{A}$ cogenerated by the subset $S = \{a, b, d\}$? The nonempty subcoalgebras of $\mathcal{A}$ contained in $S$ have universes $\{a\}, \{d\}, \{a, d\}$. Therefore, the greatest subcoalgebra contained in $S$ is $(\{a, d\}; f^A|\{a, d\})$.

Another result from Section 4.5 is that disjoint sums of $F$-coalgebras are the sums (coproducts) in the category $\mathbf{Set}_F$. We now define disjoint sums for coalgebras of type $\tau$. It can be shown that under this definition disjoint sums of coalgebras of type $\tau$ correspond to disjoint sums of $F$-coalgebras.

**Definition 9.1.10.** Let $(\mathcal{A}_j)_{j \in J}$ be a family of coalgebras of type $\tau$. The *disjoint sum* of this family is a coalgebra of type $\tau$ with the universe $\sum A_j = \{(j, a) \mid j \in J, \ a \in A_j\}$ and with the fundamental co-operations $f_i^{\sum A_j} : \sum A_j \to (\sum A_j)^{\sqcup n_i}$ for every $i \in I$ which are defined by $f_i^{\sum A_j}(k, a) := ((f_i^{\sum A_j})_1(k, a), (f_i^{\sum A_j})_2(k, a))$ with $(f_i^{\sum A_j})_1(k, a) := (f_i^{A_k})_1(a)$ and $(f_i^{\sum A_j})_2(k, a) := (k, (f_i^{A_k})_2(a))$, for $k \in J$, $a \in A_k$ and $i \in I$.

## 9.2 Congruences and Quotient Coalgebras

Congruences of $F$-coalgebras are defined as kernels of homomorphisms. For coalgebras of type $\tau$, we define a congruence to be an equivalence relation with a certain additional property. We shall see shortly that this is equivalent to a congruence being a kernel of a homomorphism in this case too.

**Definition 9.2.1.** Let $\mathcal{A} = (A; (f_i^A)_{i \in I})$ be a coalgebra of type $\tau$. A *congruence relation* $\theta$ on $\mathcal{A}$ is an equivalence relation on $A$ which satisfies the condition that $((f_i^A)_2(a), (f_i^A)_2(b)) \in \theta$ and $(f_i^A)_1(a) = (f_i^A)_1(b)$ for every pair $(a, b) \in \theta$ and for each $i \in I$.

**Example 9.2.2.** Let $A = \{a, b, c, d\}$ and let the binary co-operation $f^A$ be given by $a, c \mapsto (2, a)$, $b \mapsto (2, c)$ and $d \mapsto (2, b)$. It is obvious that the equivalence relation $\theta$ on $A$ given by the partition $\{\{a, b, c\}, \{d\}\}$ is a congruence on $\mathcal{A}$.

**Lemma 9.2.3.** *The intersection of two congruences $\theta_1$ and $\theta_2$ on a coalgebra $\mathcal{A}$ is a congruence on $\mathcal{A}$.*

**Proof:** The intersection of $\theta_1$ and $\theta_2$ is certainly an equivalence relation on $A$. Let $(a, b) \in \theta_1 \cap \theta_2$. Then $(a, b) \in \theta_j$ for $j = 1, 2$ and so $((f_i^A)_2(a), (f_i^A)_2(b)) \in \theta_j$ and $(f_i^A)_1(a) = (f_i^A)_1(b)$ for $j = 1, 2$. Therefore also $((f_i^A)_2(a), (f_i^A)_2(b)) \in \theta_1 \cap \theta_2$. ∎

This can be generalized to show that the intersection of an arbitrary family of congruence relations on $\mathcal{A}$ is a congruence on $\mathcal{A}$. This is different from the situation for $F$-coalgebras, where the intersection of congruences is in general not a congruence. For $F$-coalgebras the infimum of a family of congruences is the transitive closure of the intersection, but for coalgebras of type $\tau$ we can simply use the intersection.

For the supremum of a family of congruences, we note that the union of congruences is not in general a congruence, or even an equivalence relation. Thus we have to use the congruence relation on $\mathcal{A}$ which is generated by the union. With these infimum and supremum operators we get a lattice of congruence relations.

**Theorem 9.2.4.** *Let $\mathcal{A}$ be a coalgebra of type $\tau$. The set $Con\mathcal{A}$ of all congruence relations of $\mathcal{A}$ forms a complete lattice with respect to the operations $\wedge$ and $\vee$ defined by $\theta_1 \wedge \theta_2 := \theta_1 \cap \theta_2$ and $\theta_1 \vee \theta_2 := [\theta_1 \cup \theta_2]$. This lattice has the diagonal relation $\Delta_A$ as its least element and $\nabla_A$ as its greatest element.*

The next example shows that in general $\nabla_A$ need not be equal to $A \times A$.

**Example 9.2.5.** Let $A = \{a, b\}$ and let the binary co-operation $f^A$ be given by $a \mapsto (1, b), b \mapsto (2, a)$. Then the greatest congruence on $\mathcal{A} = (A; f^A)$ is not $A \times A$ since $(f^A)_1(a) \neq (f^A)_1(b)$. Therefore in this case $\nabla_A = \Delta_A$.

As in the algebra case, congruences can be used to produce quotient coalgebras.

**Definition 9.2.6.** Let $\mathcal{A} = (A; (f_i^A)_{i \in I})$ be a coalgebra of type $\tau$ and let $\theta$ be a congruence relation on $\mathcal{A}$. We define co-operations on the quotient set $A/\theta$ by

$$f_i^{A/\theta}([a]_\theta) = ((f_i^A)_1(a), [(f_i^A)_2(a)]_\theta),$$

for all $a \in A$. Then the coalgebra $\mathcal{A}/\theta = (A/\theta; (f_i^{A/\theta})_{i \in I})$ is called the *quotient coalgebra* of $\mathcal{A}$ by $\theta$.

The definition means that

$$(f_i^{A/\theta})_1([a]_\theta) = (f_i^A)_1(a)$$
and
$$(f_i^{A/\theta})_2([a]_\theta) = [(f_i^A)_2(a)]_\theta.$$

For this definition to be valid we have to verify that the co-operations $f_i^{A/\theta}$ defined on $A/\theta$ are well-defined. To check this, let $[a]_\theta = [b]_\theta$. Then

$$
\begin{aligned}
& f_i^{A/\theta}([a]_\theta) \\
= \ & ((f_i^A)_1(a), [(f_i^A)_2(a)]_\theta) \\
= \ & ((f_i^A)_1(b), [(f_i^A)_2(b)]_\theta) \\
= \ & f_i^{A/\theta}([b]_\theta).
\end{aligned}
$$

**Example 9.2.7.** Let $A = \{a, b, c, d\}$ and let the equivalence relation $\theta$ on $A$ be given by the partition $\{\{a, b, c\}, \{d\}\}$. Then $A/\theta = \{[a]_\theta, [d]_\theta\}$. Let the binary co-operation $f^A$ on $A$ be defined by $f^A(a) = (2, a)$, $f^A(b) = (2, c)$, $f^A(c) = (2, a)$ and $f^A(d) = (2, a)$. Then $f^{A/\theta}$ maps both classes $[a]_\theta$ and $[d]_\theta$ to $(2, [a]_\theta)$.

**Proposition 9.2.8.** *Let $\mathcal{A} = (A; (f_i^A)_{i \in I})$ be a coalgebra of type $\tau$ and let $\theta$ be a congruence on $\mathcal{A}$. Then the natural mapping $\gamma : A \to A/\theta$ defined by $a \mapsto [a]_\theta$ is a surjective homomorphism from $\mathcal{A}$ onto $\mathcal{A}/\theta$.*

**Proof:** For any $a \in A$, we have $\gamma((f_i^A)_1(a)) = (f_i^{A/\theta})_1([a]_\theta)$ $= (f_i^{A/\theta})_1(\gamma(a))$ and $\gamma((f_i^A)_2(a)) = [(f_i^A)_2(a)]_\theta = (f_i^{A/\theta})_2([a]_\theta) =$

$(f_i^{A/\theta})_2(\gamma(a))$. This shows that $\gamma$ is a homomorphism. ∎

Congruences on $F$-coalgebras were defined as kernels of homomorphisms. Now we prove that any congruence of a coalgebra of type $\tau$ corresponds to a congruence of the corresponding $F$-coalgebra.

**Theorem 9.2.9.** *Let $A$ be a coalgebra of type $\tau$. Then an equivalence relation $\theta$ on $A$ is a congruence on $A$ if and only if $\theta$ is the kernel of some homomorphism from $A$ to some coalgebra $B$.*

**Proof:** When $\theta$ is a congruence, it is clear that $\theta$ is the kernel of the natural mapping $\gamma : A \to A/\theta$ since

$$(a, b) \in \theta \Leftrightarrow [a]_\theta = [b]_\theta \Leftrightarrow \gamma(a) = \gamma(b) \Leftrightarrow (a, b) \in Ker\gamma.$$

Conversely, let $\varphi : A \to B$ be a homomorphism with $Ker\varphi$ as its kernel. Then $Ker\varphi$ is an equivalence relation on $A$. For any $(a, b) \in Ker\varphi$, we have $\varphi(a) = \varphi(b)$, so that $(f_i^B)_2(\varphi(a)) = (f_i^B)_2(\varphi(b))$ and thus $\varphi((f_i^A)_2(a)) = \varphi((f_i^A)_2(b))$ since $\varphi$ is a homomorphism. This implies that $((f_i^A)_2(a), (f_i^A)_2(b)) \in Ker\varphi$. We also have $(f_i^B)_1(\varphi(a)) = (f_i^B)_1(\varphi(b))$, that is, $(f_i^A)_1(a) = (f_i^A)_1(b)$, which makes $Ker\varphi$ a congruence relation on $A$. ∎

Thus if $\theta$ is a congruence on the coalgebra $A$ of type $\tau$, then $\theta$ is the kernel of some homomorphism of the coalgebra $A$ of type $\tau$. But this homomorphism corresponds to a homomorphism of the corresponding $F$-coalgebra and $\theta$ is the kernel of this $F$-coalgebra homomorphism and thus a congruence on the $F$-coalgebra and conversely. This shows that congruences on coalgebras of type $\tau$ correspond to congruences of the corresponding $F$-coalgebras.

The analogue of the Homomorphic Image Theorem (Theorem 1.1.13) for algebras holds for coalgebras of type $\tau$. This fact follows directly from the Diagram Lemma (Corollary 4.1.9), but we will also give a direct proof using the type $\tau$ definitions.

**Theorem 9.2.10.** *Let $A$ and $B$ be coalgebras of type $\tau$, with $\varphi$ a surjective homomorphism from $A$ onto $B$. Then $B$ is isomorphic to the quotient coalgebra $A/Ker\varphi$ and the diagram in Figure 9.3 commutes.*

**Proof:** We use the natural mapping $\gamma$ considered in Proposition 9.2.8 to define a mapping $\psi : \mathcal{B} \to \mathcal{A}/Ker\varphi$ by $\psi(b) = \gamma(a)$ for $b = \varphi(a)$ and any $a \in A$. This mapping is well-defined, since $\varphi(c) = b = \varphi(a) \in B$ implies that $(c,a) \in Ker\varphi$ and so $[c]_{Ker\varphi} = [a]_{Ker\varphi}$. It is clear that $\psi$ is onto, and we show that $\psi$ is one-to-one. If $\psi(b_1) = \psi(b_2)$ for some $b_1, b_2 \in B$, then there are elements $a_1, a_2 \in A$ with $b_1 = \varphi(a_1)$ and $b_2 = \varphi(a_2)$. Since $[a_1]_{Ker\varphi} = [a_2]_{Ker\varphi}$, that is $(a_1, a_2) \in Ker\varphi$, we get $b_1 = b_2$. Also, since $\psi(\varphi(a)) = \psi(b) = \gamma(a)$ for all $a \in A$, the diagram in Figure 9.3 commutes.

Using the fact that $\gamma$ is a homomorphism, we can show that $\psi$ is also a homomorphism. We have $\psi((f_i^B)_2(b)) = \gamma((f_i^A)_2(a)) = [(f_i^A)_2(a)]_{Ker\varphi} = (f_i^{A/\theta})_2([a]_{Ker\varphi}) = (f_i^{A/\theta})_2(\gamma(a)) = (f_i^{A/\theta})_2(\psi(b))$ since with $b = \varphi(a)$ we have $(f_i^B)_2(b) = (f_i^B)_2(\varphi(a)) = \varphi((f_i^A)_2(a))$. Moreover, $(f_i^B)_1(b) = (f_i^B)_1(\varphi(a)) = (f_i^A)_1(a) = (f_i^{A/Ker\varphi})_1(\gamma(a)) = (f_i^{A/Ker\varphi})_1(\psi(b))$. ∎

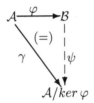

Fig. 9.3    Homomorphic Image Theorem

Let $\varphi : \mathcal{A} \to \mathcal{B}$ be a homomorphism. Two standard properties of homomorphisms are that the image $\varphi(\mathcal{C})$ of a subcoalgebra $\mathcal{C}$ of $\mathcal{A}$ should be a subcoalgebra of $\mathcal{B}$, and the pre-image $\varphi^{-1}(\mathcal{D})$ of a subcoalgebra of $\mathcal{B}$ should be a subcoalgebra of $\mathcal{A}$. For $F$-coalgebras, the proof of the latter fact requires that the functor $F$ preserves weak pullbacks (see [75]). As we mentioned earlier, weak pullbacks are defined in much the same way as pullbacks, but without the uniqueness requirement. It can be proved that the functor $F$ we are using for coalgebras of type $\tau$, defined by $X \mapsto \prod_{j \in J} X^{\sqcup n_j}$ for every set $X$, preserves pullbacks and thus weak pullbacks. Thus this pre-image fact will also hold for coalgebras of type $\tau$. However we can also prove this fact directly, in a simpler fashion.

**Theorem 9.2.11.** *Let $\mathcal{A}$ and $\mathcal{B}$ be coalgebras of type $\tau$ and let $\varphi : \mathcal{A} \to \mathcal{B}$ be a homomorphism.*

(i) *If $\mathcal{C} \leq \mathcal{A}$, then $\varphi(\mathcal{C}) \leq \mathcal{B}$.*
(ii) *If $\mathcal{D} \leq \mathcal{B}$, then $\varphi^{-1}(\mathcal{D}) \leq \mathcal{A}$.*

**Proof:** (i) We know that $\varphi(C) \subseteq B$ and we want to show that $(\varphi(C); (f_i^B|\varphi(C))_{i \in I})$ is a subcoalgebra of $(B; (f_i^B)_{i \in I})$. Assume that $c \in \varphi(C)$. Then there is an element $a \in C$ such that $c = \varphi(a)$. So $(f_i^B)_1(c) = (f_i^B)_1(\varphi(a)) = (f_i^A)_1(a) = (f_i^C)_1(a)$ and $(f_i^B)_2(c) = (f_i^B)_2(\varphi(a)) = \varphi((f_i^A)_2(a)) = \varphi((f_i^C)_2(a)) \in \varphi(C)$. Therefore $f_i^B(c) = ((f_i^C)_1(a), \varphi((f_i^A)_2(a))) \in \varphi(C)^{\sqcup n_i}$ for all $i \in I$. By the subcoalgebra criterion of Lemma 9.1.5 we have $\varphi(C) \leq B$.

(ii) Again it is clear that $\varphi^{-1}(D) \subseteq A$, and we need to show that $(\varphi^{-1}(D); (f_i^A|\varphi^{-1}(D))_{i \in I})$ is a subcoalgebra of $(A; (f_i^A)_{i \in I})$. Let $a \in \varphi^{-1}(D)$, so that $\varphi(a) \in D$. Since $\varphi : A \to B$ is a homomorphism and $D \leq B$, we have $(f_i^A)_1(a) = (f_i^B)_1(\varphi(a)) = (f_i^D)_1(\varphi(a))$ and $\varphi((f_i^A)_2(a)) = (f_i^B)_2(\varphi(a)) = (f_i^D)_2(\varphi(a)) \in D$. Then we obtain $1 \leq (f_i^A)_1(a) \leq n_i$ and $(f_i^A)_2(a) \in \varphi^{-1}(D)$ and thus $f_i^A(A) \in \varphi^{-1}(D)^{\sqcup n_i}$. By Lemma 9.1.5 the coalgebra $(\varphi^{-1}(D); (f_i^A|\varphi^{-1}(D))_{i \in I})$ is a subcoalgebra of $(A; (f_i^A)_{i \in I})$. ∎

As we did for $F$-coalgebras in Section 4.9, we can define covarieties of coalgebras of type $\tau$ as classes which are closed under the operators $H$, $S$ and $\Sigma$. It can be shown that the operators $HS$, $H\Sigma$ and $\Sigma S$ are closure operators and that

$$HS(K) \subseteq SH(K), \quad H\Sigma(K) \subseteq \Sigma H(K) \quad \text{and} \quad \Sigma S(K) \subseteq S\Sigma(K).$$

These inclusions can be used to prove the following version of Tarski's Theorem for coalgebras of type $\tau$.

**Theorem 9.2.12.** *Let $K$ be a class of coalgebras of the same type $\tau$. Then $SH\Sigma(K)$ is the smallest covariety to contain $K$.*

## 9.3 Bisimulations

In this section we define bisimulations for coalgebras of type $\tau$, and show that our definition here coincides with the definition of a bisimulation for the corresponding $F$-coalgebras, as in Section 4.6.

**Definition 9.3.1.** Let $\mathcal{A} = (A; (f_i^A)_{i \in I})$ and $\mathcal{B} = (B; (f_i^B)_{i \in I})$ be coalgebras of type $\tau$, and let $R \subseteq A \times B$ be a binary relation. Then $R$ is said to be a *bisimulation* between $\mathcal{A}$ and $\mathcal{B}$ if for all $i \in I$, $(f_i^A)_1(a) = (f_i^A)_1(b)$ and $((f_i^A)_2(a), (f_i^A)_2(b)) \in R$ whenever $(a, b) \in R$.

Now we prove that for coalgebras of type $\tau$ this definition is equivalent to Definition 4.6.1.

**Theorem 9.3.2.** *Let* $\mathcal{A} = (A; (f_i^A)_{i \in I})$ *and* $\mathcal{B} = (B; (f_i^B)_{i \in I})$ *be coalgebras of type* $\tau$ *and let* $R \subseteq A \times B$ *be a binary relation. Then* $R$ *is a bisimulation between* $\mathcal{A}$ *and* $\mathcal{B}$ *if and only if* $R$ *is a bisimulation between the corresponding F-coalgebras.*

**Proof:** First let $R$ be a bisimulation between the $F$-coalgebras $\mathcal{A}$ and $\mathcal{B}$. Then for each $i \in I$ there are co-operations $f_i^R : R \to R^{\sqcup n_i}$ on $R$ such that the projections $\pi_A : R \to A$ and $\pi_B : R \to B$ are homomorphisms. Thus for each pair $(a, b) \in R$ we have $\pi_A((f_i^R)_2(a, b)) = (f_i^A)_2(\pi_A(a, b)) = (f_i^A)_2(a)$ and $\pi_B((f_i^R)_2(a, b)) = (f_i^A)_2(\pi_B(a, b)) = (f_i^A)_2(b)$. This means that $(f_i^R)_2(a, b) = ((f_i^A)_2(a), (f_i^B)_2(b)) \in R$. Moreover, we have $(f_i^R)_1(a, b) = (f_i^A)_1(\pi_A(a, b)) = (f_i^A)_1(a)$ and $(f_i^R)_1(a, b) = (f_i^A)_1(\pi_B(a, b)) = (f_i^A)_1(b)$, showing that $(f_i^A)_1(a) = (f_i^B)_1(b)$ for all $i \in I$. This shows that $R$ is a bisimulation between the coalgebras $(A; (f_i^A)_{i \in I})$ and $(B; (f_i^B)_{i \in I})$.

Conversely, suppose that $R$ is a bisimulation between the coalgebras $\mathcal{A} = (A; (f_i^A)_{i \in I})$ and $\mathcal{B} = (B; (f_i^B)_{i \in I})$. By definition we have $(f_i^A)_1(a) = (f_i^A)_1(b)$ and $((f_i^A)_2(a), (f_i^A)_2(b)) \in R$, for each $i \in I$ and each $(a, b) \in R$. For each $i \in I$ we define $n_i$-ary co-operations on $R$ by $(f_i^R)_1(a, b) = (f_i^A)_1(a) = (f_i^B)_1(b)$ and $(f_i^R)_2(a, b) = ((f_i^A)_2(a), (f_i^B)_2(b))$ for all $(a, b) \in R$. Then $\mathcal{R} = (R; (f_i^R)_{i \in I})$ is a coalgebra of type $\tau$, and it suffices to show that the projections $\pi_A : R \to A$ and $\pi_B : R \to B$ are homomorphisms. For any $(a, b) \in R$, $(f_i^R)_1(a, b) = (f_i^A)_1(a) = (f_i^A)_1(\pi_A(a, b))$ and $\pi_A((f_i^R)_2(a, b)) = \pi_A((f_i^A)_2(a), (f_i^B)_2(b)) = (f_i^A)_2(a) = (f_i^A)_2(\pi_A(a, b))$. This shows that $\pi_A$ is a homomorphism, and the proof for $\pi_B$ is similar.  ∎

This equivalence of our definition of bisimulation for coalgebras of type $\tau$ and the definition for $F$-coalgebras means that all the results from Sections 4.6 and 4.8 are also valid for bisimulations of coalgebras of type $\tau$.

## 9.4   Terminal Coalgebras of Type $\tau$

We define terminal coalgebras of type $\tau$ as terminal $F$-coalgebras for the functor $F$ mapping any set $X$ to $\prod_{i \in I} X^{\sqcup n_i}$ described at the beginning of this chapter. This functor transforms coalgebras of type $\tau$ into $F$-coalgebras. It turns out that finite terminal coalgebras of type $\tau$ exist for the type $\tau = (n_i)_{i \in I}$ only if $n_i = 1$ for all $i \in I$.

**Theorem 9.4.1.** *The category* $\mathbf{Coalg}(\tau)$ *has finite terminal coalgebras if and only if* $n_i = 1$ *for all* $i \in I$.

**Proof:** Assume that there is a terminal coalgebra $(T; (f_i^T)_{i \in I})$ of type $\tau$ and that $n_j > 1$ for some $j \in I$. Since $f_j^T : T \to T^{\sqcup n_i}$ is a co-operation and since $T$ is finite, there are an integer $k_j \in \{1, \ldots, n_j\}$ and an element $t_j \in T$ such that $f_j^T(t) \neq (k_j, t_j)$ for all $t \in T$. Now we consider an element $(a_i)_{i \in I} \in \prod_{i \in I} T^{\sqcup n_i}$ where $a_j = (k_j, t_j)$. Corresponding to the definition of the functor $F$ the structural mapping $\alpha_T := (f_i^T)_{i \in I}$ maps $t$ to $(f_i^T(t))_{i \in I}$ for all $t \in T$. By our construction we have $(f_i^T)_{i \in I}(t) \neq (a_i)_{i \in I}$ for all $t \in T$. Therefore the structural mapping $\alpha_T$ is not bijective which contradicts Lambek's Lemma (Lemma 7.3.1).

Conversely, if $n_i = 1$ for all $i \in I$ then any one-element coalgebra $\mathcal{T} = (\{t\}; (f_i^T)_{i \in I})$ is terminal.

∎

## 9.5 Coterms

We saw in Chapter 1 that there are two main approaches to the study of algebras of a fixed type $\tau$. The algebraic approach, by the study of homomorphic images, subalgebras and product algebras, led to the definition of a variety as a class closed under the operators $H$, $S$ and $P$. The second approach is equational: using terms and identities we formed equational classes of algebras. The Galois connection $(Id, Mod)$ between sets of identities and classes of algebras led to the Main Theorem of Equational Theory, that varieties and equational classes are the same classes of algebras.

In this section we extend these ideas to the coalgebra setting. We have already considered the algebraic approach, forming covarieties as classes closed under the operators $H$, $S$ and $\Sigma$. Now we need to introduce an equational approach. For this we first define an analogue of terms called *coterms*, and use them to produce coequations and coidentities. We will prove that the model classes of sets of coidentities are in fact covarieties. These covarieties form a sublattice of the lattice of all covarieties of coalgebras of type $\tau$ which was considered for $F$-coalgebras in Section 8.2.

**Definition 9.5.1.** Let $\tau = (n_i)_{i \in I}$ be a type of coalgebras, that is, an indexed set of natural numbers, $n_i \geq 1$, for all $i \in I$, with corresponding indexed set $(f_i)_{i \in I}$ of co-operation symbols. We say that symbol $f_i$ has arity $n_i$, for $i \in I$. Let $\bigcup \{e_j^n \mid n \geq 1, \ n \in \mathbb{N}, \ 1 \leq j \leq n\}$ be a set of symbols which is disjoint from the set $\{f_i \mid i \in I\}$. We assign to each $e_j^n$ the positive integer $n$ as its arity. Then coterms of type $\tau$ are defined as follows:

(i) For every $i \in I$ the co-operation symbol $f_i$ is an $n_i$-ary coterm of type $\tau$.

(ii) For every $n \geq 1$ and $1 \leq j \leq n$ the symbol $e_j^n$ is an $n$-ary coterm of type $\tau$.

(iii) If $t_1, \ldots, t_{n_i}$ are $n$-ary coterms of type $\tau$, then $f_i[t_1, \ldots, t_{n_i}]$ is an $n$-ary coterm of type $\tau$ and $e_j^n[t_1, \ldots, t_n]$ is an $n$-ary coterm of type $\tau$, for every $i \in I$ and $n \geq 1$ and $1 \leq j \leq n$.

Let $cT_\tau^{(n)}$ be the set of all $n$-ary coterms of type $\tau$ and let $cT_\tau := \bigcup_{n \geq 1} cT_\tau^{(n)}$ be the set of all (finitary) coterms of type $\tau$.

## 9.6 Superposition of Co-operations

Returning to our study of co-operations, we now describe some ways to generate new co-operations from existing ones. The following superposition of co-operations was introduced in [15]. If $f^A$ is an $n$-ary co-operation defined on $A$ and if $g_1^A, \ldots, g_n^A$ are $k$-ary co-operations defined on set $A$, then we define $f^A[g_1^A, \ldots, g_n^A] : A \to A^{\sqcup k}$ by

$$a \mapsto ((g_{(f^A)_1(a)}^A)_1((f^A)_2(a)), (g_{(f^A)_1(a)}^A)_2((f^A)_2(a)))$$

for all $a \in A$. The co-operation $f^A[g_1^A, \ldots, g_n^A]$ is called the *superposition* of $f^A$ and $g_1^A, \ldots, g_n^A$.

The *injections* $\iota_i^{n,A}$ are special co-operations which are defined by $\iota_i^{n,A} : A \to A^{\sqcup n}$ with $a \mapsto (i, a)$ for $1 \leq i \leq n$.

Let $\mathcal{A} = (A; (f_i^A)_{i \in I})$ be a coalgebra of type $\tau$. Each coterm $t$ of type $\tau$ induces a co-operation $t^A$ on $\mathcal{A}$, by the following inductive definition:

(i) If $f_i$ is an $n_i$-ary co-operation symbol, then $f_i^A$ is the induced $n_i$-ary co-operation on $A$.

(ii) $(e_j^n)^A := \iota_j^{n,A}$ for every $n \geq 1$ and $j \leq n$, where $\iota_j^{n,A}$ is an $n$-ary injection.

(iii) If $f_i[g_1, \ldots, g_{n_i}]$ is a coterm and we inductively assume that the induced co-operations $g_1^{\mathcal{A}}, \ldots, g_{n_i}^{\mathcal{A}}$ are defined, then $(f_i[g_1, \ldots, g_{n_i}])^{\mathcal{A}} = f_i^{\mathcal{A}}[g_1^{\mathcal{A}}, \ldots, g_{n_i}^{\mathcal{A}}]$.

(iv) If the induced co-operations $g_1^{\mathcal{A}}, \ldots, g_n^{\mathcal{A}}$ are assumed to be known, then we define $(e_j^n[g_1, \ldots, g_n])^{\mathcal{A}} = g_j^{\mathcal{A}}$ for $1 \leq j \leq n$.

## 9.7 Coidentities and Covarieties

Now we have everything we need to define the concept of a coidentity satisfied by a coalgebra $\mathcal{A}$ of type $\tau$.

**Definition 9.7.1.** Let $\mathcal{A} = (A; (f_i^{\mathcal{A}})_{i \in I})$ be a coalgebra of type $\tau$ and let $s \approx t$ be a pair of coterms of type $\tau$. Then $s \approx t$ is called a coidentity in $\mathcal{A}$ if $s^{\mathcal{A}} = t^{\mathcal{A}}$ for the induced co-operations. In this case we write $\mathcal{A} \models_{coid} s \approx t$.

Let $K$ be a class of coalgebras of type $\tau$ and let $\Sigma$ be a set of pairs of coterms of type $\tau$. Then we write $K \models_{coid} \Sigma$ if every pair $s \approx t$ from $\Sigma$ is a coidentity in every coalgebra $\mathcal{A}$ from $K$.

This allows us to define a pair of operators $Coid : K \longmapsto CoidK$, and $Comod : \Sigma \longmapsto Comod\Sigma$, where for any class $K$ of coalgebras and any set $\Sigma$ of coidentities,

$$CoidK = \{s \approx t \mid s, t \in cT_\tau \text{ and } \forall \mathcal{A} \in K(\mathcal{A} \models_{coid} s \approx t)\}$$

and $\quad Comod\Sigma = \{\mathcal{A} \mid \mathcal{A} \text{ a coalgebra of type } \tau \text{ and}$
$$\forall s \approx t \in \Sigma(\mathcal{A} \models_{coid} s \approx t)\}.$$

This pair forms a Galois connection $(Coid, Comod)$, since the following conditions are satisfied:

$(G_1) \quad K_1 \subseteq K_2 \Rightarrow Coid\, K_2 \subseteq Coid\, K_1$
$\qquad \Sigma_1 \subseteq \Sigma_2 \Rightarrow Coid\, \Sigma_2 \subseteq Coid\, \Sigma_1$

and

$(G_2) \quad K \subseteq Comod\, Coid\, K$
$\qquad \Sigma \subseteq Coid\, Comod\, \Sigma.$

Classes of the form $Comod\, \Sigma$ for some set $\Sigma$ of coidentities are the (Galois-)closed sets under the closure operator $ComodCoid$, while sets of the form $Coid\, K$ for some class $K$ of coalgebras are the (Galois-) closed sets under $CoidComod$. Classes of the form $Comod\, \Sigma$ are called *co-equational classes* of coalgebras, and the collection of such classes forms a complete lattice $\mathcal{L}_{comod}(\tau)$. Dually, classes of the form $Coid\, K$ are called *co-equational theories*, and the collection of all such classes forms a complete lattice $\mathcal{E}_{coid}(\tau)$. The two lattices $\mathcal{L}_{comod}(\tau)$ and $\mathcal{E}_{coid}(\tau)$ are dually isomorphic to each other.

Our goal now is to prove that co-equational classes of the form $Comod\Sigma$ for sets $\Sigma$ of coequations of type $\tau$ are covarieties. To do this we first need three technical lemmas which extend the properties of subcoalgebras, homomorphic images and disjoint sums from fundamental co-operations to coterms.

**Lemma 9.7.2.** *Let $\mathcal{A}$ and $\mathcal{B}$ be coalgebras of type $\tau$, and let $\varphi : \mathcal{A} \to \mathcal{B}$ be a homomorphism. Then $\varphi((t^A)_2(a)) = (t^B)_2(\varphi(a))$ and $(t^A)_1(a) = (t^B)_1(\varphi(a))$ for any coterm $t \in cT_\tau$ and any $a \in A$.*

**Proof:** We give a proof by induction on the complexity of coterms. First let $t = f_i$ be a co-operation symbol. Then by Definition 9.1.1 we have $\varphi(f_i^A(a)) = \varphi((f_i^A)_1(a)), (f_i^A)_2(a)) = ((f_i^A)_1(a), \varphi((f_i^A)_2(a))) = ((f_i^A)_1(\varphi(a)), (f_i^B)_2(\varphi(a))) = f_i^B(\varphi(a))$; and then $(f_i^A)_1(a) = (f_i^B)_1(\varphi(a))$ and $\varphi((f_i^A)_2(a)) = (f_i^B)_2(\varphi(a))$.

If $t = e_i^n$ for $1 \leq i \leq n$, then $\varphi((e_i^{n,A})_2(a)) = \varphi((\iota_i^{n,A})_2(a)) = \varphi(a) = (\iota_i^{n,B})_2(\varphi(a)) = (e_i^{n,B})_2(\varphi(a))$, and $(e_i^{n,A})_1(a) = (\iota_i^{n,A})_1(a) = i = (\iota_i^{n,B})_1(\varphi(a)) = (e_i^{n,B})_1(\varphi(a))$.

Now we consider $t = f_i[t_1, \ldots, t_{n_i}]$, for some coterms $t_j$ such that $\varphi((t_j^A)_2(a)) = (t_j^B)_2(\varphi(a))$ and $(t_j^A)_1(a) = (t_j^B)_1(\varphi(a))$ for all $j = 1, \ldots, n_i$. Then

$$
\begin{aligned}
&\varphi((t^A)_2(a)) \\
={}& \varphi((f_i[t_1, \ldots, t_{n_i}]^A)_2(a)) \\
={}& \varphi((f_i^A[t_1^A, \ldots, t_{n_i}^A])_2(a)) \\
={}& \varphi((t_{(f_i^A)_1(a)}^A)_2((f_i^A)_2(a))) \\
={}& \varphi((t_{(f_i^B)_1(\varphi(a))}^A)_2((f_i^A)_2(a))) \\
={}& (t_{(f_i^B)_1(\varphi(a))}^B)_2(\varphi((f_i^A)_2(a))) \\
={}& (f_i^B[t_1^B, \ldots, t_{n_i}^B])_2(\varphi(a)) \\
={}& (t^B)_2(\varphi(a)),
\end{aligned}
$$

and

$$
\begin{aligned}
&(t^A)_1(a) \\
={}& (f_i[t_1, \ldots, t_{n_i}]^A)_1(a) \\
={}& (f_i^A[t_1^A, \ldots, t_{n_i}^A])_1(a) \\
={}& (t_{(f_i^A)_1(a)}^A)_1((f_i^A)_2(a))
\end{aligned}
$$

$$\begin{aligned}
&= (t^A_{(f^B_i)_1((\varphi(a))})_1((f^A_i)_2(a)) \\
&= (t^B_{(f^B_i)_1(\varphi(a))})_1((f^A_i)_2(a)) \\
&= (t^B_{(f^B_i)_1(\varphi(a))})_1((f^A_i)_2(\varphi(a))) \\
&= (f^B_i[t^B_1,\ldots,t_{n_i}])_1(\varphi(a)) \\
&= (t^B)_1(\varphi(a)).
\end{aligned}$$

Therefore $\varphi((t^A)_2(a)) = (t^B)_2(\varphi(a))$ and $(t^A)_1(a) = (t^B)_1(\varphi(a))$ for all coterms $t \in cT_\tau$. ∎

A similar extension can be made for subcoalgebras.

**Lemma 9.7.3.** *Let $\mathcal{A}$ and $\mathcal{B}$ be coalgebras of type $\tau$ and let $\mathcal{B} \leq \mathcal{A}$ be a subcoalgebra. Then $t^B = t^A|B$ for any coterm $t \in cT_\tau$.*

**Proof:** Again our proof is by induction on the complexity of coterms. If $t$ is a co-operation symbol $f_i$, then $t^B = f^B_i = f^A_i|B = t^A|B$ by the definition of a subcoalgebra of type $\tau$. If $t = e^n_i$ for $1 \leq i \leq n$, then $t^B = (e^n_i)^B = \iota^{n,B}_i = \iota^{n,A}_i|B = (e^n_i)^A|B = t^A|B$.

Now let $t = f_i[t_1,\ldots,t_{n_i}]$ for some coterms $t_j$ for which $t^B_j = t^A_j|B$ for all $j = 1,\ldots,n_i$. Then

$$\begin{aligned}
t^B(b) &= (f_i[t_1,\ldots,t_{n_i}])^B(b) \\
&= (f^B_i[t^B_1,\ldots,t^B_{n_i}])(b) \\
&= ((t^B_{(f^B_i)_1(b)})_1((f^B_i)_2(b)), (t^B_{(f^B_i)_1(b)})_2((f^B_i)_2(b))) \\
&= (((t^A|B)_{(f^A_i|B)_1(b)})_1((f^A_i|B)_2(b)), \\
&\quad ((t^A|B)_{(f^A_i|B)_2(b)})_2((f^A_i|B)_2(b))) \\
&= ((t^A_{(f^A_i)_1(b)})_1((f^A_i)_2(b))|B, (t^A_{(f^A_i)_1(b)})_2((f^A_i)_2(b))|B) \\
&= ((t^A_{(f^A_i)_1(b)})_1((f^A_i)_2(b)), (t^A_{(f^A_i)_1(b)})_2((f^A_i)_2(b)))|B) \\
&= ((f^A_i[t^A_1,\ldots,t^A_{n_i-1}])|B)(b) \\
&= (t^A|B)(b).
\end{aligned}$$

Therefore $t^B = t^A|B$ for all coterms $t \in cT_\tau$. ∎

Now we prove a similar proposition for disjoint sums. We will restrict ourselves to disjoint sums of two coalgebras of type $\tau$, but the proof can easily be generalized to disjoint sums of arbitrary families of coalgebras of type $\tau$.

**Lemma 9.7.4.** *Let $\mathcal{A}_1$ and $\mathcal{A}_2$ be coalgebras of type $\tau$ and let $\mathcal{A}_1 + \mathcal{A}_2$ be their disjoint sum. Then for any coterm $t \in cT_\tau$ we have $(t^{\mathcal{A}_1+\mathcal{A}_2})_1(k,a) = (t^{\mathcal{A}_k})_1(a)$ and $(t^{\mathcal{A}_1+\mathcal{A}_2})_2(k,a) = (k, (t^{\mathcal{A}_k})_2(a))$ for any $a \in \mathcal{A}_1 + \mathcal{A}_2$ and for $k = 1,2$.*

**Proof:** If $t = f_i$, then $(t^{A_1+A_2})_1(k,a) = (f_i^{A_1+A_2})_1(k,a) = (f_i^{A_k})_1(a) = (t^{A_k})_1(a)$ for $k = 1,2$ by the definition of a disjoint sum. By the same definition we also have $(t^{A_1+A_2})_2(k,a) = (f_i^{A_1+A_2})_2(k,a) = (k,(f_i^{A_k})_2(a)) = (k,(t^{A_k})_2(a))$ for $k = 1,2$.

If $t = e_i^n$, then

$$(t^{A_1+\ A_2})_1(k,a)$$
$$= (e_i^{n,A_1+A_2})_1(k,a)$$
$$= (\iota_i^{n,A_1+A_2})_1(k,a)$$
$$= i$$
$$= (\iota_i^{n,A_k})_1(k,a)$$
$$= (e_i^{n,A_k})_1(k,a)$$
$$= (t^{A_k})_1(k,a)$$

and

$$(t^{A_1+\ A_2})_2(k,a)$$
$$= (e_i^{n,A_1+A_2})_2(k,a)$$
$$= (\iota_i^{n,A_1+A_2})_2(k,a)$$
$$= (k,a)$$
$$= (k,(\iota_i^{n,A_k})_2(a))$$
$$= (k,(e_i^{n,A_k})_2(a))$$
$$= (k,(t^{A_k})_2(a))$$

for $k = 1,2$ and any $a \in A_1 + A_2$.

Inductively, let $t = f_i[t_1,\ldots,t_{n_i}]$ for coterms $t_j$ with $(t_j^{A_1+A_2})_1(k,a) = (t_j^{A_k})_1(a)$ and $(t_j^{A_1+A_2})_2(k,a) = (k,(t_j^{A_k})_2(a))$ for all $j = 1,\ldots,n_i$, and for all $a \in A_1 + A_2$, with $k = 1,2$. Then

$$(t^{A_1+A_2})_1(k,a)$$
$$= (f_i[t_1,\ldots,t_{n_i}]^{A_1+A_2})_1(k,a)$$
$$= (f_i^{A_1+A_2}[t_1^{A_1+A_2},\ldots,t_{n_i}^{A_1+A_2}])_1(k,a)$$
$$= (t_{(f_i^{A_1+A_2})_1(k,a)}^{A_1+A_2})_1((f_i^{A_1+A_2})_2(k,a))$$
$$= (t_{(f_i^{A_k})_1(a)}^{A_1+A_2})_1(k,(f_i^{A_k})_2(a))$$
$$= (t_{(f_i^{A_k})_1(a)}^{A_1+A_2})_1((f_i^{A_k})_2(a))$$
$$= (f_i^{A_k}[t_1^{A_k},\ldots,t_{n_i}^{A_k}])_1(a)$$
$$= ((f_i[t_1,\ldots,t_{n_i}])^{A_k})_1(a)$$
$$= (t^{A_k})_1(a)$$

and

$$(t^{A_1+A_2})_2(k,a)$$
$$= (f_i[t_1,\ldots,t_{n_i}]^{A_1+A_2})_2(k,a)$$
$$= (f_i^{A_1+A_2}[t_1^{A_1+A_2},\ldots,t_{n_i}^{A_1+A_2}])_2(k,a)$$

$$
\begin{aligned}
&= && (t^{A_1+A_2}_{(f_i^{A_1+A_2})_1(k,a)})_2((f_i^{A_1+A_2})_2(k,a)) \\
&= && (t^{A_1+A_2}_{(f_i^{A_k})_1(a)})_2(k,(f_i^{A_k})_2(a)) \\
&= && (k, t^{A_k}_{(f_i^{A_k})_1(a)}((f_i^{A_k})_2(a))) \\
&= && (k, (f_i^{A_k}[t_1^{A_k},\ldots,t_{n_i}^{A_k}])_2(a)) \\
&= && ((k, (f_i[t_1,\ldots,t_{n_i}])^{A_k})_2(a) \\
&= && (k, (t^{A_k})_2(a))
\end{aligned}
$$

for all $i \in I$, for all $a \in A_1 + A_2$ and for $k = 1,2$.     ∎

Now we are ready to prove our theorem, that co-equational classes are also covarieties.

**Theorem 9.7.5.** *Let $\Sigma$ be a set of coequations of type $\tau$. Then the co-equational class $Comod\Sigma$ is a covariety.*

**Proof:** We need to show that $Comod\Sigma$ is closed under the formation of subcoalgebras, homomorphic images and disjoint sums. For subcoalgebras first, let $\mathcal{A} = (A; (f_i^A)_{i \in I})$ be a coalgebra of type $\tau$ and let $\mathcal{B}$ be a subcoalgebra of $\mathcal{A}$. If $\mathcal{A} \models_{coid} s \approx t$, then $s^A = t^A$ for the co-operations induced on $\mathcal{A}$. Then $s^A|B = t^A|B$, and so by Lemma 9.7.3 we also have $s^B = t^B$, showing that $\mathcal{B} \models_{coid} s \approx t$.

Next, let $\mathcal{A}$ and $\mathcal{B}$ be coalgebras and let $\varphi : \mathcal{A} \to \mathcal{B}$ be a surjective homomorphism. Suppose that $\mathcal{A}$ satisfies $s \approx t$. For any $b \in B$, there is an element $a \in A$ such that $b = \varphi(a)$. Then using Lemma 9.7.2 we have $(s^B)_2(b) = (s^B)_2(\varphi(a)) = \varphi((s^A)_2(a)) = \varphi((t^A)_2(a)) = (t^B)_2(\varphi(a)) = (t^B)_2(b)$ and $(s^B)_1(b) = (s^B)_1(\varphi(a)) = (s^A)_1(a) = (t^A)_1(a) = (t^B)_1(\varphi(a)) = (t^B)_1(b)$. This means that $s^B = ((s^B)_1, (s^B)_2) = ((t^B)_1, (t^B)_2) = t^B$, so $\mathcal{B}$ satisfies the coidentity $s \approx t$.

Finally, let $\mathcal{A}_1$ and $\mathcal{A}_2$ be coalgebras in $Comod\Sigma$. We will give the proof only in the case of a sum of two coalgebras; the proof for arbitrary sums follows similarly. Then for all $s \approx t$ in $\Sigma$ the coalgebras $\mathcal{A}_1$ and $\mathcal{A}_2$ satisfy $s \approx t$, making $s^{A_1} = t^{A_1}$ and $s^{A_2} = t^{A_2}$. Then by Lemma 9.7.4 we have $(s^{A_1+A_2})_1(k,a) = (s^{A_k})_1(a) = (t^{A_k})_1(a) = (t^{A_1+A_2})_1(k,a)$ and $(s^{A_1+A_2})_2(k,a) = (k,(s^{A_k})_2(a)) = (k,(t^{A_k})_2(a)) = (t^{A_1+A_2})_2(k,a)$ for $k = 1,2$ and all $a \in A_k$. Then $s^{A_1+A_2} = t^{A_1+A_2}$, showing that $\mathcal{A}_1 + \mathcal{A}_2$ satisfies $s \approx t$ and that $\mathcal{A}_1 + \mathcal{A}_2$ is in $Comod\Sigma$. Altogether we have shown that $Comod\Sigma$ is a covariety.     ∎

## 9.8 Exercises for Chapter 9

1. Verify directly from Definition 9.1.1, without using the equivalence of type $\tau$ coalgebras with $F$-coalgebras, that the identity mapping $id_A$ is a homomorphism on any type $\tau$ coalgebra $\mathcal{A}$ and that the composition of two homomorphisms is a homomorphism (where defined).

2. Let $\mathcal{A}$, $\mathcal{B}$ and $\mathcal{C}$ be coalgebras of type $\tau$. Prove the following:
a) If $\mathcal{A} \leq \mathcal{B}$ and $\mathcal{B} \leq \mathcal{C}$, then $\mathcal{A} \leq \mathcal{C}$.
b) If $\mathcal{A} \subseteq \mathcal{B} \subseteq \mathcal{C}$ and $\mathcal{A} \leq \mathcal{C}$ and $\mathcal{B} \leq \mathcal{C}$, then $\mathcal{A} \leq \mathcal{B}$.

3. Prove that for any coalgebra $\mathcal{A}$ of type $\tau$, both $\emptyset$ and $\mathcal{A}$ itself are sub-coalgebras of $\mathcal{A}$.

4. For $\Sigma$ a set of coequations of type $\tau$, let $(\mathcal{A}_j)_{j \in J}$ be a family of coalgebras of type $\tau$ in $Comod\Sigma$. Prove that $\prod_{j \in J} \mathcal{A}_j$ is also in $Comod\Sigma$.

5. Prove that the functor $F$ which defines coalgebras of type $\tau$ as $F$-coalgebras preserves weak pullbacks.

# Chapter 10

# Clones of Operations and Co-operations

The concepts of composition or superposition of operations and of clones of operations were defined in Chapter 1 (Section 1.5). In this chapter we consider these topics in more detail, and extend them to composition and clones of co-operations.

Operations are among the most fundamental objects in mathematics. For any base set $A$, we have used the notation $O^n(A)$ for the set of all $n$-ary operations defined on $A$, and $O(A) = \bigcup\limits_{n=1}^{\infty} O^n(A)$ for the set of all finitary operations on $A$. A particularly important case occurs when $A$ is the two-element set $\{0, 1\}$, in which case the operations are usually called Boolean operations or Boolean functions. This simple case of operations on a two-element set nevertheless plays an important role in many fields, including switching circuits, binary logic, discrete optimization, operations research, combinatorics, algorithms, and universal algebra.

Boolean operations are connected to algebra and binary logic by viewing the two elements 0 and 1 as the classical truth values "false" and "true" respectively. Then the Boolean operations are precisely the term operations of the two-element Boolean algebra $\underline{2} = (\{0, 1\}, \wedge, \vee, \neg, 0, 1)$ introduced in Section 1.1, where $\wedge, \vee$ and $\neg$ denote the logical operators of conjunction, disjunction and negation, respectively. The picture in Figure 10.1 shows how we may connect this Boolean logic approach to circuit design. The picture represents an elementary switching circuit (automaton without memory) realizing the usual logical operation of conjunction. Then the switching circuit in Figure 10.2 realizes the composite Boolean operation $x_1 \wedge (x_1 \wedge (x_1 \wedge x_2))$.

The connection of Boolean operations to binary logic can be extended, when the base set $A$ is finite but of size greater than two, to multi-valued

Fig. 10.1   Elementary Switching Circuit

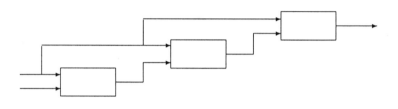

Fig. 10.2   Compound Switching Circuit

logic, specifically $|A|$-valued logic. It can be shown (see [63]) that clones of $|A|$-valued set logic functions, that is, functions mapping $n$-tuples of subsets of $A$ into subsets of $A$, are isomorphic to clones of functions defined on sets $B$ with $|B| = 2^{|A|}$. Completeness and approximation properties for such multi-valued logic operations have applications in the design of optical, biological, and electronic circuitry. Set logic systems are based on multiplex computing or logic-values multiplexing; the simultaneous transmission of logic values enables the construction of superchips with no interference problems.

In Section 1 of this chapter we recall some definitions and notation relating to superposition and clones of operations, and describe several ways to model the composition of operations by algebraic structures. Section 2 describes the lattice of all clones of Boolean functions, while Section 3 discusses the functional completeness problem and maximal clones. In Section 4 we consider some clones which occur in universal algebra, the clones of algebras and varieties. In Sections 5 and 6 we turn to the coalgebra point of view: we define composition of co-operations and study clones of coalgebras. In Section 7 we describe the lattice of all clones of Boolean co-operations.

## 10.1 Clones as Algebraic Structures

In this section we recall some basic information about composition and clones of operations, and discuss several ways in which clones can be modeled as algebraic structures.

**Definition 10.1.1.** Let $A$ be a non-empty base set. For each $n, m \geq 1$, we define an operation

$$S_m^{n,A} : O^n(A) \times (O^m(A))^n \to O^m(A),$$

by setting $S_m^{n,A}(f^A, g_1^A, \ldots, g_n^A)(a_1, \ldots, a_m)$ equal to
$f^A(g_1^A(a_1, \ldots, a_m), \ldots, g_n^A(a_1, \ldots, a_m))$, for all $a_1, \ldots, a_m \in A$.

The operations $S_m^{n,A}$ are called superposition or composition operations on $O(A)$, and $O(A)$ is closed under these operations. For each $n \geq 1$, there are $n$ *projection operations* $e_i^{n,A}$ of arity $n$, defined on set $A$ by $e_i^{n,A}(a_1, \ldots, a_n) = a_i$, for $1 \leq i \leq n$. A *clone of operations* on the set $A$ is any set of operations on $A$ which is closed under composition and which contains all the projections $e_i^{n,A}$ for all $n \geq 1$ and $1 \leq i \leq n$.

It is easy to see that the intersection of any number of clones on $A$ is again a clone on $A$, and that the set of all clones on $A$ can be ordered by set inclusion. The largest clone on $A$ is the whole set $O(A)$, while the smallest is the set $J(A)$ of all the projection operations on $A$. In fact the collection of all clones on $A$, with intersection as meet, forms a lattice. This lattice will be discussed further, in the special case that $A = \{0, 1\}$, in Section 10.2.

**Example 10.1.2.** Another example of a clone on the base set $A = \{0, 1\}$ is the set of all zero-preserving Boolean operations. An operation $f^A$ is said to *preserve* 0 if $f^A(0, 0, \ldots, 0) = 0$, and it is easy to check that the set of all such operations contains the projections and is closed under composition of operations. Similarly, the set of all one-preserving Boolean operations is also a Boolean clone.

There are several ways in which to regard clones as algebraic structures, that is, to model them using sets of objects and operations defined on these sets. One such approach was given by Mal'cev in [59]: he defined a clone as the base set of a subalgebra of the so-called *full iterative algebra* $\mathcal{O}(A) = (O(A); *, \zeta, \tau, \Delta, e_1^2)$. This algebra has one binary operation $*$, three unary operations $\zeta$, $\tau$ and $\Delta$, and one nullary operation $e_1^2$, defined by the following list of properties:

$$(f^A * g^A)(x_1, \ldots, x_{m+n-1}) := f^A(g^A(x_1, \ldots, x_m), x_{m+1}, \ldots, x_{m+n-1}),$$
$$\text{for any } f^A \in O^n(A), g^A \in O^m(A);$$
$$(\tau f^A)(x_1, \ldots, x_n) := f^A(x_2, x_3, \ldots, x_n, x_1);$$
$$(\zeta f^A)(x_1, \ldots, x_n) := f^A(x_2, x_1, x_3, \ldots, x_n);$$
$$(\Delta f^A)(x_1, \ldots, x_{n-1}) := f^A(x_1, x_1, \ldots, x_{n-1}), \text{ if } f^A \in O^n(A) \text{ with } n > 1;$$
and
$$(\tau f^A)(x_1) = (\zeta f^A)(x_1) = (\Delta f^A)(x_1) = f^A(x_1), \text{ if } f^A \text{ is a unary function.}$$

It is easy to verify that the binary operation $*$ is associative, so that a Mal'cev clone or iterative algebra is a semigroup with three additional unary operations and a nullary operation. It is also not hard to prove that universes of subalgebras of the iterative algebra $\mathcal{O}(A)$ are clones in the sense defined above and conversely. Such universes have also been called closed classes or superposition-closed classes, and in this terminology the composition operation is usually called superposition.

Another way to give an algebraic structure to clones is to use heterogeneous (or many-sorted, or multi-based) algebras, as mentioned in Chapter 1 ([46], [10]). In this approach we use

$$\mathcal{O}(A) = ((O^n(A))_{n \in \mathbb{N}^+}; (S_m^n)_{m,n \in \mathbb{N}^+}, (e_i^{n,A})_{n \in \mathbb{N}^+, 1 \leq i \leq n}),$$

where $\mathbb{N}^+$ denotes the set of positive natural numbers. This is the heterogeneous clone of all operations defined on a base set $A$, and any clone on $A$ is a subalgebra of this algebra.

Recalling from Chapter 1 the equivalence of varieties and equational classes, we note that all such heterogeneous clones belong to a variety $K_0$ of heterogeneous algebras, the variety determined by the following three *clone identities*:

(C1)  $S_m^p(z, S_m^n(y_1, x_1, \ldots, x_n), \ldots, S_m^n(y_p, x_1, \ldots, x_n)) \approx$
      $S_m^n(S_n^p(z, y_1, \ldots, y_p), x_1, \ldots x_n), \quad$ for all $m, n, p \in \mathbb{N}^+$,

(C2)  $S_m^n(e_i^n, x_1, \ldots, x_n) \approx x_i, \quad$ for all $m \in \mathbb{N}^+$, $1 \leq i \leq n$,

(C3)  $S_n^n(y, e_1^n, \ldots e_n^n) \approx y, \quad$ for all $n \in \mathbb{N}^+$.

Up to isomorphism, the elements of the variety $K_0$ are exactly the clones regarded as heterogeneous algebras (see for instance [80]). Our heterogeneous clones correspond to algebraic theories, or particular categories, in the sense of F. W. Lawvere ([58]).

Finally, clones can also be described using relations. Extending the usual notion of a binary relation on $A$, we define an $h$-ary *relation* on the set $A$, for any positive integer $h$, to be any subset $\varrho$ of $A^h$. In other words, $h$-ary relations on $A$ are sets of $h$-tuples consisting of elements of $A$. To connect relations with operations, we say that an $n$-ary operation $f^A \in O^n(A)$ *preserves* an $h$-ary relation $\varrho$ if for every $n$ $h$-tuples

$$(a_{11}, \ldots, a_{1h}), \quad (a_{21}, \ldots, a_{2h}), \ldots, (a_{n1}, \ldots, a_{nh})$$

all in $\varrho$, the $h$-tuple

$$(f^A(a_{11}, a_{21}, \ldots, a_{n1}), f^A(a_{12}, a_{22}, \ldots, a_{n2}), \ldots, f^A(a_{1h}, a_{2h}, \ldots, a_{nh}))$$

is also in $\varrho$.

This formulation can also be condensed using matrix notation: $f^A$ preserves $\varrho$ iff for every $h \times n$ matrix $Y = (y_{ij})$ whose column vectors $\underline{y_1}, \ldots, \underline{y_n}$ all belong to $\varrho$, the vector $(f^A(\underline{y}^1), \ldots, f^A(\underline{y}^n))$ also belongs to $\varrho$.

An operation $f^A$ preserving a relation $\varrho$ is also often expressed as $\varrho$ is *invariant* under $f^A$, $f^A$ is a *polymorphism* of $\varrho$, or $f^A$ is *compatible* with $\varrho$. This relationship of preservation establishes a Galois connection between sets of operations and sets of relations on a base set $A$, as follows. For any relation $\varrho$, we define $Pol\varrho$ to be the set of all operations which preserve $\varrho$; and conversely for any operation $f^A$ we can consider the set of all relations which are preserved by $f^A$. This extends to sets of relations and of operations, with $PolR$ defined to be the intersection $\bigcap\{Pol\varrho \mid \varrho \in R\}$ for any set $R$ of relations. Using the machinery of Galois connections, it can be shown that any set of operations on set $A$ which has the form $PolR$ for some set $R$ of relations on $A$ is a clone, and conversely any clone can be expressed as $PolR$ for some such set $R$. For more information on this Galois connection see [67].

We conclude this section with examples of some important Boolean clones which are defined by relations.

**Example 10.1.3.** Taking $A$ to be the Boolean base set $\{0,1\}$, we consider the binary relation $\varrho = \{(0,1), (1,0)\}$. Then a $2 \times n$ matrix $Y = (y_{ij})$ has all columns in $\varrho$ iff $y_{2j} = 1 - y_{1j} = \neg y_{1j}$ holds for all $1 \leq j \leq n$, where the symbols $-$ and $\neg$ denote the Boolean operations of subtraction modulo 2 and negation respectively. For an operation $f^A$ to preserve $\varrho$, it is necessary to have

$$(f^A(y_{11}, \ldots, y_{1n}), f^A(\neg y_{11}, \ldots, \neg y_{1n})) \in \varrho$$

hold for all $y_{11}, \ldots, y_{1n} \in \{0, 1\}$, so that

$$\neg f^A(y_{11}, \ldots, y_{1n}) = f^A(\neg y_{11}, \ldots, \neg y_{1n})$$

or

$$f^A(y_{11}, \ldots, y_{1n}) = \neg(f^A(\neg y_{11}, \ldots, \neg y_{1n}))$$

for all $y_{11}, \ldots, y_{1n} \in \{0, 1\}$.

The *dual* of a Boolean operation $f^A$ is the operation $(f^A)^*$ defined by $(f^A)^*(y_{11}, \ldots, y_{1n}) = \neg(f^A(\neg y_{11}, \ldots, \neg y_{1n}))$. A *self-dual* Boolean operation is one with $(f^A)^* = f^A$. Using Post's notation from ([69]), we denote the set of all self-dual Boolean operations by $D_3$.

**Example 10.1.4.** Again in the Boolean case, let us consider the binary relation $\varrho := \{(0, 0), (0, 1), (1, 1)\}$. Here $(x, y) \in \varrho$ iff $x \leq y$, and a $2 \times n$ matrix has all columns in $\rho$ whenever $y_{1j} \leq y_{2j}$ for all $j = 1, \ldots, n$. Thus an operation $f^A$ preserves $\varrho$ if

$$y_{11} \leq y_{21}, \ldots, y_{1n} \leq y_{2n} \quad \text{implies} \quad f^A(y_{11}, \ldots, y_{1n}) \leq f^A(y_{21}, \ldots, y_{2n}).$$

This is the standard definition of a *monotone* Boolean operation. The set of all monotone Boolean operations is usually denoted by $A_1$.

Unary relations form a special case here, since a *unary relation* is simply a subset of the base set $\{0, 1\}$. When our relation $\varrho$ is the singleton set $\{0\}$, there is a single $1 \times n$ matrix with all columns in $\{0\}$, namely $Y = (0, \ldots, 0)$. A Boolean operation $f^A$ preserves $\{0\}$ iff $f^A(0, \ldots, 0) = 0$, which gives us (the Boolean case of) the clone of *zero-preserving* operations mentioned previously. The usual notation is to use $C_2$ for the set of all one-preserving Boolean operations and $C_3$ for the set of all zero-preserving Boolean operations.

**Example 10.1.5.** As our last example consider the 4-ary relation $\varrho := \{(y_1, y_2, y_3, y_4) \in \{0, 1\}^4 | y_1 + y_2 = y_3 + y_4\}$, where $+$ denotes addition modulo 2. A Boolean operation $f^A$ is *linear* if there are elements $c_0, c_1, \ldots, c_n \in \{0, 1\}$ such that $f^A(x_1, \ldots, x_n) = c_0 + c_1 x_1 + \cdots + c_n x_n$. It can be shown that a Boolean operation is linear iff it preserves the relation $\varrho$. The set of all linear Boolean operations is denoted by $L_1$.

## 10.2 The Lattice of All Boolean Clones

We saw in Section 10.1 that for any given base set $A$, the collection of all clones of operations on $A$ forms a lattice under set inclusion. We can also see this lattice result from the algebraic viewpoint of iterative algebras. In this view, clones of operations defined on $A$ are universes of subalgebras of the algebra $\mathcal{O}(A) = (O(A); *, \zeta, \tau, \Delta, e_1^2)$. Since the collection of universes of all subalgebras of a given algebra always forms a complete lattice with respect to set inclusion, our collection of all clones of operations defined on a given set $A$ forms a complete lattice $\mathcal{L}(A)$.

This lattice $\mathcal{L}(A)$ has been the object of much study. E. L. Post first proved that in the Boolean case $A = \{0, 1\}$, the lattice is countably infinite (announced in [68] and published in [69]). The structure of the lattice in this case is completely known, as we shall describe shortly, and the lattice of all Boolean clones is often called *Post's lattice*. But when the cardinality of $A$ is three or more, the corresponding lattice of all clones on $A$ has $2^{\aleph_0}$ elements, and the lattice is much more difficult to study.

Post's research on the lattice of Boolean clones was motivated by the question of whether a Boolean function (a truth function) is expressible via other given truth functions. Later applications such as switching circuits in computer science have increased the significance of this approach. Post's original proof is rather complicated, but simpler methods have since been found; see for instance [71].

To describe Post's lattice, we need some notation for operations and clones. We use *et* and *vel* to denote the usual Boolean-logic operations of conjunction and disjunction respectively, as well as the notation $et(x_1, x_2) = x_1 \wedge x_2$ and $vel(x_1, x_2) = x_1 \vee x_2$. To simplify our notation, we shall omit the superscript $A$ or $\{0, 1\}$ on operations, writing $f$ instead of $f^A$. We use $c_0^1$ and $c_1^1$ for the *unary constant operations* with value 0 and 1 respectively, $e_1^1$ for the unary projection which is thus the identity operation on $A$, and $N$ to denote the negation operation. The *dual* of a clone $C$ is the clone $C^*$ consisting of all operations dual to the operations in $C$. With this notation we can give the following list of all Boolean clones:

$C_1 := O_{\{0,1\}}$,
$C_3 := Pol\{0\}$, with dual clone $C_2 = Pol\{1\}$,
$C_4 := C_2 \cap C_3$,
$A_1 := Pol \leq$, where $\leq := \{(00), (01), (11)\}$, (the clone of monotone Boolean operations),

$A_3 := A_1 \cap C_3$, with dual clone $A_2 := A_1 \cap C_2$,

$A_4 := A_1 \cap C_4$,

$D_3 := Pol\overline{N}$, where $\overline{N} := \{(01),(10)\}$, (the clone of self-dual operations),

$D_1 := D_3 \cap C_4$,

$D_2 := D_3 \cap A_1$,

$L_1 := Pol\varrho_G$, where $\varrho_G := \{(x,y,z,u) \in \{0,1\}^4 : x+y = z+u\}$, (the clone of linear Boolean operations),

$L_3 := L_1 \cap C_3$, with dual clone $L_2 := L_1 \cap C_2$,

$L_4 := L_1 \cap C_4$,

$L_5 := L_1 \cap D_3$,

$F_8^\mu := PolD_\mu$, where $D_\mu := \{0,1\}^\mu \setminus \{(1,\ldots,1)\}$, for $\mu \geq 2$ with dual

$F_4^\mu := PolD'_\mu$ where $D'_\mu := \{0,1\}^\mu \setminus \{(0,\ldots,0)\}$,

$F_7^\mu := F_8^\mu \cap A_1$, with dual clone $F_3^\mu := F_4^\mu \cap A_1$,

$F_6^\mu := F_8^\mu \cap A_4$, with dual clone $F_2^\mu := F_4^\mu \cap A_4$,

$F_5^\mu := F_8^\mu \cap C_4$, with dual clone $F_1^\mu := F_4^\mu \cap C_4$,

$F_8^\infty := \bigcap_{\mu=2}^{\infty} PolD_\mu$, with dual clone $F_4^\infty := \bigcap_{\mu=2}^{\infty} PolD'_\mu$,

$F_7^\infty := F_8^\infty \cap A_1$, with dual clone $F_3^\infty$,

$F_6^\infty := F_8^\infty \cap A_4$, with dual clone $F_2^\infty$,

$F_5^\infty := F_8^\infty \cap C_4$, with dual clone $F_1^\infty$,

$P_1 := \langle\{et\}\rangle$, with dual clone $S_1 := \langle\{vel\}\rangle$,

$P_3 := \langle\{et,c_0^1\}\rangle$, with dual clone $S_3 := \langle\{vel,c_1^1\}\rangle$,

$P_5 := \langle\{et,c_1^1\}\rangle$, with dual clone $S_5 := \langle\{vel,c_0^1\}\rangle$,

$P_6 := \langle\{et,c_0^1,c_1^1\}\rangle$, with dual clone $S_6 := \langle\{vel,c_0^1,c_1^1\}\rangle$,

$O_9 := \langle\{e_1^1,c_0^1,c_1^1,N\}\rangle = \langle\{N,c_0^1\}\rangle$,

$O_8 := \langle\{e_1^1,c_0^1,c_1^1\}\rangle$,

$O_6 := \langle\{e_1^1,c_0^1\}\rangle$, with dual clone $O_5 := \langle\{e_1^1,c_1^1\}\rangle$,

$O_4 := \langle\{e_1^1,N\}\rangle = \langle\{N\}\rangle$,

$O_1 := \langle\{e_1^1\}\rangle$.

Figure 10.3 shows the Hasse diagram for the countably infinite lattice of all Boolean clones (Post's lattice).

Dual clones are symmetric to each other across the vertical center of the lattice. The lattice can be seen to be both atomic and dually atomic, having finitely many atoms and dual atoms. Note that the dual atoms are the five clones $C_2$, $C_3$, $A_1$, $L_1$ and $D_3$ already described in the examples

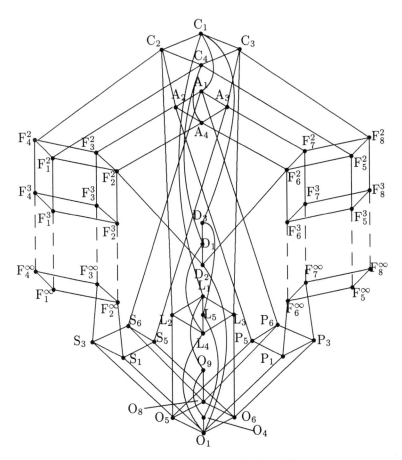

Fig. 10.3   Post's Lattice of all Boolean Clones

in Section 10.1. Another interesting result is that every Boolean clone is finitely generated.

It is well-known that any Boolean operation can be expressed as a composition of the conjunction, disjunction and negation operations; and in fact since each of conjunction and disjunction can be expressed by means of the other and the negation operation, we only need a composition of conjunction and negation, or of disjunction and negation. This fact gives rise to some *normal forms* for operations, as follows. For $x \in \{0,1\}$ let $x^1 := x$ and $x^0 := Nx = 1 - x$. A *clause* over the variables $x_1, \ldots, x_n$ is the Boolean function given by

$$x_{i_1}^{\sigma_1} \vee \cdots \vee x_{i_l}^{\sigma_l}$$

with $1 \leq l \leq n$, $1 \leq i_1 \leq \cdots \leq i_l \leq n$, and $\sigma_1, \ldots, \sigma_l \in \{0, 1\}$.

Historically, the problem of deciding whether for every $n$-tuple $(a_1, \ldots, a_n) \in \{0, 1\}^n$ at least one clause vanishes on the $n$-tuple opened the way for $NP$-completeness problems in computational complexity, an area which is very important today in theoretical computer science, combinatorics, and the theory of algorithms. Here we show that every $n$-ary Boolean operation $f$, other than the $n$-ary constant operation $c_0^n$ with value 0, may be represented by a conjunction of clauses. For each $(a_1, \ldots, a_n) \in f^{-1}(0)$ we use the clause $x_1^{1-a_1} \vee \cdots \vee x_n^{1-a_n}$: this clause takes the value 0 iff $x_1^{1-a_1} = \cdots = x_n^{1-a_n} = 0$, that is, exactly on $(a_1, \ldots, a_n)$. The conjunction of all such clauses, for $n$-tuples mapped by $f$ to 0, forms an operation called the (complete) *conjunctive normal form* of $f$. The *disjunctive normal form* of an operation $f$ is defined dually.

**Example 10.2.1.** Let $f$ be the ternary Boolean function given by the truth table in Figure 10.4.

| $x_1$ | 0 | 0 | 0 | 0 | 1 | 1 | 1 | 1 |
|---|---|---|---|---|---|---|---|---|
| $x_2$ | 0 | 0 | 1 | 1 | 0 | 0 | 1 | 1 |
| $x_3$ | 0 | 1 | 0 | 1 | 0 | 1 | 0 | 1 |
| $f(x_1, x_2, x_3)$ | 0 | 0 | 1 | 0 | 1 | 1 | 0 | 0 |

Fig. 10.4  Table for Operation $f$ from Example 10.2.1

This operation takes the value 0 on five different 3-tuples, giving us the five clauses $x_1 \vee x_2 \vee x_3$, $x_1 \vee x_2 \vee Nx_3$, $x_1 \vee Nx_2 \vee Nx_3$, $Nx_1 \vee Nx_2 \vee x_3$ and $Nx_1 \vee Nx_2 \vee Nx_3$. The conjunction of these five gives the conjunctive normal form of $f$:

$$f(x_1, x_2, x_3) = (x_1 \vee x_2 \vee x_3) \wedge (x_1 \vee x_2 \vee Nx_3) \wedge (x_1 \vee Nx_2 \vee Nx_3)$$
$$\wedge (Nx_1 \vee Nx_2 \vee x_3) \wedge (Nx_1 \vee Nx_2 \vee Nx_3).$$

This process gives a way to express any operation using only the three operations of conjunction, disjunction and negation. (Note that the remaining case, the constant $c_0^n$, can also be obtained, as $x \wedge Nx$.) We say that the set consisting of these three operations is *functionally complete*, meaning that every operation in $O(A)$ can be generated from it by composition or superposition. Since the disjunction operation can also be produced using the other two, the two-element set $\{\wedge, N\}$ is also functionally complete.

An interesting question in Boolean logic is whether there is a one-element functionally complete set, that is, whether there is a single operation from which all Boolean operations can be generated. It is well known that the *Sheffer function*, the operation $sh^A$ defined by $sh^A(x_1, x_2) = Nx_1 \wedge Nx_2$ forms a one-element generating system for Boolean operations. In general, the *functional completeness problem* is to decide for a given set $A$ and a subset $F \subseteq O(A)$ whether $F$ is functionally complete. In the next section we present the solution to this problem for an arbitrary finite set $A$ of cardinality at least two.

## 10.3 The Functional Completeness Problem

The functional completeness problem is the problem of deciding for a given set $F$ of operations on a base set $A$ whether $F$ generates the whole clone of operations on that set. The solution to this problem can be given by means of the concept of a maximal subalgebra of a given algebra. The following criteria for maximality of a subalgebra is a straightforward consequence of the definition, and we leave its proof as an exercise.

**Definition 10.3.1.** Let $\mathcal{A}$ and $\mathcal{B}$ be algebras of the same type, with $\mathcal{B}$ a subalgebra of $\mathcal{A}$. Then $\mathcal{B}$ is called a *maximal subalgebra* of $\mathcal{A}$ if there is no subalgebra $\mathcal{C}$ of $\mathcal{A}$ such that $\mathcal{B} \subset \mathcal{C} \subset \mathcal{A}$.

**Proposition 10.3.2.** *Let $\mathcal{B}$ be a subalgebra of $\mathcal{A}$. Then $\mathcal{B}$ is a maximal subalgebra of $\mathcal{A}$ iff for all $q \in A \setminus B$ the set $B \cup \{q\}$ generates the algebra $\mathcal{A}$.*

As an example of the use of this characterization to describe maximal clones, let us consider the Boolean clone $C_3$, of all zero-preserving operations on the two-element base set $A = \{0, 1\}$. We can show that this clone is maximal in $O(A)$ by showing that for any operation $f^A$ not in $C_3$, the set $C_3 \cup \{f^A\}$ generates all of $O(A)$. We start with the fact that $f^A(0, \ldots, 0) = 1$. Using superposition, we obtain from $f^A$ a unary Boolean function $f'^A$ with $f'^A(0) = 1$. But there are only four unary Boolean functions, and the only two of these which have $f'^A(0) = 1$ are the negation operation $N$ and the unary constant $c_1^1$. If $f'^A$ is the negation operation, we know that we can combine it with the conjunction operation already in $C_3$ to generate all of $O(A)$. Similarly, if $f'^A$ is the unary constant $c_1^1$, then since addition modulo 2 is already in $C_3$ we can generate the negation by $Nx = x + c_1^1$; and once we have both the conjunction and the negation we can generate all of $O(A)$. This shows that $C_3$ is a maximal clone.

The importance of maximal subalgebras of an algebra $\mathcal{A}$ arises from the fact that every subalgebra of a finitely generated algebra can be extended to a maximal one. This is straightforward to prove using Zorn's Lemma (see for instance [36]).

**Theorem 10.3.3.** *Every proper subalgebra of a finitely generated algebra $\mathcal{A}$ can be extended to a maximal subalgebra of $\mathcal{A}$.*

These results can now be combined to give a general completeness criterion for finitely generated algebras.

**Theorem 10.3.4.** *Let $\mathcal{A}$ be a finitely generated algebra. A set $X \subseteq A$ is a generating set for $\mathcal{A}$ iff no universe of a maximal subalgebra of $\mathcal{A}$ contains the set $X$.*

**Proof**: For the only if direction, we proceed by contrapositive. Suppose that there is some maximal subalgebra $\mathcal{M}$ of $\mathcal{A}$ whose universe $M$ contains $X$. Then by the definition of the subalgebra generated by $X$ as the intersection of all the subalgebras of $\mathcal{A}$ which contain $X$, we see that the subalgebra generated by $X$ is contained in $\mathcal{M}$, and so cannot be all of $\mathcal{A}$. Conversely, again by contrapositive, if $X$ does not generate all of $\mathcal{A}$, then by Theorem 10.3.3 the proper subalgebra it generates can be extended to some maximal subalgebra of $\mathcal{A}$, and the universe of this subalgebra then contains $X$. $\blacksquare$

We now have all the tools needed to answer the functional completeness question in the Boolean case. It is well known that for $A = \{0,1\}$ the clone $O(A)$ is generated by $\{\vee, N\}$ and is thus finitely generated. Post's results show that there are exactly five maximal subclones of the full clone: $C_2$, $C_3$, $A_1$, $D_3$, and $L_1$. By Theorem 10.3.4 a set $F$ of Boolean functions is functionally complete iff for each of these five maximal subclones there exists a function $f \in F$ such that $f$ is not included in the maximal subclone.

To use this same approach for the case of a finite base set $A$ of cardinality greater than two, we need to know that $O(A)$ is finitely generated and then to determine all maximal subclones of $O(A)$. In the case of a finite set $A$, the first condition is accomplished by the following theorem going back to E. L. Post.

**Theorem 10.3.5.** *For any finite set $A$ of size at least two, there is a finite generating system for the clone $O(A)$.*

**Proof:** Let $A$ be a finite set and let 0 and 1 be two different elements in $A$. Let $+$ and $\cdot$ be two binary operations from $A^2$ to $A$ with the properties that $0 + x \approx x \approx x + 0$ and $x \cdot 1 \approx x$ and $x \cdot 0 \approx 0$. For each $a \in A$, define a unary operation $\chi_a : A \to A$ with

$$\chi_a(x) = \begin{cases} 1 \text{ if } x = a \\ 0 \text{ otherwise} \end{cases}.$$

If $f^A : A^n \to A$ is an arbitrary $n$-ary operation and $(a_{1j}, \ldots, a_{nj})$ is an arbitrary $n$-tuple on $A$ then $f^A(a_{1j}, \ldots, a_{nj})$ is equal to $f^A(a_{1j}, \ldots, a_{nj}) \chi_{a_{nj}}(a_{nj})$. This shows that $f^A$ can be obtained as $f^A(x_1, \ldots, x_n) = \sum\limits_{(a_1, \ldots, a_n) \in A^n} f^A(a_1, \ldots, a_n) \cdot \prod\limits_{i \leq 1} \chi_{a_i}(x_i)$, where $\Sigma$ and $\Pi$ denote the extended sum and product respectively. Thus we have a generating set for $O(A)$ consisting only of the operations $+$, $\cdot$ and $(\chi_a)_{a \in A}$, and since $A$ is finite, this is a finite generating system. ∎

We note that this proof does include the Boolean case as well, and the formulation $f^A(x_1, \ldots, x_n) = \sum\limits_{(a_1, \ldots, a_n) \in A^n} f^A(a_1, \ldots, a_n) \cdot \prod\limits_{i \leq 1} \chi_{a_i}(x_i)$ corresponds to the conjunctive normal form described earlier, by taking $+$ to be the disjunction operator, $\cdot$ to be conjunction, and $\chi_0(x) = Nx = x$ and $\chi_1(x) = x$.

The maximal clones of operations on any finite set $A$ are known, with a deep theorem explicitly describing these clones due to I. G. Rosenberg ([73], [74]). The special cases $|A| = 2, 3, 4$ were solved earlier by E. L. Post ([69]), S. V. Jablonski ([51]), and A. I. Mal'cev (unpublished). New proofs of Rosenberg's Theorem were given by R. W. Quackenbush ([70]) and D. Lau ([56]). Every maximal clone on a finite set $A$ is of the form $Pol\rho$ for some $h$-ary relation $\rho$. Thus the description of the maximal clones amounts to the determination of the corresponding $h$-ary relations $\rho \subseteq A^h$; Rosenberg's Theorem produces a list of all such relations.

Using this description of the maximal clones on a finite set, along with Theorems 10.3.4 and 10.3.5, gives us the following answer to the functional completeness problem in such cases.

**Corollary 10.3.6.** *Let $A$ be a finite set, and let $F$ be a subset of $O(A)$. Then $F$ generates $O(A)$ if and only if $F$ is not contained in one of the maximal subclones of $O(A)$.* ∎

## 10.4  Clones of Algebras and Varieties

In this section we show briefly how a clone may be associated to any algebra or variety. Let $\mathcal{A} = (A; (f_i^{\mathcal{A}})_{i \in I})$ be an algebra, with $f_i^{\mathcal{A}}$ an $n_i$-ary operation for each $i \in I$. The *term clone* of $\mathcal{A}$, or simply the *clone* of $\mathcal{A}$, is the clone on the base set $A$ generated by the family $(f_i^{\mathcal{A}})_{i \in I}$ of its fundamental operations. The universe of this clone is the set of all the term operations of the algebra $\mathcal{A}$.

Two algebras with the same term clone are said to be *rationally equivalent* to each other. Our work on generating sets of Boolean operations in the previous section provides an example of two rationally equivalent clones. Consider the two-element *Boolean algebra* $2_B = (\{0,1\}; \wedge, \vee, \Rightarrow, N, 0, 1)$, where the fundamental operations are conjunction, disjunction, implication, negation, and the constants 0 and 1 respectively. Since the set $\{\wedge, N\}$ generates all operations on the base set $\{0,1\}$, we see that the Boolean algebra $2_B$ is rationally equivalent to the algebra $(\{0,1\}; \wedge, N)$. In fact these two algebras each have the whole clone $O(\{0,1\})$ as their term clone. Any algebra $\mathcal{A}$ whose term clone is the whole clone $O(A)$ is called a *primal* algebra. The properties of primal algebras were first studied in [32]. Algebras whose term clone is a maximal subclone of $O(A)$ are also important: such algebras are called *preprimal*, and have been studied in [17] and [52].

The term operations of an algebra $\mathcal{A}$ can also be generated in another way. We can consider the set $W_\tau(X)$ of all terms of the type $\tau = (n_i)_{i \in I}$, and the set $W_\tau(X)^{\mathcal{A}}$ of the corresponding term operations induced on $\mathcal{A}$. This set is also the universe of a clone, and it is not difficult to prove that this gives the term clone of $\mathcal{A}$.

For the clone of a variety, we start with any variety $V$ of algebras of type $\tau$. As we saw in Chapter 1, the set $Id V$ of all identities satisfied by all algebras in $V$ is a fully invariant congruence on the absolutely free algebra $\mathcal{F}_\tau(X)$. Then the quotient algebra $\mathcal{F}_V(X) := \mathcal{F}_\tau(X)/_{Id V}$ is known as the free algebra with respect to the variety $V$, or the $V$-free algebra, freely generated by the set $X$. The clone of this algebra $\mathcal{F}_V(X)$ is called the *clone* of the variety $V$, and is denoted by $clone V$. The universe of the clone of a variety $V$ consists of equivalence classes of terms with respect to the equivalence relation $Id V$.

There are close connections between algebraic properties of clones of varieties, regarded as heterogeneous algebras as in Section 10.1, and relationships between varieties. For instance, if $V$ is a subvariety of a variety $W$, then the clone of $V$ is a homomorphic image (as a heterogeneous algebra) of the clone of $W$. Subclones correspond to reducts of varieties, and direct products of clones correspond to non-indexed products of varieties. Minimal varieties, those whose only proper subvariety is the trivial variety of one-element algebras, correspond to simple clones. Details of these connections may be found in [82].

## 10.5  Superposition and Clones of Co-operations

In the remaining sections of this chapter we consider clones of co-operations, beginning here with the definition of superposition of co-operations and the verification that this superposition does give a clone. We recall from Chapter 4 that for any non-empty base set $A$, the $n$-th copower $A^{\sqcup n}$ of $A$ is the union of $n$ disjoint copies of $A$; formally, we define $A^{\sqcup n}$ as the cartesian product $A^{\sqcup n} := \underline{n} \times A$, where $\underline{n} := \{1, \ldots, n\}$. An element $(i, a)$ in this copower corresponds to the element $a$ in the $i$-th copy of $A$, for $1 \leq i \leq n$. A co-operation on $A$ is a mapping $f^A : A \to A^{\sqcup n}$ for some $n \geq 1$; the natural number $n$ is called the arity of the co-operation $f^A$. We also need to recall that any $n$-ary co-operation $f^A$ on set $A$ can be uniquely expressed as a pair $(f_1^A, f_2^A)$ of mappings, $f_1^A : A \to \{1, \ldots, n\}$ and $f_2^A : A \to A$; the first mapping gives the labeling used by $f^A$ in mapping elements to copies of $A$, and the second mapping tells us what element of $A$ any element is mapped to.

We shall denote by $cO^n(A) = \{f^A : A \to A^{\sqcup n}\}$ the set of all $n$-ary co-operations defined on $A$, and by $cO(A) := \cup_{n \geq 1} cO^n(A)$ the set of all finitary co-operations defined on $A$. In Chapter 9 we also defined the superposition of co-operations, as introduced in [15]. If $f^A \in cO^n(A)$ and $g_1^A, \ldots, g_n^A \in cO^{(k)}(A)$ then we define a $k$-ary co-operation $f^A[g_1^A, \ldots, g_n^A] : A \to A^{\sqcup k}$ by

$$a \mapsto ((g_{(f^A)_1(a)}^A)_1((f^A)_2(a)), (g_{(f^A)_1(a)}^A)_2((f^A)_2(a)))$$

for all $a \in A$.

The co-operation $f^A[g_1^A, \ldots, g_n^A]$ is called the *superposition* of $f^A$ and $g_1^A, \ldots, g_n^A$. It will also be denoted by $comp_k^n(f^A, g_1^A, \ldots, g_n^A)$.

**Example 10.5.1.** As with operations, we shall refer to the case of a two-element base set $A = \{0, 1\}$ as the Boolean case for co-operations. It is clear that the unary Boolean co-operations are simply the four different unary operations on $A$. There are exactly 16 different binary Boolean co-operations, as shown in the table in Figure 10.5 below. To simplify our notation, we shall omit the superscript on the co-operation name for these 16 operations, for the remainder of this section.

| | $i_1^2$ | $h_1$ | $h_2$ | $w$ | $d$ | $g_1$ | $h_4$ | $g_2$ |
|---|---|---|---|---|---|---|---|---|
| 0 | (0,0) | (0,0) | (0,1) | (0,1) | (0,0) | (0,0) | (0,1) | (0,1) |
| 1 | (0,1) | (0,0) | (0,1) | (0,0) | (1,1) | (1,0) | (1,1) | (1,0) |

| | $i_2^2$ | $h_5$ | $h_6$ | $h_7$ | $h_8$ | $h$ | $h_9$ | $h_{10}$ |
|---|---|---|---|---|---|---|---|---|
| 0 | (1,0) | (1,0) | (1,1) | (1,1) | (1,0) | (1,0) | (1,1) | (1,1) |
| 1 | (1,1) | (1,0) | (1,1) | (1,0) | (0,1) | (0,0) | (0,1) | (0,0) |

Fig. 10.5    The 16 Binary Boolean Co-operations

The co-operation $d$ from Figure 10.5 is called the *Boolean binary diagonal co-operation*. (Diagonal co-operations can of course be defined for arbitrary arities $n \geq 1$.) The co-operations $h_1, h_2, h_5, h_6$ are called the *constant Boolean co-operations*. In general, there are exactly $(2n)^2$ Boolean co-operations of a given arity $n$.

To illustrate the superposition definition, let us calculate the superposition of the binary Boolean co-operations $w$, $g_1$ and $g_2$ from the table in Figure 10.5.    By our definition $w[g_1, g_2]$ maps 0 to $((g_1)_1(w_2(1)), (g_1)_2(w_2(1))) = (1,0)$, and similarly it maps 1 to $(0,0)$. Therefore $w[g_1, g_2] = h$.

Now we can use the superposition operation defined on $cO(A)$ to make a clone. As special elements we need the *injection co-operations*:    for any $n \geq 1$ and any $1 \leq i \leq n$, the $i$-th $n$-ary injection $i_i^{n,A} : A \rightarrow A^{\sqcup n}$ is defined by $a \mapsto (i, a)$ for all $a \in A$. Using these we get a multi-based algebra

$$((cO^n(A))_{n \geq 1}, (comp_k^n)_{k,n \geq 1}, (i_i^{n,A})_{1 \leq i \leq n}),$$

called the *clone of co-operations* on $A$. It was shown in [15] that this structure is indeed a clone, since it satisfies the three clone axioms (C1), (C2), and (C3).

**Theorem 10.5.2.** *The algebra*

$$((cO^n(A))_{n \geq 1}, (comp_k^n)_{k,n \geq 1}, (i_i^{n,A})_{1 \leq i \leq n})$$

*is a clone.*

**Proof:** We verify that this structure satisfies the three clone axioms. To prove (C1) for co-operations $f^A \in cO^{(p)}(A)$ and $g_1^A, \ldots, g_p^A \in cO^n(A)$ and

$h_1^A, \ldots, h_k^A \in cO_A^{(m)}$, we compare the two sides of the equation

$$(f^A[g_1^A, \ldots, g_p^A])[h_1^A, \ldots, h_n^A] \;=\; f^A[g_1^A[h_1^A, \ldots, h_n^A], \ldots,$$
$$g_p^A[h_1^A, \ldots, h_n^A]]. \qquad (*)$$

Since $h_1^A, \ldots, h_n^A$ are $m$-ary co-operations, both sides result in an $m$-ary co-operation. For each $a \in A$ we have $f^A[g_1^A, \ldots, g_p^A](a) = (g_1^A{}_{f^A(a)}(f_2^A(a)), g_2^A{}_{f^A(a)}(f_2^A(a)))$. Therefore, $(f^A[g_1^A, \ldots, g_p^A])_1(a) = g_1^A{}_{f^A(a)}(f_2^A(a))$ and $(f^A[g_1^A, \ldots, g_p^A])_2(a) = g_2^A{}_{f^A(a)}(f_2^A(a))$. Thus on the left side of (*) we have

$$(f^A[g_1^A, \ldots, g_p^A])[h_1^A, \ldots, h_n^A](a)$$
$$= \;(h_1^A{}_{(f^A[g_1^A, \ldots, g_p^A])_1(a)}((f^A[g_1^A, \ldots, g_p^A])_2(a)),$$
$$h_2^A{}_{(f^A[g_1^A, \ldots, g_p^A])_1(a)}((f^A[g_1^A, \ldots, g_p^A])_2(a)))$$
$$= \;(h_1^A{}_{g_1^A{}_{f^A(a)}(f_2^A(a))}(g_2^A{}_{f_1^A(a)}(f_2^A(a))),$$
$$h_2^A{}_{g_1^A{}_{f^A(a)}(f_2^A(a))}(g_2^A{}_{f_1^A(a)}(f_2^A(a)))).$$

For the right side of equation (*) we define $R_i := g_i^A[h_1^A, \ldots, h_n^A]$ for $i = 1, \ldots, p$. Then

$$f^A[g_1^A[h_1^A, \ldots, h_n^A], \ldots, g_p^A[h_1^A, \ldots, h_n^A]](a)$$
$$= \;f^A[R_1, \ldots, R_p](a)$$
$$= \;(R_{f_1^A(a)})_1(f_2^A(a)), (R_{f_1^A(a)})_2(f_2^A(a)))$$
$$= \;R_{f_1^A(a)}(f_2^A(a))$$
$$= \;g_{f_1^A(a)}[h_1^A, \ldots, h_n^A](f_2^A(a))$$
$$= \;((h^A{}_{(g_{f_1^A(a)})_1^A(f_2^A(a))})_1(g_{f_1^A(a)})_2^A(f_2^A(a)),$$
$$((h^A{}_{(g_{f_1^A(a)})_1^A(f_2^A(a))})_2(g_{f_1^A(a)})_2^A(f_2^A(a)))$$
$$= \;(h_1^A{}_{g_1^A{}_{f_1^A(a)}(f_2^A(a))}(g_2^A{}_{f_1^A(a)}(f_2^A(a))),$$
$$h_2^A{}_{g_1^A{}_{f_1^A(a)}(f_2^A(a))}(g_2^A{}_{f_1^A(a)}(f_2^A(a)))).$$

This shows that the two sides are equal for all $a \in A$, and (C1) is satisfied.

For (C2) we have to prove that $\iota_i^{n,A}[g_1^A, \ldots, g_n^A] = g_i^A$ for $1 \leq i \leq n$, where $\iota_i^{n,A}$ is an injection and $g_1^A, \ldots, g_n^A \in cO^n(A)$. For any $a \in A$ we have $(\iota_i^{n,A})_1(a) = i$ and $(\iota_i^{n,A})_2(a) = a$. Therefore for all $a \in A$,

$$\iota_i^{n,A}[g_1^A, \ldots, g_n^A](a)$$
$$= \left(g_1^A{}_{(\iota_i^{n,A})_1(a)}((\iota_i^{n,A})_2(a)), g_2^A{}_{(\iota_i^{n,A})_1(a)}((\iota_i^{n,A})_2(a))\right)$$
$$= \left((g_i)^A{}_1(a), (g_i)^A{}_2(a)\right)$$
$$= g_i^A(a).$$

To prove (C3) we show that $g^A[\iota_1^{n,A}, \ldots, \iota_n^{n,A}](a) = g^A(a)$ for every $a \in A$, where $g^A \in cO^n(A)$ and $\iota_1^{n,A}, \ldots, \iota_n^{n,A}$ are the $n$-ary injections. Here we have

$$g^A[\iota_1^{n,A}, \ldots, \iota_n^{n,A}](a)$$
$$= \left((\iota^{n,A})_{1\,g_1^A(a)}(g_2^A(a)), (\iota^{n,A})_{2\,g_1^A(a)}(g_2^A(a))\right)$$
$$= \left(g_1^A(a), g_2^A(a)\right) = g^A(a).$$

This finishes the proof.                                                    ∎

As in the case of operations, there is a one-based variant of this algebra. Using only $n$-ary co-operations, for a fixed $n \geq 1$, and the operation $comp^n := comp_n^n$, we form the structure $(cO^n(A), comp^n, \iota_1^A, \ldots, \iota_n^A)$. This is a unitary Menger algebra of rank $n$. As an example, the set of the 16 binary Boolean co-operations, with the operation $comp^2$ and the two injections $\iota_1^{2,A}, \iota_2^{2,A}$ forms a unitary Menger algebra of rank 2.

Just as in Section 10.4 we defined the (term) clone of any algebra, we can produce a clone determined by any coalgebra. To do this we use the concept of an indexed coalgebra, as introduced in Chapter 9. An indexed coalgebra, or a coalgebra of type $\tau$, is a structure of the form $\mathcal{A} = (A; (f_i^A)_{i \in I})$ consisting of a non-empty set $A$ and a family of finitary co-operations $f_i$ of arity $n_i$. The type $\tau$ is then $(n_i)_{i \in I}$.

The clone we are looking for is the clone generated by the set of fundamental co-operations of the coalgebra, the set $F^A := \{f_i^A \mid i \in I\}$. This set $F^A$ can also be expressed as the union of the sets $F^{A(n)}$, of $n$-ary co-operations, for $n \geq 1$.

The clone generated by the fundamental co-operations of a coalgebra can also be defined using coterms and induced coterm operations, both of which were defined in Chapter 9. We shall use the notation $cT_\tau^{(n)}$ for the set of all $n$-ary coterms of type $\tau$, for $n \geq 1$, and $cT_\tau := \cup_{n \geq 1} cT_\tau^{(n)}$ for the set of all coterms of type $\tau$.

We let $cT_\tau^{A(n)}$ be the set of all $n$-ary co-operations induced on $\mathcal{A}$ by $n$-ary coterms of type $\tau$, and let $cT_\tau^A := \cup_{n \geq 1} cT_\tau^{A(n)}$. We now show that these sets form the universes of a subclone of the clone of all co-operations on $A$. The next two theorems are easy to prove using induction on the complexity of a coterm.

**Theorem 10.5.3.** *The sequence* $(cT^{A(n)})_{n \geq 1}$ *is the universe of a subalgebra of the clone* $((cO^n(A))_{n \geq 1}; (comp_m^n)_{m,n \geq 1}, (\imath_i^{n,A})_{1 \leq i \leq n})$ *of all co-operations defined on* $A$.

**Theorem 10.5.4.** *Let* $\mathcal{A} = (A; (f_i^{\mathcal{A}})_{i \in I})$ *be an indexed coalgebra of type* $\tau$, *and let* $cT_\tau$ *be the set of all coterms of type* $\tau$. *Then* $cT_\tau^A$ *is the clone generated by the fundamental co-operations of* $\mathcal{A}$.

## 10.6  Clones of Boolean Co-operations

As we remarked in Section 10.5, clones of co-operations on the two-element set $A = \{0, 1\}$ are called clones of Boolean co-operations. In this final section of this chapter we shall describe the lattice of all clones of Boolean co-operations. Our treatment is based on [23], which used the concept of a *sharp* co-operation from [15].

**Definition 10.6.1.** Let $A$ be any finite set. Let $\imath_1^{n,A}, \ldots, \imath_n^{n,A}$ be the injections on $A$ and let $f^A \in cO^n(A)$. Then we say that $f^A$ *depends essentially* on its $i$-th input if there exists a co-operation $g^A \in cO^n(A)$ such that $f^A[\imath_1^{n,A}, \ldots, \imath_{i-1}^{n,A}, g, \imath_{i+1}^{n,A}, \ldots, \imath_n^{n,A}] \neq f$. We say that $f^A \in cO^n(A)$ is *essentially* $k$-*ary* if there are exactly $k$ inputs upon which $f^A$ essentially depends. A co-operation $f^A$ is called *sharp* if there is a $k \in \mathbb{N}$ such that $f^A$ is both $k$-ary and essentially $k$-ary.

We recall that any $n$-ary co-operation $f^A$ on set $A$ can be uniquely expressed as a pair $(f_1^A, f_2^A)$ of mappings, $f_1^A : A \to \{1, \ldots, n\}$ and $f_2^A : A \to A$. The first mapping gives the labeling used by $f^A$ in mapping elements to copies of $A$, and the second mapping tells us what element of $A$ any element is mapped to. With this notation, we see that an $n$-ary co-operation $f^A$ on $A$ depends on its $i$-th input iff $f_1^A$ maps at least one element of $A$ to $i$, and is essentially $k$-ary iff $f_1^A$ has a $k$-element range.

In addition, it is easy to see that if a co-operation on an $m$-element set is essentially $k$-ary, then $k \leq m$. This tells us that in the Boolean case, any essentially $k$-ary Boolean co-operation can only be unary or binary. There are exactly 16 binary Boolean co-operations, as listed in the table in Figure 10.5, and of these the eight shown in the table in Figure 10.6 are sharp.

(As in Section 5, we omit the superscript $A$ on the names of these Boolean co-operations.)

|   | $d$ | $g_1$ | $h_4$ | $g_2$ | $h_8$ | $h$ | $h_9$ | $h_{10}$ |
|---|-----|-----|-----|-----|-----|-----|-----|-----|
| 0 | (0,0) | (0,0) | (0,1) | (0,1) | (1,0) | (1,0) | (1,1) | (1,1) |
| 1 | (1,1) | (1,0) | (1,1) | (1,0) | (0,1) | (0,0) | (0,1) | (0,0) |

Fig. 10.6   Sharp Binary Boolean Co-operations

Unary co-operations on $A = \{0,1\}$ are functions from $A$ to $\{1\} \times A$, and there are four such functions, all of which are sharp. In a slight abuse of notation, we label them with the names used in the binary table from Section 10.5, according to their behaviour on 0 and 1, as shown in Figure 10.7.

|   | $\iota_1^2$ | $h_1$ | $h_2$ | $w$ |
|---|-----|-----|-----|-----|
| 0 | (0,0) | (0,0) | (0,1) | (0,1) |
| 1 | (0,1) | (0,0) | (0,1) | (0,0) |

Fig. 10.7   Sharp Unary Boolean Co-operations

Csákány proved in [15] that any essentially $k$-ary co-operation is a superposition of a $k$-ary co-operation and some injections, and conversely that any $n$-ary co-operation can be obtained as the superposition of an essentially $k$-ary co-operation for some $k \leq n$ and some injections. This means that any clone of co-operations is uniquely determined by its sharp co-operations.

**Definition 10.6.2.** A co-operation $f^A$ on $A$ is called *Sheffer* if it generates (by superposition) the whole clone $cO(A)$ of all co-operations on $A$.

To describe the characterization from [15] of all Sheffer co-operations, we need some notation and another definition. For any partition $\pi$ on $A$, let $\theta_\pi$ be the corresponding equivalence relation.

**Definition 10.6.3.** A co-operation $f^A = (f_1^A, f_2^A)$ is said to *preserve* a partition $\pi$ on $A$ if $\theta_\pi \subseteq Ker f_1^A$ and $\theta_\pi$ is a congruence on the unary algebra $(A; f_2^A)$. A set $C \subseteq cO(A)$ preserves a partition $\pi$ on $A$ if every $f^A \in C$ preserves $\pi$. A co-operation $f^A = (f_1^A, f_2^A)$ preserves a subset $S \subseteq A$ if $f_2^A$ preserves $S$.

**Proposition 10.6.4.** ([15]) *A co-operation $f^A$ on a finite set $A$ is Sheffer iff $f^A$ preserves neither partitions different from the least one nor non-empty proper subsets of $A$.*

**Corollary 10.6.5.** ([15]) *Let $n$ be a prime number. A co-operation $f^A$ on $A = \{0, 1, 2, \ldots, n-1\}$ is Sheffer iff it is essentially at least binary and $f_2^A$ is a non-identity cyclic permutation on $A$.*

This Corollary tells us that there are exactly two sharp Boolean Sheffer co-operations, namely $g_2$ and $h_{10}$. These two co-operations thus each generate the whole clone $cO(\{0,1\})$.

Now we define a binary relation $\sim$ on the set of all sharp co-operations on $A$: we set $f^A \sim g^A$ iff $f^A$ and $g^A$ generate the same clone of co-operations. This relation is obviously an equivalence relation on the set of sharp co-operations, and to describe the lattice of all Boolean clones of co-operations we need to study sharp co-operations only up to this equivalence.

It is clear that co-operations $f^A$ and $f^A[\imath_1^{2,A}, \imath_2^{2,A}]$ are equivalent for each $f^A$, and we have just shown that $g_2$ and $h_{10}$ are equivalent. We can show that the following pairs of sharp co-operations are also equivalent: $g_1$ and $h$; $d$ and $h_8$; $h_4$ and $h_9$. This reduces our considerations to the sharp co-operations $\imath_1^2$, $h_1$, $h_2$, $w$, $g_2$, $g_1$, $h_4$ and $d$.

The injection $\imath_1^2$ generates the smallest clone of Boolean co-operations, the one consisting of all the injections. Next we look for the minimal clones in the lattice. In order to be minimal a clone must be generated by one element, and have the property that any non-injection co-operation in the clone generates the clone. Since $g_1[\imath_1^2, \imath_1^2] = h_1$ and $h_4[\imath_1^2, \imath_1^2] = h_2$, the clones generated by $g_1$ and $h_4$ respectively are not minimal. However it can be verified that the clones generated by each of $h_1$, $h_2$, $w$ and $d$ are minimal. There are thus exactly four minimal clones of Boolean co-operations. Moreover, the clone $\langle g_1 \rangle$ contains $\langle h_1 \rangle$ as a subclone and similarly $\langle h_2 \rangle \subseteq \langle h_4 \rangle$.

**Lemma 10.6.6.** *The following relationships hold between Boolean co-operations:*

(i) $h_1 = g_1[\imath_1^2, \imath_1^2] = h_2[w, w]$.

(ii) $h_2 = h_1[w, w] = h_4[\imath_1^2, \imath_1^2]$.

(iii) $h_4 = d[h_2, d]$.

(iv) $g_1 = d[\imath_1^2, h_1[\imath_2^2, \imath_1^2]]$, $g_2 = g_1[w, g_1[\imath_2^2, \imath_1^2]] = g_1[h_4, g_1[\imath_2^2, \imath_1^2]] = g_1[h_2, \imath_2^2]$
$= d[w, w[\imath_2^2, \imath_1^2]] = h_4[\imath_1^2, h_1[\imath_2^2, \imath_1^2]] = h_4[h_4[\imath_2^2, \imath_1^2], w[\imath_2^2, \imath_1^2]]$.

**Proof**: The proofs of (i), (ii) and (iii) are all similar, and we give here only the proof of the first equality in (i). Since $g_1[\iota_1^2, \iota_1^2](0) = ((\iota_1^2)_1(0), (\iota_1^2)_2(0)) = (0,0) = h_1(0)$ and $g_1[\iota_1^2, \iota_1^2](1) = ((\iota_1^2)_1(0), (\iota_1^2)_2(0)) = (0,0) = h_1(1)$, we see that $g_1[\iota_1^2, \iota_1^2]$ is the same as $h_1$.

(iv) Since $h_1[\iota_2^2, \iota_1^2](1) = ((\iota_2^2)_1(0), (\iota_1^2)_2(0)) = (1,0)$, we get $d[\iota_1^2, h_1[\iota_2^2, \iota_1^2]](1) = ((h_1[\iota_2^2, \iota_1^2])_1(1), (h_1[\iota_2^2, \iota_1^2])_2(1)) = (1,0) = g_1(1)$ and $d[\iota_1^2, h_1[\iota_2^2, \iota_1^2]](0) = ((\iota_1^2)_1(0), (\iota_1^2)_2(0)) = (0,0) = g_1(0)$. This shows that $d[\iota_1^2, h_1[\iota_2^2, \iota_1^2]]$ is the same as $g_1$. Similarly, $g_2 = g_1[w, g_1[\iota_2^2, \iota_1^2]] = g_1[h_4, g_1[\iota_2^2, \iota_1^2]] = g_1[h_2, \iota_2^2] = d[w, w[\iota_2^2, \iota_1^2]] = h_4[\iota_1^2, h_1[\iota_2^2, \iota_1^2]] = h_4[h_4[\iota_2^2, \iota_1^2], w[\iota_2^2, \iota_1^2]]$. ∎

Lemma 10.6.6 allows us to conclude that the clones $\langle d, h_2 \rangle = \langle d, w \rangle = \langle g_1, h_4 \rangle = \langle g_1, h_2 \rangle = \langle g_1, w \rangle = \langle h_4, h_1 \rangle = \langle h_4, w \rangle$ all equal the clone generated by $g_2$, which is the whole clone $cO(\{0,1\})$. We can also get $\langle d, h_1 \rangle = \langle d, g_1 \rangle$ and $\langle h_1, w \rangle = \langle h_2, w \rangle$ from the lemma.

From the six sharp Boolean co-operations $d$, $g_1$, $h_4$, $h_1$, $h_2$ and $w$, we can make 15 different pairs. The discussion above means that we need consider only the pairs $(d, g_1)$, $(d, h_4)$, $(h_1, h_2)$ and $(h_1, w)$, each of which generates a clone of Boolean co-operations. To consider any larger subsets of the set of the six sharp co-operations, we need only consider cases where the subset contains both of the co-operations from one of these pairs.

It can be shown that the following list of clones are all equal to the whole clone of co-operations:

$\langle d, g_1, h_4 \rangle$, $\quad \langle h, g_1, h_2 \rangle$, $\quad \langle d, g_1, w \rangle$, $\quad \langle d, h_4, h_1 \rangle$, $\quad \langle d, h_4, w \rangle$, $\quad \langle h_1, h_2, h_4 \rangle$, $\langle h_1, h_2, g_1 \rangle$, $\quad \langle h_1, h_2, d \rangle$, $\quad \langle h_1, h_4, w \rangle$, $\quad \langle h_1, w, g_1 \rangle$, $\quad \langle h_1, w, d \rangle$;

and that $\langle h_1, w, h_2 \rangle = \langle h_1, w \rangle$ and $\langle d, g_1, h_1 \rangle = \langle d, g_1 \rangle$ and $\langle d, h_4, h_2 \rangle = \langle d, h_2 \rangle$.

These calculations show that no new clones are obtained, giving us at most the following 12 clones of Boolean co-operations:

$\langle g_2 \rangle = cO_{\{0,1\}}$, $\langle \iota_1^2 \rangle$ $\langle h_1 \rangle$, $\quad \langle d \rangle$, $\quad \langle h_2 \rangle$, $\quad \langle w \rangle$, $\quad \langle g_1 \rangle$, $\quad \langle h_4 \rangle$, $\quad \langle h_1, h_2 \rangle$, $\langle d, g_1 \rangle$, $\quad \langle h_1, w \rangle$ and $\langle d, h_4 \rangle$.

Finally we need to show that these 12 clones are indeed all distinct. Following the notation from the Boolean operation case, we let $\neg$ denote the negation and $+$ the addition operation on $\{0,1\}$, and let $\leq$ be the usual order on $\{0,1\}$.

**Proposition 10.6.7.** *The following sets of Boolean co-operations are clones:*

$C_{0c} := \cup_{n \geq 1} \{f^A \mid f^A \in cO^n(\{0,1\})(\exists i \in \{1, \ldots, n\}(f^A(0) = (i,0)))\}$,

$C_{1c} := \cup_{n \geq 1} \{f^A \mid f^A \in cO^n(\{0,1\})(\exists i \in \{1, \ldots, n\}(f^A(1) = (i,1)))\}$,

$D_c := \cup_{n \geq 1} \{f^A \mid f^A \in cO^n(\{0,1\})(f_1^A(a) = f_1^A(\neg a)$ and $f_2^A(a)$
$= \neg f_2^A(\neg a), a \in \{0,1\})\}$,

$M_c := \cup_{n \geq 1} \{f^A \mid f^A \in cO^n(\{0,1\})(a_1 \leq a_2 \Rightarrow f_1^A(a_1) = f_1^A(a_2), f_2^A(a_1)$
$\leq f_2^A(a_2)$ and $a_1, a_2 \in \{0,1\})\}$,

$L_c := \cup_{n \geq 1} \{f^A \mid f^A \in cO^n(\{0,1\})(\exists i \in \{1, \ldots, n\} \, \exists a_0, a_1 \in \{0,1\}(\forall x$
$\in \{0,1\}(f^A(x) = (i, a_0 + a_1 x))))\}$,

$C_2 := C_{0c} \cap C_{1c} = \cup_{n \geq 1} \{f^A \mid f^A \in cO^n_{\{0,1\}}(\exists i,j \in \{1, \ldots, n\}(f^A(0)$
$= (i,0), f^A(1) = (j,1)))\}$,

$C_3 := \cup_{n \geq 1} \{f^A \mid f^A \in cO^n_{\{0,1\}}(\exists i,j \in \{1, \ldots, n\}(f^A(0) = (i,0), f^A(1)$
$= (j,0)))\} \cup (\cup_{n \geq 1}\{i_i^n \mid 1 \leq i \leq n\})$,

$C_4 := \cup_{n \geq 1} \{f^A \mid f^A \in cO^n(\{0,1\})(\exists i,j \in \{1, \ldots, n\}(f^A(0)$
$= (i,1), f^A(1) = (j,1)))\} \cup (\cup_{n \geq 1}\{i_i^n \mid 1 \leq i \leq n\})$,

$M_0 := C_3 \cap M_c = \cup_{n \geq 1} \{f^A \mid f^A \in cO^n(\{0,1\})(\exists i \in \{1, \ldots, n\}(f^A(0)$
$= (i,0), f^A(1) = (i,0)))\} \cup (\cup_{n \geq 1}\{i_i^n \mid 1 \leq i \leq n\})$ and

$M_1 := C_4 \cap M_c = \cup_{n \geq 1} \{f^A \mid f^A \in cO^n(\{0,1\})(\exists i \in \{1, \ldots, n\}(f^A(0)$
$= (i,1), f^A(1) = (i,1)))\} \cup (\cup_{n \geq 1}\{i_i^n \mid 1 \leq i \leq n\})$.

**Proof:** Starting with the set $C_{0c}$, we note first that all the injections $i_i^{n,A}$ are in the set, since $i_i^{n,A}(0) = (i,0)$ for all $i$ with $1 \leq i \leq n$. For closure under superposition, let $f^A \in C_{0c}^{(n)}$ and $g_1^A, \ldots, g_n^A \in C_{0c}^{(m)}$. Then $f^A[g_1^A, \ldots, g_n^A](0)$
$= ((g_{f_1^A(0)}^A)_1(f^A(0)), (g_{f_1^A(0)}^A)_2(f^A(0))) = ((g_j^A)_1(0), (g_j^A)_2(0)) = (k,0)$, for some $j,k$ with $1 \leq j \leq n, 1 \leq k \leq m$. Therefore, $f^A[g_1^A, \ldots, g_n^A] = comp_m^n(f^A, g_1^A, \ldots, g_n^A)$ is in $C_{0c}$. It can be shown similarly that $C_{1c}$ is also a clone.

To show that $D_c$ is a clone, we note again that since $(i_i^{n,A})_1(a) = i = (i_i^{n,A})_1(\neg a)$ and $(i_i^{n,A})_2(a) = a = \neg(i_i^{n,A})_2(\neg a)$ for all $a \in \{0,1\}$ and $1 \leq i \leq n$, all of the injections $i_i^{n,A}$ are in $D_c$. For any $f^A \in D_c^{(n)}$ and $g_1^A, \ldots, g_n^A \in D_c^{(m)}$, we have

$$(f^A[g_1^A, \ldots, g_n^A])_1(a)$$
$$= (g_{f_1^A(a)}^A)_1(f_2^A(a))$$
$$= (g_{f_1^A(\neg a)}^A)_1(\neg f_2^A(\neg a))$$
$$= (g_{f_1^A(\neg a)}^A)_1(f_2^A(\neg a))$$
$$= (f^A[g_1^A, \ldots, g_n]^A)_1(\neg a) \qquad \text{and}$$

$$(f^A[g_1^A, \ldots, g_n^A])_2(a)$$
$$= (g_{f_1^A(a)}^A)_2(f_2^A(a))$$
$$= \neg(g_{f_1^A(a)}^A)_2(\neg(f_2^A(a)))$$
$$= \neg(g_{f_1^A(a)}^A)_2(\neg(\neg f_2^A(\neg a)))$$
$$= \neg(g_{f_2^A(a)}^A)_2(f_2^A(\neg a))$$
$$= \neg(g_{f_1^A(\neg a)}^A)_2(f_2^A(\neg a))$$
$$= \neg f^A[g_1^A, \ldots, g_n^A](\neg a).$$

This shows closure under superposition, and makes $D_c$ a clone of Boolean co-operations. The proofs for $M_c$, $L_c$, $C_3$ and $C_4$ are very similar to the previous ones, and we leave them as exercises. That $C_2$, $M_0$ and $M_1$ are clones follows from the fact that the intersection of clones is a clone, and our proof is complete. ∎

The ten clones of Proposition 10.6.7 are easily seen to be pairwise distinct, since for each pair of these clones a co-operation can be found which is in one but not the other. We can also list for each of the ten clones which of the eight (up to equivalence) sharp Boolean co-operations are in the clone:

$$d, g_1, i_1^2, h_1 \in C_{0c}, \qquad h_4, d, i_1^2, h_2 \in C_{1c},$$
$$i_1^2, w \in D_c, \qquad i_1^2, h_1, h_2 \in M_c,$$
$$i_1^2, h_1, h_2, w \in L_c, \qquad i_1^2, d \in C_2,$$
$$i_1^2, g_1, h_1 \in C_3, \qquad i_1^2, h_2, h_4 \in C_4,$$
$$i_1^2, h_1 \in M_0, \qquad i_1^2, h_2 \in M_1.$$

Since any clone is uniquely determined by its sharp members, we can write each clone in terms of its sharp generators: $C_{0c} = \langle d, g_1 \rangle$, $C_{1c} = \langle d, h_4 \rangle$, $D_c = \langle w \rangle$, $M_c = \langle h_1, h_2 \rangle$, $L_c = \langle h_1, h_2, w \rangle$, $C_2 = \langle d \rangle$, $C_3 = \langle g_1 \rangle$, $C_4 = \langle h_4 \rangle$, $M_0 = \langle h_1 \rangle$ and $M_1 = \langle h_2 \rangle$.

We also have the four minimal clones $\langle h_1 \rangle$, $\langle h_2 \rangle$, $\langle d \rangle$ and $\langle w \rangle$. The list of ten clones from Proposition 10.6.7, along with the smallest clone of all the injections and the largest clone $cO(A)$, gives us exactly 12 clones of Boolean co-operations. The lattice of all clones of Boolean co-operations is given by the Hasse diagram in Figure 10.8.

Perhaps the most significant feature of this lattice of all clones of Boolean co-operations is that it is finite. This is a strong difference from

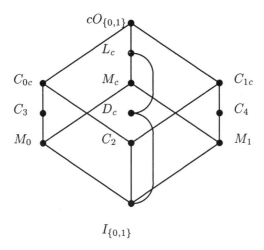

Fig. 10.8   Lattice of all Clones of Boolean Co-operations

the lattice of clones of operations on a finite set, which is countably infinite for the base set $A = \{0,1\}$ but uncountably infinite otherwise.

Two coalgebras, possibly of different types, are said to be *rationally equivalent* if they have the same universe and the same clone of induced term co-operations. The following list shows, up to rational equivalence, all the two-element indexed coalgebras:

$$(\{0,1\}; g_1), \quad (\{0,1\}; g_2), \quad (\{0,1\}; h_1), \quad (\{0,1\}; h_2),$$
$$(\{0,1\}; h_4), \quad (\{0,1\}; d), \quad (\{0,1\}; w), \quad (\{0,1\}; h_1, h_2),$$
$$(\{0,1\}; d, g_1), \quad (\{0,1\}; h_1, w), \quad (\{0,1\}; d, h_4), \quad (\{0,1\}; i_1^2).$$

## 10.7   Exercises for Chapter 10

1. Let $A = \{0,1\}$. Prove that any composition of one-preserving operations on $A$ is also one-preserving.

2. Prove that the binary operation $*$ of Mal'cev's full iterative algebra (from Section 10.1) is associative.

3. Prove that universes of subalgebras of Mal'cev's full iterative algebra

are clones, in the sense that they contain all projection operations and are closed under composition of operations, and conversely that clones are universes of subalgebras of the full iterative algebra.

4. Let $A = \{0, 1\}$, and let $\varrho$ be the binary relation on $A^4$ defined by

$$\rho = \{(y_1, y_2, y_3, y_4) \in A^4 \mid y_1 + y_2 = y_3 + y_4\}.$$

Prove that a Boolean operation $f^A$ is linear iff it preserves $\varrho$.

5. Prove Proposition 10.3.2: For $\mathcal{B}$ a subalgebra of an algebra $\mathcal{A}$, $\mathcal{B}$ is a maximal subalgebra of $\mathcal{A}$ iff $B \cup \{q\}$ generates $\mathcal{A}$ for any element $q$ in $A \setminus B$.

6. Use Zorn's Lemma to prove Theorem 10.3.3, that every proper subalgebra of a finitely generated algebra $\mathcal{A}$ can be extended to a maximal subalgebra of $\mathcal{A}$.

7. Prove that there are exactly $(2n)^2$ Boolean co-operations of a given arity $n \geq 1$.

8. Complete the proof of Theorem 10.6.6.

9. Prove that the sets $M_c$, $L_c$, $C_3$ and $C_4$ from Proposition 10.6.7 are clones of Boolean co-operations.

# Chapter 11

# Semigroups of Operations and Co-operations

Throughout this chapter we look at operations and co-operations defined over an arbitrary finite set $A$. In addition to the usual clone operations discussed in the previous chapter, we here define a binary operation $+$ on the set $O^n(A)$ of all $n$-ary operations on $A$. This operation is associative, giving us a semigroup $(O^n(A); +)$, and in Sections 1 to 6 we examine various properties of this semigroup. In the remaining sections of this chapter we look at semigroup properties for the analogous semigroup $(cO^n(A); +)$ for co-operations. The results of this chapter are taken from [23], [13] and [24].

## 11.1  Basic Definitions and Properties

We shall assume throughout this chapter that $A$ is a finite base set with $|A| \geq 2$, and that $n \geq 1$ is a fixed natural number. As before, we denote by $O^n(A)$ the set of all $n$-ary operations on $A$. For simplicity of notation we shall drop the superscript for operations $f^A$, denoting a typical operation simply as $f$.

Using the $n$-ary composition (or superposition) operation $S^{n,A}$ we define a binary operation $+$ on $O^n(A)$ by

$$f + g \; := \; S^{n,A}(f, g, \ldots, g) \; = \; f(g, \ldots, g).$$

A similar binary operation can be defined on the set $cO^n(A)$ of all $n$-ary co-operations on $A$, by

$$f + g \; := \; comp_n^n(f, g, \ldots, g).$$

It can be verified using the clone axioms (see Exercise 11.1) that both these operations are associative, giving us two semigroups to consider. Semigroups have been extensively studied as algebraic structures, and there are a number of interesting properties we can look at for our two semigroups. In the remainder of this section we define some of these properties.

Let $\mathcal{S} = (S; \circ)$ be any semigroup, so that $\circ$ is an associative binary operation on set $S$. An element $a$ of $S$ is called an *idempotent* if $a \circ a = a$, and is called *regular* if there is an element $b \in S$ such that $a \circ b \circ a = a$. By definition every idempotent element is regular, but the converse is not generally true. A semigroup is said to be *regular* if all its elements are regular, and a semigroup in which every element is idempotent is called a *band*.

We shall be particularly interested in the following kinds of bands. A semigroup $(S; \circ)$ is called a *left-zero semigroup* if $x \circ y = x$ for all $x, y \in S$. Dually, a *right-zero semigroup* is one in which $x \circ y \approx y$ is an identity. A semigroup $\mathcal{S}$ is called a *rectangular band* if the identities $x \circ x \approx x$ and $x \circ y \circ z \approx x \circ z$ are both satisfied by $\mathcal{S}$. It is obvious that any rectangular band is a regular semigroup. A semigroup $(S; \circ)$ is called a *normal band* if it is idempotent and $x \circ y \circ u \circ v = x \circ u \circ y \circ v$ for all $x, y, u, v \in S$, and $\mathcal{S}$ is called a *regular band* if $x \circ y \circ z \circ x = x \circ y \circ x \circ z \circ x$ for all $x, y, z \in S$. *Semilattices* are commutative and idempotent semigroups.

A *zero semigroup* or *constant semigroup* is a semigroup in which the binary operation is constant: the identity $x \circ y \approx u \circ v$ is satisfied. In this case there exists some element $a \in S$ such that $x \circ y = a$ for all $x$ and $y$ in $S$, and we call $a$ the *constant value* of $\mathcal{S}$.

If $a$ is an element in a semigroup $\mathcal{S} = (S; \circ)$, it is convenient to use power notation to denote successive applications of $\circ$ to $a$: that is, we set $a^1 = a$, $a^2 = a \circ a$, and inductively $a^{n+1} = a \circ a^n$. Then the subsemigroup generated by $a$ is the set $< a > = \{a, a^2, a^3, \ldots\}$. The *order* of element $a$ is defined to be the cardinality of this subsemigroup. Since in this chapter we shall deal only with finite semigroups of operations, we consider here only the case of an element with finite order. When $a$ has finite order, there exist two positive integers, the *index* $m$ and the *period* $r$ of $a$, with the properties that $a^{r+m} = a^m$ and $< a > = \{a, a^2, \ldots, a^{r+m-1}\}$. In this case, the order of $a$ is the number $r + m - 1$.

A slightly different notation is used in the special case of the semigroup $\mathcal{H}_n$ of all mappings on the $n$-element set $\{1, 2, \ldots, n\}$, under the usual composition of mappings. Let $f$ be any such mapping. If $f$ is a bijection, its order is its usual order as a permutation. If $f$ is not a bijection, then we let $\lambda(f)$ be the least non-negative integer $m$ such that $Im f^m = Im f^{m+1}$. The order of $f$ is then the order of the permutation obtained by restricting $f$ to the set $Im f^{\lambda(f)}$. Again we call $\lambda(f)$ the index of $f$, and $r$ the period.

*Green's relations* are five equivalence relations definable on any semigroup, which tell us about the structure of the semigroup. We start with the left and right Green's relations $\mathcal{L}$ and $\mathcal{R}$ on a semigroup $S = (S; \circ)$. For any elements $a$ and $b$ of $S$, we say that $a\mathcal{L}b$ if $a = b$ or there exist some $c$ and $d$ in $S$ such that $c \circ a = b$ and $d \circ b = a$; dually $a\mathcal{R}b$ if $a = b$ or there exist some $f$ and $g$ in $S$ such that $a \circ f = b$ and $b \circ g = a$. It follows easily from these definitions that $\mathcal{L}$ is always a right congruence on $S$, while $\mathcal{R}$ is always a left congruence. Two more Green's relations are defined in terms of these: $\mathcal{H} = \mathcal{L} \cap \mathcal{R}$ and $\mathcal{D} = \mathcal{L} \circ \mathcal{R}$ (the usual composition of relations). The relation $\mathcal{J}$ is defined by $a\mathcal{J}b$ iff $a = b$ or there exist elements $p$, $q$, $r$ and $s$ in $S$ such that $a = p \circ b \circ q$ and $b = r \circ a \circ s$. If $\mathcal{T}$ denotes any of these five relations, we use the notation $T_a$ for the equivalence class containing $a$ in the corresponding relation.

More detail about any of these semigroup properties may be found in [48].

## 11.2 Unary Operations

We begin our investigation of the semigroups $(O^n(A); +)$ with the special case that $n = 1$. In this case our semigroup becomes the *full transformation semigroup* on the set $A$, the semigroup of all mappings $f : A \to A$. Our operation $+$ on such mappings is then the ordinary composition of mappings, and we follow the usual convention of denoting this operation by $\circ$ instead of $+$.

Our first lemma lists some well-known properties of the transformation semigroup on $A$; again, more detail may be found in [48]. We see in this lemma that the kernel and image of a transformation $f$ determine its $\mathcal{R}$- and $\mathcal{L}$-class respectively. We recall the notation $Ker f$ for the kernel of $f$, that is the set $\{(a, b) \mid f(a) = f(b)\}$, and $Im\ f$ for the image of $f$.

**Proposition 11.2.1.** *Let $A$ be a finite set.*

(i) *The operation $f \in O^1(A)$ is idempotent iff $f(a) = a$ for all $a \in Imf$; that is, iff the restriction of $f$ to the set $Im\ f$ is the identity function on this set.*

(ii) *$(O^1(A); \circ)$ is a regular semigroup.*

(iii) *For any $f$ and $g$ in $O^1(A)$, $(f, g) \in \mathcal{R}$ iff $Imf = Img$.*

(iv) *For any $f$ and $g$ in $O^1(A)$, $(f, g) \in \mathcal{L}$ iff $Kerf = Kerg$.*

(v) *A subgroup $S \subseteq (O^1(A); \circ)$ is a right-zero semigroup iff $Imf = Img$,* *for all $f, g \in S$.*

Our first goal is to characterize constant subsemigroups of $(O^1(A); \circ)$. If $S$ is a constant semigroup, then there is a constant value $f$ in $S$ such that $g \circ h = f$ for all $g$ and $h$ in $S$. This element $f$ is an idempotent element, and it is the only idempotent element in $S$. We introduce the following binary relation on $O^1(A)$.

**Definition 11.2.2.** The binary relation $\sim$ is defined on $O^1(A)$ by the rule that $g \sim h$ iff $g^2 = h^2$ and $g^3 = h^3$.

Notice that $g^3 = h^3$ means that $g$ and $h$ agree on $Im\ g^2 = Im\ h^2$. It is easy to see that $\sim$ is an equivalence relation on $O^1(A)$, and we can describe its equivalence classes or blocks. The proof of the following lemma is a straightforward application of the definitions and the properties of bijection, and we leave it as an exercise.

**Lemma 11.2.3.** *Let $A$ be a finite set.*

(i) *For any $g \in O^1(A)$, the equivalence class $[g]_\sim$ contains at most one idempotent element.*

(ii) *If $f$ is idempotent, then for all $g \in [f]_\sim$ we have $Ker g \subseteq Ker f$ and $(g \circ f)|Im f = (f \circ g)|Im f = f|Im f$.*

(iii) *If $Im f = A$, then $[f]_\sim = \{f\}$.*

(iv) *If $f$ is idempotent and $|Im f| = |A| - 1$, then $[f]_\sim = \{f\}$.*

If $f$ is idempotent and the cardinality of its image is exactly $|A| - 1$, then there exists a uniquely determined element $x \in A$ such that $Im\ f = A \setminus \{x\}$. Moreover then $f(y) = y$ for all $y \neq x$. This means that there is exactly one equivalence class $B$ with respect to $Ker f$ which contains two elements, while all the other classes contain only one element. In fact the restriction of $f$ to $A \setminus B$ is the identity function on this set.

The following lemma contains parts (iii) and (iv) of Lemma 11.2.3 as special cases.

**Lemma 11.2.4.** *Let $f$ be an idempotent element of $(O^1(A); \circ)$. Then $[f]_\sim = \{f\}$ iff every equivalence class of $Ker f$ has size at most two.*

**Proof:** To prove the forward direction by contradiction, we first assume that $[f]_\sim = \{f\}$ but there is an equivalence class $B$ with $|B| > 2$. Let $x$, $y$ and $z$ be three distinct elements in block $B$, and note that by definition we have $f(x) = f(y) = f(z)$. Also since $f$ is idempotent, and hence equal to the identity function when restricted to $Im\ f$, we may assume that $f(x) = x$. Now we use this information to produce an operation $g \neq f$ which is in $[f]_\sim$, for our contradiction. We define $g : A \to A$ by setting $g(z) = y$, $g(l) = x$ for any $l$ in $B \setminus \{z\}$, and $g(t) = f(t)$ for any $t \notin B$.

It is clear that $f \neq g$, since they disagree on the input $z$. To show that $g \in [f]_\sim$, we have to show that $g^2 = f^2$ and $g^3 = f^3$. For the first of these, we consider the three different cases for $g^2$. On the input $z$, we have $g^2(z) = g(y) = x = f(z) = f^2(z)$. For any input $l \in B \setminus \{z\}$ we have $g^2(l) = g(g(l)) = g(x) = x = f(l) = f^2(l)$. And finally for any input $t \notin B$, we note first that we must also have $f(t) \notin B$; for if not, then $f(t) = f(f(t)) = x$ would imply $t \in B$. Then we have $g^2(t) = g(g(t)) = g(f(t)) = f(f(t)) = f^2(t)$. Thus we have shown that $f^2 = g^2$.

To show that $g^3 = f^3$, we first show that $g \circ f = f$, again by considering three types of inputs. For any $t \notin B$, since also $f(t) \notin B$, we have $(g \circ f)(t) = g(f(t)) = f(f(t)) = f(t)$. For any $l \in B \setminus \{z\}$, we have $(g \circ f)(l) = g(x) = x = f(l)$, and finally also $(g \circ f)(z) = f(z)$. Now using $g^2 = f^2$ and $g \circ f = f$, we see that $g^3 = g \circ g^2 = g \circ f^2 = g \circ f \circ f = f^2 = f = f^3$. Overall we have $g \neq f$ but $g \in [f]_\sim$, which is our contradiction.

For the converse, again proceeding by contradiction we assume that no equivalence class has more than two elements, but there is an element $g \neq f$ such that $g \in [f]_\sim$. Since $f \neq g$, there is an element $x \in A$ with $f(x) \neq g(x)$. We will show that the elements $x$, $f(x)$ and $g(x)$ give three distinct elements all in the same equivalence class of $Ker f$, which will contradict our assumption.

We already know that $f(x) \neq g(x)$. To show that $x$ is also distinct from these two elements, we note that by Lemma 11.2.3(ii) we have $(g \circ f)|Imf = (f \circ g)|Imf = f|Imf$. Therefore for any element $t \in Im\ f$, we can write $t = f(a)$ for some $a \in A$, and we have $g(t) = g(f(a)) = (g \circ f)(a) = f(a) = t$. This shows that $g$ is the identity function on the set $Im\ f$. Can $x$ be in $Im\ f$? If so, we would have $x = f(a)$ for some $a \in A$, and then $f(x) = f(f(a)) = f(a) = x$ and also $g(x) = x$, by the previous remark; but this contradicts our choice of $x$. Therefore we have $x \notin Im\ f$, and in particular we see that $f(x)$ cannot equal $x$. Similarly, if $g(x) = x$ then we would have $g^2(x) = x$ but also $g^2(x) = f^2(x) = f(x)$, resulting in the contradiction $f(x) = g(x) = x$. This shows that the three elements $x$, $f(x)$ and $g(x)$ are all distinct, and it is clear that they are all in the same class under $Ker f$. ∎

If $S$ is a constant semigroup with idempotent constant value $f$, then any subset of $S$ which contains $f$ is also a constant semigroup with value $f$. We can now characterize the largest semigroup which is constant with value $f$.

**Theorem 11.2.5.** *Let $f$ be an idempotent element of $(O^1(A); \circ)$. Then the following are equivalent:*

(i) *The largest constant semigroup with value $f$ is $\{f\}$.*
(ii) $[f]_\sim = \{f\}$.
(iii) *Every equivalence class of $Ker f$ has size at most two.*

**Proof**: The equivalence of (ii) and (iii) was proved in Lemma 11.2.4, and we now show the equivalence of (i) and (ii). First, suppose that $\{f\}$ is the largest constant semigroup with value $f$. Then for any $g \in [f]_\sim$, we have $g^2 = f^2 = f$ and $g^3 = f^3 = f$ by definition, as well as $(g \circ f)|Imf = (f \circ g)|Imf = f|Imf$ by Lemma 11.2.3(ii). This makes $\{g, f\}$ the universe of a constant semigroup with constant value $f$, which forces $f = g$ and so $[f]_\sim = \{f\}$. Conversely, let $[f]_\sim = \{f\}$, and let $S$ be any constant semigroup containing $f$ as its constant value. Then for any $g$ in $S$, we have $g^2 = f^2 = f$ and $g^3 = f^3 = f$, so that $g \in [f]_\sim$, hence $g = f$. Therefore $S$ must equal $\{f\}$ only.                                  ∎

The next lemma gives one more property of the cardinality of $[f]_\sim$.

**Lemma 11.2.6.** *Let $f$ be an idempotent element of $(O^1(A); \circ)$. If the class $[f]_\sim$ contains more than one element, then it contains at least three elements.*

**Proof**: Assume that $[f]_\sim$ contains more than one element. Then by Lemma 11.2.4 there is a block $B \in A/Ker f$ with size at least two, and as in the proof of that lemma we can construct an element $g \neq f$ with $g \in [f]_\sim$. Using the notation from that proof, a third element $h$ can be defined by $h(y) = z$, $h(l) = x$ for $l \in B \setminus \{y\}$, and $h(t) = f(t)$ for all $t \notin B$. One can check that $h$ is also in $[f]_\sim$, but different from $f$ and $g$.                                  ∎

We can also use another binary relation on $O^1(A)$ to characterize constant semigroups. We define the relation $\varrho$ by

$$g \varrho h \iff g \circ h = h^2 = g^2 = h \circ g,$$

for any $g, h \in O^1(A)$. This relation is obviously reflexive and symmetric, but in general is not transitive. It has the following additional properties.

**Lemma 11.2.7.** *Let $f$ be an idempotent element of $O^1(A)$.*

(i) $\varrho \subseteq \sim$.

(ii) $g\varrho f$ *for all* $g \in [f]_\sim$.

(iii) *The restriction of the relation* $\sim$ *to the class* $[f]_\sim$ *is the transitive closure of the restriction of* $\varrho$ *to* $[f]_\sim$.

(iv) *The restriction of the relation* $\sim$ *to* $[f]_\sim$ *is transitive on* $[f]_\sim$ *if no block* $B \in A/Kerf$ *contains more than two elements.*

**Proof:** (i) It is straightforward to show that if $g\varrho h$, then $g^3 = h^3$ as well and so $g \sim h$.

(ii) This follows from Lemma 11.2.3, part (ii).

(iii) Using part (i), we see that the restriction of $\varrho$ to the set $[f]_\sim$ is contained in the restriction of $\sim$ to this set; and since the latter relation is transitive, it also contains the transitive closure of $\varrho$ restricted to $[f]_\sim$. For the opposite inclusion, suppose that the pair $(g, h)$ is in the restriction of $\sim$ to the set $[f]_\sim$. Then $f \circ g = g \circ f = f = f^2 = g^2$, and $h \circ f = f \circ h = f = f^2 = h^2$. Thus $g\varrho f$ and $f\varrho h$, and $(g, h)$ is in the transitive closure of the restriction of $\varrho$ to $[f]_\sim$.

(iv) We shall prove that the restriction of $\varrho$ to the set $[f]_\sim$ is transitive when all blocks of the kernel of $f$ have size at most two; the conclusion then follows by (iii). Let $(g, h)$ and $(h, l)$ be pairs in $\varrho$, with $g$, $h$ and $l$ in $[f]_\sim$. Since $g, l \in [f]_\sim$, we have $g^2 = l^2 = f^2 = f$.

Every element $a$ of $A$ is in some block $B$ of the kernel of $f$. Suppose first that this block $B$ has size one, so $B = \{a\}$. Then the fact that $f$ is idempotent forces $f(a) = a$, and hence $f(g(a)) = f(a) = a$ implies also that $g(a) = a$. Similarly we must also have $l(a) = a$. Now we have

$$
\begin{aligned}
(g \circ l)(a) &= g(l(a)) &= g(a) &= a &= f(a), \quad \text{and} \\
(l \circ g)(a) &= l(g(a)) &= l(a) &= a &= f(a).
\end{aligned}
$$

If however block $B$ has size two, we can write it as $B = \{a, b\}$ for some element $b \in A$, with $f(a) = f(b) = a$. Then

$$
\begin{aligned}
(g \circ f)(a) &= g(f(a)) &= g(a) &= f(a) &= a, \quad \text{and} \\
(f \circ g)(a) &= f(g(b)) &= f(b) &= a,
\end{aligned}
$$

and therefore either $g(b) = a$ or $g(b) = b$. Suppose that $g(b) = b$. Then $g(g(b)) = g(b) = b \neq a = (f \circ f)(a)$, a contradiction since $g \in [f]_\sim$. Thus

$g(b) = a$. Similarly, $l(b) = a$. Now the same calculation as in the previous case shows that $(g \circ l)(a) = (l \circ g)(a) = f(a)$. Altogether we have $g \circ l = g^2 = f^2 = f = l^2 = l \circ g$, making $(g, l)$ in $\varrho$. ∎

We shall say that two mappings $g$ and $h$ *agree* on a subset $A'$ of our base set $A$ if the restrictions of $g$ and $h$ onto $A'$ are equal.

**Theorem 11.2.8.** *Let $f$ be an idempotent element of $(O^1(A); \circ)$ and $f \in S \subseteq O^1(A)$. Then the following are equivalent:*

(i) *$S$ is a constant semigroup with constant value $f$.*
(ii) *$S$ is a subset of $[f]_\sim$ and $\varrho$ is transitive on $S$.*
(iii) *For each $g \in S$ we have $g^2 = f$; and for all $y \in Imf$ and all $x \in [y]_{Kerf}$ we have $g(x) \in \bigcap\{[y]_{Kerh} \mid h \in S\}$.*
(iv) *For all $g, h \in S$ we have $g^2 = h^2 \in S$ and $g$ and $h$ agree on both $Imh$ and $Img$.*

**Proof:** (i) ⇒ (ii): Let $S$ be a constant semigroup with value $f$. For any element $g \in S$, we have $g^2 = g^3 = f = f^2 = f^3$, so that $g \in [f]_\sim$. Moreover, any two elements $g$ and $h$ in $S$ are $\varrho$-related, since $g \circ h = h \circ g = g^2 = f$ in $S$.

(ii) ⇒ (iii): Let $\varrho$ be transitive on the set $S \subseteq [f]_\sim$. For (iii), let $g \in S$, $y \in Imf$ and $x \in [y]_{Kerf}$. Then $f(x) = f(y) = y$, with the latter equality from the idempotence of $f$. Now let $h \in S$. Then we have $h \circ g = h^2 = f^2 = f = h \circ f$ since $f \in S$. Therefore $(g(x), h(x)) \in Kerh$ and $(h(x), f(x)) \in Kerh$ which implies $(g(x), f(x)) \in Kerh$. Since $f(x) = y$, we have $g(x) \in [y]_{Kerh}$. Therefore $g(x) \in \bigcap\{[y]_{Kerh} \mid h \in S\}$.

(iii) ⇒ (i): Let $S$ have the property from (iii), and let $h, g$ be any elements of $S$. First we have both $g^2 = h^2 = f$. For any $x \in A$, since $f(x) \in Imf$ and $f(x) = f^2(x)$ we have $x$ and $f(x)$ in the same equivalence class for the kernel of $f$. Similarly, since $h(h(x)) = f(x)$ and $g(g(x)) = f(x)$, all three of $x$, $f(x)$, $g(x)$ and $h(x)$ are in the same kernel class for $f$. From this we see that $(h \circ g)(x) = (g \circ h)(x) = f(x)$. It follows that $S$ is a constant semigroup.

(i) ⇒ (iv): If $S$ is a constant semigroup with value $f$, then for any two elements $g$ and $h$ of $S$ we have $g^2 = f = h^2$ and also $g^3 = g^2 = f = h^2 = h^3$. This implies that the restriction of $g$ to the set $Im\, g^2$ is the identity function, and similarly for $h$. Since $g^2 = h^2$, we have $Img^2 = Imh^2$ and thus the restriction of $g$ to $Imh^2$ equals the restriction of $h$ to $Imh^2$. But $g \circ h = h^2$, so that $g$ and $h$ agree on both $Img$ and $Imh$.

(iv) $\Rightarrow$ (i): Condition (iv) means that any two squares are equal in $S$, and that any two elements $h$ and $g$ of $S$ agree on $Imh$ and $Img$. Since $f \in S$ is idempotent, the first part tells us that $g^2 = f$ for all $g \in S$. From the second part we also get $g \circ h = h^2 = f$ and $h \circ g = g^2 = f$. This shows that $S$ is a constant semigroup, with constant value $f$. ∎

## 11.3　From Unary to $n$-ary

There is a useful connection between $n$-ary operations on a finite set $A$, for $n \geq 2$, and unary operations on a related set, which allows us to extend many of our results on unary operations from the previous section to the $n$-ary case.

Let $n \geq 2$. Any $n$-ary operation $f : A^n \to A$ determines a unary operation $f^{\otimes n}$ on the cartesian product set $A^n$: we define $f^{\otimes n} : A^n \to A^n$ by the rule that

$$f^{\otimes n}(a_1, \ldots, a_n) := (f(a_1, \ldots, a_n), \ldots, f(a_1, \ldots, a_n))$$

for all $(a_1, \ldots, a_n) \in A^n$.

Let $(O^n(A))^{\otimes n}$ be the set of all operations of the form $f^{\otimes n}$ for $f \in O^n(A)$. The main result which makes the carryover from unary to $n$-ary possible is the following, from [35]. As with many of the results from this section, the proof is straightforward, and we leave it as an exercise for the reader to check the details.

**Proposition 11.3.1.** *Let $A$ be a finite set of size at least two. $((O^n(A))^{\otimes n}; \circ)$ is a subsemigroup of $(O^1(A^n); \circ)$, and is isomorphic to $(O^n(A); +)$.*

The construction of $f^{\otimes n}$ from an $n$-ary operation $f$ is actually a special case of a more general construction. Let $f_1, \ldots, f_n$ be $n$-ary operations on $A$. Then the unary operation $f_1 \otimes \cdots \otimes f_n : A^n \to A^n$ is defined by

$$(a_1, \ldots, a_n) \mapsto (f_1(a_1, \ldots, a_n), \ldots, f_n(a_1, \ldots, a_n))$$

for all $(a_1, \ldots, a_n) \in A^n$.

It is clear that for any $f$, we have $f^{\otimes n} = f \otimes \cdots \otimes f$. The following equation ($*$) is also easy to check:

($*$) $\quad (f_1 \otimes \cdots \otimes f_n) \circ (g_1 \otimes \cdots \otimes g_n) = S^n(f_1, g_1, \ldots, g_n) \otimes \cdots \otimes S^n(f_n, g_1, \ldots, g_n).$

There is a close connection between properties of an $n$-ary operation $f$ and the corresponding unary $f^{\otimes n}$. First, it is obvious that for any $b \in A$, we have $b \in Imf$ if and only if $(b, \ldots, b) \in Imf^{\otimes n}$. The classes of $f$ and $f^{\otimes n}$ for the Green's relations $\mathcal{L}$ and $\mathcal{R}$ are also closely connected. We can show that two $n$-ary operations $f$ and $g$ are $\mathcal{L}$-related on $O^n(A)$ iff the induced unary operations $f^{\otimes n}$ and $g^{\otimes n}$ are related by $\mathcal{L}$ on $O^1(A^n)$. For the first direction, suppose that $(f, g) \in \mathcal{L}$ on $O^n(A)$. Then there exist $n$-ary operations $h$ and $l$ on $A$ such that $h + f = g$ and $l + g = f$. Then the isomorphism from Proposition 11.3.1 tells us immediately that $h^{\otimes n} \circ f^{\otimes n} = g^{\otimes n}$ and $l^{\otimes n} \circ g^{\otimes n} = f^{\otimes n}$, making $f^{\otimes n}$ related to $g^{\otimes n}$ by $\mathcal{L}$ on $O^1(A^n)$. Conversely, suppose that $(f^{\otimes n}, g^{\otimes n}) \in \mathcal{L}$ on $O^1(A^n)$, so that there exist unary operations $h$ and $l$ on $A^n$ with $h \circ f^{\otimes n} = g^{\otimes n}$ and $l \circ g^{\otimes n} = f^{\otimes n}$. The operation $h \in O^1(A^n)$ can be expressed by some $n$-tuple $h = (h_1, \ldots, h_n) = h_1 \otimes \cdots \otimes h_n$, and similarly for $l$. Using this formulation we have $(h_1 \otimes \cdots \otimes h_n) \circ (f \otimes \cdots \otimes f) = g \otimes \cdots \otimes g$ and $(l_1 \otimes \cdots \otimes l_n) \circ (g \otimes \cdots \otimes g) = f \otimes \cdots \otimes f$. Applying equation (*) above results in $S^n(h_1, f, \ldots, f) \otimes \cdots \otimes S^n(h_n, f, \ldots, f) = g \otimes \cdots \otimes g$ and $S^n(l_1, g, \ldots, g) \otimes \cdots \otimes S^n(l_n, g, \ldots, g) = f \otimes \cdots \otimes f$. From this we get $h_i + f = g$ and $l_i + g = f$ for all $1 \leq i \leq n$, making $f$ related to $g$ by $\mathcal{L}$ on $O^n(A)$. A similar proof can be given to show that two $n$-ary relations $f$ and $g$ are related by $\mathcal{R}$ on $O^n(A)$ iff the corresponding $n$-ary functions $f^{\otimes n}$ and $g^{\otimes n}$ are related by $\mathcal{R}$ on $O^1(A^n)$.

Using these observations about image and $\mathcal{L}$- and $\mathcal{R}$-classes, and the isomorphism from Proposition 11.3.1, the following results are easy to prove.

**Theorem 11.3.2.** *Let $A$ be a finite set of size at least two, and let $n \geq 2$. For any $n$-ary operations $f$ and $g$ on $A$, the following properties hold:*

(i) $f \in Idem(O^n(A))$ *iff for all $a \in Imf$ we have $f(a, \ldots, a) = a$.*

(ii) $(f, g) \in \mathcal{R}_{O^n(A)}$ *iff $Kerf = Kerg$.*

(iii) $(f, g) \in \mathcal{L}_{O^n(A)}$ *iff $Imf = Img$.*

(iv) $S \subseteq (O^n(A); +)$ *is a right-zero semigroup iff $S \subseteq Idem(O^n(A))$ and for all $f, g \in S$ we have $Imf = Img$.*

In analogy with our treatment of constant semigroups of operations in the unary case in Section 11.2, we now introduce some terminology and some relations. For any subset $S$ of $A$, we shall say that an $n$-ary operation $f$ is an *identity* on $S$ if $f(a, \ldots, a) = a$ for all $a \in S$. This happens iff the restriction of $f^{\otimes n}$ to the diagonal relation on $S^n$ is the identity mapping. Then the idempotent elements of $(O^n(A); +)$ are precisely the operations $f$ such that $f$ is an identity on $Imf$. Moreover, if $f$ is an idempotent element of $(O^n(A); +)$ and $g + f = f$, then $g$ is an identity on $Imf$.

**Definition 11.3.3.** Let $A$ be a finite set of size at least two.

(i) Let $f \in O^n(A)$. We call $f$ a *permutation* if for all $a$, $b$ in $A$, $f(a, \ldots, a)$ $= f(b, \ldots, b)$ implies $a = b$.
(ii) For any $S \subseteq A$, we say that two operations $g, h \in O^n(A)$ *agree on* $S$ if $g(a, \ldots, a) = h(a, \ldots, a)$ for all $a \in S$.

It is clear that an operation $f$ is a permutation in the sense of this definition iff the restriction of $f^{\otimes n}$ to the diagonal $\Delta_{A^n}$ is a permutation in the usual sense.

**Definition 11.3.4.** Let $A$ be a finite set of size at least two, and let $n \geq 2$.

(i) We define a binary relation $\sim_n$ on $O^n(A)$ by $g \sim_n h$ iff $g + g = h + h$ and $g$ and $h$ agree on the set $Im(g + g)$.
(ii) We define a binary relation $\varrho_n$ on $O^n(A)$ by $g\varrho_n h$ iff $g + g = h + h$ and $g$ and $h$ agree on $Img \cup Imh$.

It is clear from the definitions that the relation $\sim_n$ is an equivalence relation on $O^n(A)$, and that $\varrho_n$ is reflexive and symmetric. The proofs of the following results are straightforward and will be skipped.

**Lemma 11.3.5.** *Let $A$ be a finite set of size at least two, and let $n \geq 2$. Let $f$, $g$ and $h$ be in $O^n(A)$.*

(i) $f + g = f = g + f$ *iff* $f^{\otimes n} \circ g^{\otimes n} = f^{\otimes n} = g^{\otimes n} \circ f^{\otimes n}$.
(ii) $g + h = f$ *iff* $g^{\otimes n} \circ h^{\otimes n} = f^{\otimes n}$.
(iii) $g = h$ on $Img \cup Imh$ *iff* $g^{\otimes n} = h^{\otimes n}$ on $Img^{\otimes n} \cup Imh^{\otimes n}$.
(iv) $g\varrho_n h$ *iff* $g^{\otimes n}\varrho h^{\otimes n}$.
(v) $g \sim_n h$ *iff* $g^{\otimes n} \sim h^{\otimes n}$.

As we did in the unary case, we can characterize idempotents and constant semigroups using the classes of our relations. We use $[f]_{\sim_n}$ for the class or block of $f$ under the relation $\sim_n$, and define $[a]_f$ to be the set $\{b \in A \mid f(b, \ldots, b) = a\}$.

**Proposition 11.3.6.** *Let $f \in O^n(A)$. Then the block $[f]_{\sim_n}$ contains $f + f$ as its only idempotent element iff $f$ is an identity on the set $Im(f + f)$.*

**Proof:** Suppose that $[f]_{\sim_n}$ contains an idempotent element $\alpha$. Then $\alpha + \alpha$ $= \alpha = f + f$; so the only idempotent element is $f + f$, and it is related by

$\sim_n$ to $f$. By definition this means that $f + f = f$ on $Im(f+f)$, which forces $f + f = f$. Thus $f$ is an idempotent, and it is an identity on $Im(f+f)$. Conversely, let $f$ be an identity on $Im(f+f)$. Then the restriction of $f^{\otimes n}$ to the set $Im(f+f)$ is the diagonal of $Im(f+f)$. It follows that $(f^{\otimes n})^2$ is the only idempotent element in $[f^{\otimes n}]_\sim$ and hence that $f + f$ is the only idempotent in $[f]_{\sim_n}$.                                                    ∎

**Proposition 11.3.7.** *Let $A$ be a finite set of size at least two. Let $f$ be an idempotent element of $(O^n(A); +)$. If there exists an element $a \in Imf$ for which the set $[a]_f$ contains more than two elements, then the class $[f]_{\sim_n}$ contains more than two elements. In a limited converse, if the class $[f]_{\sim_n}$ contains at least two elements, then there is an element $a$ in $Imf$ for which $[a]_f$ contains at least two elements.*

**Proof:** Let us assume first that there is an element $a \in Imf$ whose class $[a]_f$ contains pairwise different elements $b$, $c$ and $d$. By definition, $f(b,\dots,b) = f(c,\dots,c) = f(d,\dots,d) = a$. It follows from Theorem 11.3.2(i) that $a$ itself is in $[a]_f$, so we may assume that $a = d$. We define an $n$-ary operation $g_1$ on $A$, with the rules that $g_1(c,\dots,c) = b$ and $g_1$ agrees with $f$ on every other input $n$-tuple. Our goal is to show that this operation $g_1$ is in $[f]_{\sim_n}$ and is different from $f$. To simplify our notation, we shall write $\underline{x}$ to mean $(x, x, \dots, x)$ for any symbol $x$. We first observe that

$$
\begin{aligned}
(g_1 + g_1)(\underline{a}) &= g_1(\underline{a}) = a = f(\underline{a}) = (f+f)(\underline{a}),\\
(g_1 + g_1)(\underline{b}) &= g_1(\underline{a}) = a = f(\underline{b}) = (f+f)(\underline{b}),\\
(g_1 + g_1)(\underline{c}) &= g_1(\underline{b}) = a = f(\underline{c}) = (f+f)(\underline{c}).
\end{aligned}
$$

For $\underline{x} \in A^n \setminus \{\underline{c}\}$ we get $(g_1 + g_2)(\underline{x}) = g_1(g_1(\underline{x})) = g_1(f(\underline{x}))$. Now we consider whether $f(\underline{x})$ is in $[a]_f$ or not. If it is, we have $f(f(\underline{x})) = a = f(\underline{a})$, and by idempotency of $f$ we get $f(\underline{x}) = a$; then $(g_1 + g_1)(\underline{x}) = g_1(f(\underline{x})) = g_1(\underline{a}) = a = f(\underline{a}) = (f + f)(\underline{x})$. If $f(\underline{x})$ is not in $[a]_f$, then $(g_1 + g_1)(\underline{x}) = g_1(f(\underline{x})) = f(f(\underline{x})) = (f + f)(\underline{x})$ since $f(\underline{x})$ is not equal to any of $\underline{a}$, $\underline{b}$ or $\underline{c}$. In either case we have $g_1 + g_1 = f + f$.

Next we will show that $f$ and $g_1$ agree on $Im(f+f)$. Let $z$ be any element in $Im(f+f)$. By idempotency of $f$ we have $f(\underline{z}) = z$. If $z \in [a]_f$, then $z = a$ and $g_1(\underline{z}) = g_1(\underline{a}) = a = f(\underline{a}) = f(\underline{z})$. If $z \notin [a]_f$, then $z$ cannot equal any of $\underline{a}$, $\underline{b}$ or $\underline{c}$ and we get $g_1(\underline{z}) = f(\underline{z})$. Again, in either case $f$ and $g_1$ agree on $Im(f+f)$, which shows that $g_1 \in [f]_{\sim_n}$.

Next we define another $n$-ary operation $g_2$ on $A$, by having $g_2(\underline{b}) = c$

but $g_2$ agreeing with $f$ on every other $n$-tuple input. In a similar fashion to the proof above, we can show that $g_2$ is also in $[f]_{\sim_n}$, and clearly $f$, $g_1$ and $g_2$ are all different operations. Thus we have at least three elements in $[f]_{\sim_n}$.

For the converse, we suppose that there are two distinct operations $g_1$ and $g_2$ in $[f]_{\sim_n}$. There must exist an $n$-tuple $\overline{x} = (x_1, \ldots, x_n) \in A^n$ such that $g_1(\overline{x}) \neq g_2(\overline{x})$. We may assume that $f(\overline{x}) = a$. We want to show that $g_1(\overline{x})$ and $g_2(\overline{x})$ are in $[a]_f$. We claim that the $n$-tuple $\overline{x}$ cannot be in $\Delta_{Imf}$. For if $\overline{x} \in \Delta_{Imf}$, then there exists $b \in Imf$ such that $\overline{x} = (b, \ldots, b)$. Since $f$ agrees with both $g_i$ on $Im(f + f)$, we have $f + f + f = g_i + f + f$ for $i = 1, 2$. This means that $f = g_i + f$ for $i = 1, 2$, and so $f(\overline{x}) = f(\underline{b}) = b = (g_i + f)(\underline{b}) = g_i(\underline{b})$. But now we have $g_i(\underline{b}) = b$ for $i = 1, 2$, and hence $g_1(\overline{x}) = g_1(\underline{b}) = g_2(\underline{b}) = g_2(\overline{x})$, which is a contradiction.

Since $f$ and both $g_i$ agree on $Im(f + f)$ and $\sim_n$ is an equivalence relation, we see that $g_i$ agrees with $f$ on $Im(g_i + g_i)$. Therefore $g_i + g_i + g_i = f + g_i + g_i$. But we have $g_i + g_i = f$, so that $g_i + g_i + g_i = f + f = f$. It follows that $f + g_i = f$, for $i = 1, 2$. Then $(f + g_1)(\overline{x}) = f(g_1(\overline{x})) = f(\overline{x}) = a = f(\underline{a})$ and $(f + g_2)(\overline{x}) = f(g_2(\overline{x})) = f(\overline{x}) = a = f(\underline{a})$. Therefore $g_1(\underline{x})$ and $g_2(\underline{x})$ are two distinct elements in $[a]_f$. ∎

The limited converse of this proposition can be tightened a little. It can be shown that if there exists an element $a \in Imf$ with exactly two elements in $[a]_f$ and there exists an element $\overline{x} = (x_1, \ldots, x_n)$ such that $\overline{x}$ is not of the form $\underline{c}$ for any $c \in [a]_f$ and $f(\overline{x}) = a$, then there are at least two elements in $[f]_{\sim_n}$.

Combining all the results of this section leads to a proof of the following characterization of constant subsemigroups $(S; +)$ of $(O^n(A); +)$.

**Theorem 11.3.8.** *Let $f \in Idem(O^n(A))$ and $f \in S \subseteq O^n(A)$. Then the following are equivalent:*

(i) *$S$ is a constant semigroup with constant value $f$.*
(ii) *$S \subseteq [f]_{\sim_n}$ and any two elements of $S$ are $\varrho$-related.*
(iii) *Any two elements $g, h \in S$ satisfy $g + g = h + h \in S$ and agree on $Imh$.*

As an alternate way to describe constant subsemigroups of $O^n(A)$, we define for any idempotent $f$ the set of operations

$$F_f := \{g \mid g \in O^n(A) \text{ and } g + g = g + f = f + g = f\}.$$

Clearly any constant semigroup with value $f$ must be contained in $F_f$. But $F_f$ itself need not be a semigroup, since it is not necessarily closed under the $+$ operation. It is easy to see that the extra condition needed for closure is that $h + g = g + h = f$ for any $g, h \in F_f$. Thus a semigroup $S$ is a constant semigroup with value $f$ iff $S \subseteq F_f$ and $h + g = g + h = f$ for all $h, g \in S$.

## 11.4 Bands of $n$-ary Operations

As remarked in Section 11.1, idempotent semigroups or bands are particularly interesting kinds of semigroups. In this section we will characterize three kinds of idempotent subsemigroups of $(O^n(A); +)$: rectangular bands, semilattices and normal bands.    Let us recall that these three kinds of bands are each defined by a single identity besides associativity and idempotence, the identities $x \circ y \circ z \approx x \circ z$, $x \circ y \approx y \circ x$ and $x \circ y \circ u \circ v \approx x \circ u \circ y \circ v$ respectively. We begin with semilattices of operations. The following preliminary result is easy to show using the commutativity law and the property from Theorem 11.3.2 that for $f$ an idempotent, $f(a, \ldots, a) = a$ for any $a \in Im f$.

**Proposition 11.4.1.** *If* $S \subseteq (O^n(A); +)$ *is a semilattice, then* $Im(f + g) = Im f \cap Im g$ *for every* $f, g \in S$.

Next we want to characterize commutativity of $+$. For two idempotent operations $f, g \in O^n(A)$ the following property is useful. If $g(a, \ldots, a) \in Im f$ for all $a \in Im f$, then $f + g$ and $g + f$ agree on the diagonal set $\Delta_{Im f}$: for any $(a, \ldots, a)$ in this set, with $g(a, \ldots, a) \in Im f$, it follows that $f(g(a, \ldots, a), \ldots, g(a, \ldots, a)) = g(a, \ldots, a) = g(f(a, \ldots, a), \ldots, f(a, \ldots, a))$. We shall again use the notation $\underline{a}$ to represent the $n$-tuple $(a, \ldots, a)$ for any element $a \in A$, and recall the notation $[a]_g = \{b \in A \mid g(b, \ldots, b) = a\}$.

**Proposition 11.4.2.** *Let* $f, g \in Idem(O^n(A))$. *Then* $f + g = g + f$ *if and only if the following two conditions are satisfied:*

(i) *$g(a, \ldots, a) \in Im f$ for all $a \in Im f$.*
(ii) *For any $n$-tuple $\overline{x} = (x_1, \ldots, x_n)$ not in $\Delta_{Im f}$, if $f(x_1, \ldots, x_n) = a$ then $g(x_1, \ldots, x_n)$ is in the class $[g(\underline{a})]_f$.*

**Proof:** Let us assume first that $f + g = g + f$. For condition (i), if $a \in Imf$, then $g(\underline{a}) = g(\underline{f(\underline{a})}) = (g + f)(\underline{a}) = (f + g)(\underline{a}) = f(g(\underline{a}))$, and thus $g(\underline{a}) \in Imf$. For (ii), let $\underline{x}$ be an $n$-tuple not in $\Delta_{Imf}$ and suppose that $f(\underline{x}) = a$. Then $(f + g)(\underline{x}) = f(g(\underline{x})) = (g + f)(\underline{x}) = g(\underline{f(\underline{x})}) = g(\underline{a})$; from this, we have $g(\underline{x}) \in [g(\underline{a})]_f$.

Conversely, it follows from (i) and the remark just above that $(f + g)(\underline{a}) = (g + f)(\underline{a})$ for any $n$-tuple $(a, \ldots, a) \in \Delta_{Imf}$. Now suppose that $f(\underline{x}) = a$ for some $\underline{x}$ not in $\Delta_{Imf}$. By condition (ii), $(f + g)(\underline{x}) = f(g(\underline{x})) = g(\underline{a}) = g(\underline{f(\underline{x})}) = (g + f)(\underline{x})$. Altogether we get $f + g = g + f$. ∎

**Theorem 11.4.3.** *Let* $S \subseteq Idem(O^n(A))$ *and let* $\langle S \rangle$ *be the subsemigroup of* $(O^n(A); +)$ *generated by* $S$. *Then* $\langle S \rangle$ *is a semilattice if and only if the following conditions are satisfied for every* $f, g \in S$:

(i) $g(a, \ldots, a) \in Imf$ *for all* $a$;
(ii) *for any* $n$-*tuple* $\underline{x}$ *not in* $\Delta_{Imf}$, *if* $f(\underline{x}) = a$ *then* $g(\underline{x}) \in [g(\underline{a})]_f$.

**Proof:** One direction has already been proved in Proposition 11.4.2. For the other direction, we will show that both idempotence and commutativity of $\langle S \rangle$ can be established from conditions (i) and (ii). Let $f, g \in \langle S \rangle$. Then we can write $f = f_1 + \cdots + f_k$ and $g = g_1 + \cdots + g_l$, for some operations $f_1, \ldots, f_k, g_1, \ldots, g_l \in S$. Repeated use of Proposition 11.4.2 then shows that $f$ and $g$ commute. Idempotency can also be proved similarly. ∎

Rectangular bands contained in $(O^n(A); +)$ have the following properties:

**Proposition 11.4.4.** *Let* $S \subseteq (O^n(A); +)$ *be a rectangular band. Then for any two elements* $f$ *and* $g$ *in* $S$:

(i) $f$ *and* $g$ *have the same size image,*
(ii) *the kernels of* $f$ *and* $g$ *agree on the set* $Imf \times Img$.

**Proof:** (i) By contradiction, suppose that there were two elements $f, g \in S$ with different size images, and without loss of generality let $|Imf| < |Img|$. Since $Im(f + g) \subseteq Imf$, we get $|Im(f + g)| \leq |Imf|$. But $Im(g + f + g) = \{g(\underline{a}) \mid a \in Im(f + g)\}$ so we also have $|Im(g + f + g)| \leq |Im(f + g)|$. But this implies $|Im(g + f + g)| < |Img|$, which contradicts $g + f + g = g$.
(ii) Let $a \in Imf$ and $b \in Img$ with $(\underline{a}, \underline{b}) \in Kerf$. Then $a = f(\underline{a}) = f(\underline{b})$ since $f$ is idempotent. Moreover, we have $(g + f + g)(\underline{b}) = (g + f)(\underline{b}) =$

$g(\underline{f(\underline{b})}) = g(\underline{b}) = b$ since $b \in Img$, $g$ is idempotent and $g + f + g = g$. It follows that $g(\underline{f(\underline{a})}) = g(\underline{a}) = b$ and $(\underline{a}, \underline{b}) \in Kerg$. The opposite inclusion is dual, and we have equality. ∎

The following lemma and theorem provide a characterization of all sub-semigroups of $O^n(A)$ which are rectangular bands. We leave the proofs as exercises; details may be found in [13].

**Lemma 11.4.5.** *Let $f, g \in Idem(O^n(A))$. Then the following are equivalent:*

(i) $f + g + f = f$.
(ii) $g(\underline{a}) \in [a]_f$ *for all $a \in Imf$.*
(iii) $(\underline{a}, g(\underline{a})) \in Kerf \cap Kerg$ *for all $a \in Imf$.*
(iv) *If $g(\underline{a}) = b$, then $f(g(\underline{a})) = a$ for all $a \in Imf$ and all $b \in Img$.*

**Theorem 11.4.6.** *Let $S \subseteq Idem(O^n(A))$. Then $\{f + g \mid f, g \in S\}$ is the universe of a rectangular band iff for all $f, g \in S$ the following conditions are satisfied:*

(i) $g(\underline{a}) \in [a]_f$ *for all $a \in Imf$.*
(ii) *For all $g_1, g_2 \in S$ and for all $a, b, c \in A$, if $f(\underline{a}) = a$ and $g_1(\underline{a}) = b$ and $g_2(\underline{a}) = c$, then $\{\underline{a}, \underline{b}, \underline{c}\}^2 \subseteq Kerf \cap Kerg_1 \cap Kerg_2$.*

Similar techniques lead to the following characterization for normal band subsemigroups of $O^n(A)$, from [13].

**Proposition 11.4.7.** *Let $S \subseteq Idem(O^n(A))$. The subsemigroup generated by $S$ is a normal band if and only if for all $f, g, h_1, h_2 \in S$ and for all $b \in Img$ we have $((h_1 + h_2)(\underline{b}), (h_2 + h_1)(\underline{b})) \in Kerf$.*

## 11.5   Semigroups of Boolean Operations

In this section we look at the special case that our base set $A = \{0, 1\}$, the Boolean case. We describe idempotent and regular subsemigroups of Boolean operations, characterize some bands of Boolean operations, and look at Green's relations on the semigroup $\mathcal{O}^n(A) = (O^n(A); +)$. Throughout this section we assume that $n$ is a fixed natural number, at least one, and that $A = \{0, 1\}$.

We begin with a basic classification of elements of $O^n(A)$. Each $n$-ary operation can be described by what it does to the two constant $n$-tuples $\underline{0}$

$= (0,\ldots,0)$ and $\underline{1} = (1,\ldots,1)$ , and with only two outputs possible in the Boolean case we have four categories of operations. We recall from Section 10.2 that $C_4^n$ is the set of those $n$-ary Boolean operations which map $\underline{0}$ to 0 and $\underline{1}$ to 1. The class $\neg C_4^n$ consists of $n$-ary operations mapping $\underline{0}$ and $\underline{1}$ to 1 and 0 respectively. We now also define two additional sets, $K_0^n$ as the set of all $n$-ary Boolean operations mapping both $\underline{0}$ and $\underline{1}$ to 0 and $\neg K_0^n$ as the negation of this set. By definition every $n$-ary Boolean operation must be contained in exactly one of the four sets $C_4^n$, $\neg C_4^n$, $K_0^n$ and $\neg K_0^n$.

The following lemma lists some elementary but useful properties of the operation + on these sets.

**Lemma 11.5.1.** *Let $f$ and $g$ be $n$-ary Boolean operations. Then*

(i) $f + g = g$ *for all $f, g \in C_4^n$.*
(ii) $f + g = \neg g$ *for all $f, g \in \neg C_4^n$.*
(iii) $f + g = c_0^n$ *for all $f, g \in K_0^n$.*
(iv) $f + g = c_1^n$ *for all $f, g \in \neg K_0^n$.*
(v) *If $f \in C_4^n$ and $g \in \neg C_4^n$, then $f + g = g$ and $g + f = \neg f$.*
(vi) *If $f \in C_4^n$ and $g \in K_0^n$, then $f + g = g$ and $g + f = c_0^n$.*
(vii) *If $f \in C_4^n$ and $g \in \neg K_0^n$, then $f + g = g$ and $g + f = c_1^n$.*
(viii) *If $f \in \neg C_4^n$ and $g \in K_0^n$, then $f + g = \neg g$ and $g + f = c_0^n$.*
(ix) *If $f \in \neg C_4^n$ and $g \in \neg K_0^n$, then $f + g = \neg g$ and $g + f = c_1^n$.*
(x) *If $f \in K_0^n$ and $g \in \neg K_0^n$, then $f + g = c_0^n$ and $g + f = c_1^n$.*
(xi) $f + f = c_0^n$ *iff $f \in K_0^n$.*
(xii) $f + f \in C_4^n$ *iff $f \in C_4^n \cup \neg C_4^n$.*

To discuss the order of elements of $O^n(A)$, we note from parts (i) through (v) of the previous lemma that any Boolean operation $f$ must satisfy either $f + f = f$ or $f + f + f = f$ or $f + f = f + f + f$. This proves the following result.

**Proposition 11.5.2.** *Any Boolean operation of arity $n$ has order 1 or 2.*

Various parts of Lemma 11.5.1 also allow us to characterize both idempotent and regular elements of the semigroup of $n$-ary Boolean operations. It follows from Lemma 11.5.1(i) that any $f \in C_4^n$ is an idempotent element, and of course the two constant operations $c_0^n, c_1^n$ are also idempotents; and similarly the two constants as well as any operation from $C_4^n \cup \neg C_4^n$ are regular. In fact these are the only idempotent and regular elements; proofs of these facts are left as exercises for the reader, and details may be found in [13].

**Theorem 11.5.3.** *Let $A = \{0, 1\}$ and let $n \geq 1$.*

(i) *A Boolean operation $f \in O^n(A)$ is idempotent if and only if $f \in C_4^n \cup \{c_0^n, c_1^n\}$.*

(ii) *A Boolean operation $f$ is a regular element of $O^n(A)$ if and only if $f \in C_4^n \cup \neg C_4^n \cup \{c_0^n, c_1^n\}$.*

It is not always the case in an arbitrary semigroup that the product of idempotent or regular elements is again idempotent or regular, respectively. However we do have such closure for regular and idempotent Boolean operations. Let us denote by $Idem(O^n(A))$ the set $C_4^n \cup \{c_0^n, c_1^n\}$ of all idempotent $n$-ary Boolean operations, and by $Reg(O^n(A))$ the set $C_4^n \cup \{c_0^n, c_1^n\} \cup \neg C_4^n$ of all regular elements of $O^n(A)$.

**Proposition 11.5.4.** *Let $A = \{0, 1\}$. Then $(Idem(O^n(A)); +)$ and $(Reg(O^n(A)); +)$ are subsemigroups of $(O^n(A); +)$, and $(Idem(O^n(A)); +)$ is a subsemigroup of $(Reg(O^n(A)); +)$.*

**Proof:** Closure of the sets of regular and idempotent elements under our addition operation can be shown in a straightforward case-by-case analysis, using Theorem 11.5.3 and the properties from Lemma 11.5.1. ∎

Our next proposition gives some characterizations as to when a subsemigroup of $(O^n(A); +)$ is one of various kinds of bands.

**Proposition 11.5.5.** *Let $A = \{0, 1\}$ and let $n \geq 1$. Let $S = (S; +)$ be a subsemigroup of $(O^n(A); +)$.*

(i) *$S$ is a right-zero semigroup iff $S \subseteq C_4^n$ or $S = \{c_0^n\}$ or $S = \{c_1^n\}$.*

(ii) *$S$ is a left-zero semigroup iff $S \subseteq \{c_0^n, c_1^n\}$ or $S = \{f\}$ for some $f \in C_4^n$.*

(iii) *$S$ is a semilattice iff $S \subseteq \{c_0^n, f\}$ or $S \subseteq \{c_1^n, f\}$ for some $f \in C_4^n$.*

(iv) *$S$ is a rectangular band iff $S$ is a left-zero or a right-zero semigroup.*

(v) *$S$ is an idempotent semigroup iff $S \subseteq C_4^n \cup \{c_0^n, c_1^n\}$.*

**Proof:** (i) It follows from Lemma 11.5.1(i) that any subset of $C_4^n$ is the universe of a right-zero semigroup. The one-element subsemigroups $(\{c_0^n\}; +)$ and $(\{c_1^n\}; +)$ are also right-zero semigroups. Conversely, suppose that $S$ is a right-zero semigroup. Then $S$ is also idempotent, so by Theorem 11.5.3(i) we have $S \subseteq C_4^n \cup \{c_0^n, c_1^n\}$. But since $c_0^n + f = c_0^n$ and $c_1^n + f = c_1^n$ for any operation $f$, it follows that $S$ must be a subset of $C_4^n$ or equal to one of $\{c_0^n\}$ or $\{c_1^n\}$.

(ii) Any non-empty subset of $\{c_0^n, c_1^n\}$ is the universe of a left-zero semigroup, as is any set $\{f\}$ for $f \in C_4^n$. Suppose conversely that $S$ is a left-zero semigroup. Again the idempotence of $S$ forces $S \subseteq C_4^n \cup \{c_0^n, c_1^n\}$, and it is easy to show that only the cases given are possible.

(iii) It is easy to see that for any $f \in C_4^n$, any non-empty subset of either $\{c_0^n, f\}$ or $\{c_1^n, f\}$ is the universe of a semilattice. That these are the only possible semilattices follows from the need for both idempotence and commutativity, and the properties from Lemma 11.5.1.

(iv) Any left-zero or right-zero semigroup is a rectangular band. In the other direction, any rectangular band $S$ is idempotent and so $S \subseteq C_4^n \cup \{c_0^n, c_1^n\}$. But the identity $x + y + x = x$ for rectangular bands prevents us from having both a constant and a non-constant element in $S$, so either $S \subseteq \{c_0^n, c_1^n\}$ or $S \subseteq C_4^n$. Then it follows from (i) and (ii) that $S$ is a left-zero or right-zero semigroup.

(v) One direction follows immediately from Theorem 11.5.3(i), and it is straightforward to show conversely that any subset of $C_4^n \cup \{c_0^n, c_1^n\}$ is an idempotent semigroup. ∎

Using similar techniques, the following description of commutative and constant subsemigroups of $O^n(A)$ can be given; see [13] for details.

**Proposition 11.5.6.** *Let* $A = \{0, 1\}$ *and* $n \geq 1$. *Let* $S = (S; +)$ *be a subsemigroup of* $(O^n(A); +)$.

(i) $S$ *is commutative if and only if* $c_0^n \in S^n \subseteq K_0^n$ *or* $c_1^n \in S^n \subseteq \neg K_0^n$ *or* $S^n = \{c_0^n, f\}$ *or* $S^n = \{c_1^n, f\}$ *or* $S^n = \{f\}$ *or* $S^n = \{f, \neg f\}$ *for some* $f \in C_4^n$.

(ii) $S$ *is a constant semigroup if and only if* $c_0^n \in S^n \subseteq K_0^n$ *or* $c_1^n \in S^n \subseteq \neg K_0^n$ *or* $S^n = \{f\}$ *for some* $f \in C_4^n$.

Our characterizations so far of various subsemigroups of $\mathcal{O}^n(A)$ can be extended in a natural way to clones. We say that a clone $C$ of operations on set $A$ has a particular semigroup property if for every $n \geq 1$ the $n$-clone $C^n$ of all $n$-ary operations in $C$ is the universe of a semigroup $(C^n; +)$ with the property. The previous results then characterize such clones in the Boolean case. We note also that the intersection of clones having a semigroup property also has that property. For the join of two clones $C_1$ and $C_2$, we have $(C_1 \vee C_2)^n = \; < C_1 \cup C_2 >^n = (\bigcap \{C \mid C$ is a subclone of $O(A)$ and $C_1 \cup C_2 \subseteq C\})^n = \bigcap \{C^n \mid C$ is a subclone of $O(A)$ and $C_1^n \cup C_2^n \subseteq C^n\} = C_1^n \vee C_2^n$. Thus the join of two clones with a semigroup property also has the same property. This means that the family of all such clones forms a sublattice of the lattice of all Boolean clones.

We turn now to the Green's relations on the semigroup $(O^n(A); +)$ of Boolean operations. The next two theorems describe all the $\mathcal{R}$- and $\mathcal{L}$-classes of elements of $O^n(A)$.

**Theorem 11.5.7.** *Let* $A = \{0, 1\}$ *and let* $n \geq 1$. *Let* $f$ *be any* $n$-*ary Boolean operation from* $O^n(A)$. *Then the* $\mathcal{R}$-*class of* $f$ *is either* $\{f\}$ *or the set* $C_4^n \cup \neg C_4^n$.

**Proof**: We use the decomposition of the set $O^n(A)$ into the disjoint union of the four sets $C_4^n$, $\neg C_4^n$, $K_0^n$ and $\neg K_0^n$, along with the properties from Lemma 11.5.1. By property (i), any two elements of $C_4^n$ are $\mathcal{R}$-related, since $f + g = g$ and $g + f = f$ for any elements $f, g \in C_4^n$. Next we consider two elements $f$ and $g$ from $\neg C_4^n$. In this case, $\neg f$ and $\neg g$ are both in $C_4^n$, and hence are $\mathcal{R}$-related. Also by properties (ii) and (v) we have $g + g = \neg g$ and $\neg g + g = g$, making $g$ and $\neg g$ related by $\mathcal{R}$, and similarly for $f$ and $\neg f$. It then follows that all elements of $C_4^n \cup \neg C_4^n$ are $\mathcal{R}$-related to each other.

Now we look at elements of $K_0^n$ and dually $\neg K_0^n$. Since $c_0^n + h = c_0^n$ for all $h \in O^n(A)$, we see that $c_0^n$ can only be $\mathcal{R}$-related to itself. For any non-constant $f$ in $K_0^n$, again $f + h = c_0^n$ for all $h \in O^n(A)$ means that $f$ can only be $\mathcal{R}$-related to itself. The proof for elements of $\neg K_0^n$ is similar.

∎

**Theorem 11.5.8.** *Let* $A = \{0, 1\}$ *and let* $n \geq 1$. *Let* $f$ *be any* $n$-*ary Boolean operation from* $O^n(A)$. *Then the* $\mathcal{L}$-*class of* $f$ *is either* $\{f\}$ *or* $\{f, \neg f\}$.

**Proof**: Again we use the decomposition of $O^n(A)$ into four sets, and the properties from Lemma 11.5.1. First, if $f \in C_4^n$, then $h + f$ can only equal $f$, $\neg f$, $c_0^n$ or $c_1^n$, depending on whether $h$ is in $C_4^n$, $\neg C_4^n$, $K_0^n$ or $\neg K_0^n$ respectively. Thus the only three things other than itself that $f$ can be related to are $\neg f$, $c_0^n$ and $c_1^n$. Since $\neg f + f = \neg f$ and $\neg f + \neg f = f$, we see that $f$ is $\mathcal{L}$-related to $\neg f$. However, since $h + c_0^n$ is equal to either $c_0^n$ or $c_1^n$ for any $h \in O^n(A)$, we cannot make $h + c_0^n = f$, and so $c_0^n$ is not $\mathcal{L}$-related to $f$. Similarly, $f$ cannot be $\mathcal{L}$-related to $c_1^n$ either, so that the $\mathcal{L}$-class of $f$ is simply $\{f, \neg f\}$. A similar argument yields the same result if $f$ is in $\neg C_4^n$.

Next we consider elements of $K_0^n$. Since $h + c_0^n$ always equals either $c_0^n$ or $c_1^n$ for any $h \in O^n(A)$, the constant $c_0^n$ can only be $\mathcal{L}$-related to one of the two constants. It is then easy to see that $\{c_0^n, c_1^n\}$ forms an $\mathcal{L}$-class. For any non-constant $f \in K_0^n$, the properties of the sums $h + f$ show that $f$ can only be $\mathcal{L}$-related to one of $f$, $\neg f$, or $c_0^n$ or $c_1^n$. As before, the two constants are related only to each other, so that the $\mathcal{L}$-class of $f$ is again $\{f, \neg f\}$. The final case, that $f \in \neg K_0^n$, leads in a similar way to the classes $\{c_0^n, c_1^n\}$ and $\{f, \neg f\}$ for any non-constant $f$. ∎

These descriptions of the $\mathcal{R}$- and $\mathcal{L}$-classes of the semigroup $\mathcal{O}^n(A)$ in the Boolean case can also be specialized to describe the classes of any subsemigroup $\mathcal{S}$ of $\mathcal{O}^n(A)$.

In the next proposition we consider the question of when a union of subsemigroups of $\mathcal{O}^n(A)$ is again a subsemigroup.

**Proposition 11.5.9.** *Let $(\mathcal{S}_j)_{j \in J}$ be a family of subsemigroups of $\mathcal{O}^n(A)$, and let $S = \bigcup_{j \in J} S_j$. Then $S = (S; +)$ is a subsemigroup of $\mathcal{O}^n(A)$ iff either $S$ contains no elements of $\neg C_4^n$ or $S$ is closed under the $\neg$ operation.*

**Proof**: From Lemma 11.5.1, we see that for any $f$ and $g$ in the union set $S$, we have

$$f + g = \begin{cases} g & \text{if } f \in C_4^n \\ \neg g & \text{if } f \in \neg C_4^n \\ c_0^n & \text{if } f \in K_0^n \\ c_1^n & \text{if } f \in \neg K_0^n. \end{cases}$$

Moreover by Lemma 11.5.1 (iii) and (iv), if $S$ contains any element of $K_0^n$ then it contains $c_0^n$, and similarly if there is any element of $\neg K_0^n$ in $S$ then $c_1^n$ is in $S$. Thus $(S; +)$ is a semigroup if it contains no element of $\neg C_4^n$, and otherwise $S$ is a semigroup if it is closed under $\neg$.

Conversely, if $(S; +)$ is a semigroup which contains an element $f$ of $\neg C_4^n$, then for any $g \in S$ we have $f + g = \neg g$ in $S$ as well, so $S$ is closed under $\neg$ in this case. ∎

Since the universe of any right-zero semigroup or any constant semigroup contains no elements of $\neg C_4^n$, any union of right-zero semigroups or of constant semigroups is again a semigroup of Boolean operations.

In the remainder of this section we describe the classification by Butkote, Denecke and Ratanaprasert ([13]) of all semigroups of $n$-ary Boolean operations into six types, with some information about possible sizes in each case. We know of course that there are exactly $2^{2^n}$ Boolean operations of arity $n$, so any such semigroup is finite with maximum size $2^{2^n}$. Our classification will depend upon whether or not a semigroup $S$ contains an element of $\neg C_4^n$. If there is such an element $f$ in $S$, then for any $g \in S$ we also have $f + g = \neg g$ in $S$, and $S$ is closed under $\neg$ as well as under $+$. This gives us the useful fact that the cardinality of any semigroup containing an element of $\neg C_4^n$ must be an even natural number. We can extend this to see that such $S$ must contain an equal number of elements from $C_4^n$ and $\neg C_4^n$, and an equal number of elements from $K_0^n$ and $\neg K_1^n$.

Our classification of semigroups of Boolean operations will use the following new types of semigroups.

**Definition 11.5.10.** A semigroup $S = (S; +)$ is called a *right-zero-constant semigroup* if $S$ can be written as $S_1 \cup S_2$ for two disjoint subsets $S_1$ and $S_2$ and there is a fixed element $b^*$ in $S_2$ such that

$$a + b = \begin{cases} b & \text{if } a \in S_1 \\ b^* & \text{if } a \in S_2. \end{cases}$$

A semigroup $S = (S; +)$ is called a *two-constant semigroup* if $S$ can be written as $S_1 \cup S_2$ for two disjoint subsets $S_1$ and $S_2$ and there are two fixed elements $b^*$ in $S_1$ and $b^{**}$ in $S_2$ such that

$$a + b = \begin{cases} b^* & \text{if } a \in S_1 \\ b^{**} & \text{if } a \in S_2. \end{cases}$$

A semigroup $S = (S; +)$ is called a *right-zero-two-constant semigroup* if $S$ is the union of three disjoint sets $S_1$, $S_2$ and $S_3$, and there are two fixed elements $b^* \in S_2$ and $b^{**} \in S_3$ such that

$$a + b = \begin{cases} b & \text{if } a \in S_1 \\ b^* & \text{if } a \in S_2 \\ b^{**} & \text{if } a \in S_3. \end{cases}$$

**Proposition 11.5.11.** *Let $A = \{0, 1\}$ and let $n \geq 1$. Any subsemigroup $S$ of $\mathcal{O}^n(A)$ which does not contain any operation from $\neg C_4^n$ fits one of the following five conditions:*

(i) $S$ *is a right-zero semigroup and* $|S| \leq 2^{2^n - 2}$.
(ii) $S$ *is a constant semigroup and* $|S| \leq 2^{2^n - 2}$.
(iii) $S$ *is a right-zero-constant semigroup and* $|S| \leq 2^{2^n - 1}$.
(iv) $S$ *is a two-constant semigroup and* $|S| \leq 2^{2^n - 1}$.

(v) $S$ *is a right-zero-two-constant semigroup and* $|S| \leq 3 \cdot 2^{2^n - 2}$.

*Conversely, any semigroup matching one of these five conditions is isomorphic to a subsemigroup of* $\mathcal{O}^n(A)$.

**Proof:** Our analysis of cases is based on the fact that $S \subseteq C_4^n \cup K_0^n \cup \neg K_0^n$. First, if $S$ is totally contained in $C_4^n$, then $S$ is a right-zero semigroup, with size at most $2^{2^n - 2}$. Conversely, any right-zero semigroup of size at most $2^{2^n - 2}$ is isomorphic to a subsemigroup of $\mathcal{O}^n(A)$, since any two right-zero semigroups of the same cardinality are isomorphic. Next, if $S$ is totally contained in $K_0^n$, then $S$ is a constant semigroup with constant value $c_0^n$, and a counting argument shows that this cardinality is at most $2^{2^n - 2}$. In addition, any subset of $K_0^n$ containing $c_0^n$ is a subsemigroup, and so any semigroup of this cardinality is isomorphic to a subsemigroup of $\mathcal{O}^n(A)$. A similar argument holds for any subset of $\neg K_0^n$, with constant value $c_1^n$.

Next we consider any subsemigroup $S$ for which $S$ contains an element in two of the three sets $C_4^n$, $K_0^n$ and $\neg K_0^n$. If $S$ contains an element from $C_4^n$ and an element from $K_0^n$, then $S$ is a right-zero-constant semigroup with special element $b^* = c_0^n$, and has cardinality at most $2^{2^n - 1}$. Conversely, any subset of $C_4^n \cup K_0^n \cup \neg K_0^n$ which contains $c_0^n$ is a right-zero-constant semigroup. The same result holds if $S$ contains an element from $C_4^n$ and an element from $\neg K_0^n$, this time with right-zero-constant element $c_1^n$. If $S$ contains elements from both $K_0^n$ and $\neg K_0^n$, then it contains both constants $c_0^n$ and $c_1^n$ and hence is a two-constant semigroup. Finally, if $S$ contains elements from all three of the partition sets, then $S$ is a right-zero-two-constant semigroup.

We also note that any two-constant semigroups are isomorphic, if they have the same cardinality for their corresponding disjoint sets. This means that any such semigroup is isomorphic to a subsemigroup of $\mathcal{O}^n(A)$. The analogous result is also true for right-zero-two-constant semigroups, completing our proof. ∎

To finish our classification of the subsemigroups of $\mathcal{O}^n(A)$, we turn now to subsemigroups $S$ which contain an element of $\neg C_4^n$. In this case we obtain a structure called a *four-part semigroup*, which we now define. We use four finite and pairwise disjoint sets with elements denoted as follows:

$$S_1 = \{a_{11}, a_{12}, \ldots, a_{1n_r}\},$$
$$S_2 = \{a_{21}, a_{22}, \ldots, a_{2n_r}\},$$
$$S_3 = \{a_{31}, a_{32}, \ldots, a_{3n_s}\},$$

$$S_4 = \{a_{41}, a_{42}, \ldots, a_{4n_s}\}.$$

We define a binary operation $*$ on $S = S_1 \cup S_2 \cup S_3 \cup S_4$ by

$$a_{ij} * a_{lk} = \begin{cases} a_{lk} & \text{if } a_{ij} \in S_1 \\ a_{tk} & \text{if } a_{ij} \in S_2 \text{ where } t = \begin{cases} 1 \text{ if } l = 2 \\ 2 \text{ if } l = 1 \\ 3 \text{ if } l = 4 \\ 4 \text{ if } l = 3 \end{cases} \\ a^* \in S_3 \text{ if } a_{ij} \in S_3 \\ a^{**} \in S_4 \text{ if } a_{ij} \in S_4. \end{cases}$$

The binary operation $*$ is well-defined, and it can be checked that it is associative, giving us a semigroup $(S; *)$ called a four-part semigroup. Note that since sets $S_1$ and $S_2$ have the same cardinality, as do $S_3$ and $S_4$, our four-part semigroup has even cardinality.

**Theorem 11.5.12.** *Let $A = \{0, 1\}$ and let $n \geq 1$. Any subsemigroup $S$ of $\mathcal{O}^n(A)$ which contains an operation from $\neg C_4^n$ is a four-part semigroup with cardinality at most $2^{2^n}$. Conversely any four-part semigroup of cardinality at most $2^{2^n}$ is isomorphic to a subsemigroup of $\mathcal{O}^n(A)$.*

**Proof:** Let $S$ be a subsemigroup of $\mathcal{O}^n(A)$ which contains an operation from $\neg C_4^n$. Property (ii) of Lemma 11.5.1 tells us that $g + g = \neg g$ for any $g \in \neg C_4^n$, which means that $S$ also contains an element of $C_4^n$ as well. Now we use the basic four-part partition of $\mathcal{O}^n(A)$ to partition the set $S$ into the four subsets $S \cap C_4^n$, $S \cap \neg C_4^n$, $S \cap K_0^n$ and $S \cap \neg K_0^n$. The properties from Lemma 11.5.1 show that for any $f \in S$,

$$f + g = \begin{cases} g & \text{if } f \in C_4^n \\ \neg g & \text{if } f \in \neg C_4^n \\ c_0^n & \text{if } f \in K_0^n \\ c_1^n & \text{if } f \in \neg K_0^n; \end{cases}$$

which proves that $S$ is indeed a four-part semigroup. Its cardinality is of course at most the cardinality of $\mathcal{O}^n(A)$, which is $2^{2^n}$.

In the opposite direction, suppose that $S$ is any four-part semigroup of size at most $2^{2^n}$. We construct a subsemigroup of $\mathcal{O}^n(A)$ isomorphic to $S$, as follows. First we choose any $n_r$ elements of $C_4^n$, where $n_r$ is the size of the set $S_1$ from $S$. Let $S_1'$ be the set of these elements, and let $S_2' = \neg S_1'$. If

$S_3$ and $S_4$ are empty, we use the empty set for each of $S_3'$ and $S_4'$; otherwise we form $S_3'$ by choosing $n_s$ elements from $K_0^n$, including $c_0^n$, where $n_s$ is the size of set $S_3$. Again we take $S_4' = \neg S_3'$, so that $c_1^n \in S_4'$. Let $S'$ be the union of these four sets. Then Lemma 11.5.1 guarantees that the operation $+$ on this set satisfies

$$f + g = \begin{cases} g & \text{if } f \in S_1' \\ \neg g & \text{if } f \in S_2' \\ c_0^n & \text{if } f \in S_3' \\ c_1^n & \text{if } f \in S_4'. \end{cases}$$

It is then clear that $(S'; +)$ is a semigroup isomorphic to the original four-part semigroup $\mathcal{S}$. ∎

## 11.6 Semigroups of Boolean Co-operations

In the remaining sections of this chapter we turn to the study of semigroups of co-operations. Throughout $A$ will denote a finite base set, and $n \geq 1$ will be a natural number. The set $cO^n(A)$ of all $n$-ary co-operations on $A$ forms the universe of a semigroup, under the operation $+$ defined by

$$f + g \; := \; comp_n^n(f, g, \ldots, g).$$

We begin our study in this section with the special case of Boolean co-operations, taking $A$ to be the two-element set $\{0, 1\}$. Most of the results on Boolean operations in the previous section were based on the properties of such operations under addition, as recorded in Lemma 11.5.1, which stemmed from the partition of the set of all Boolean operations into the four subclasses $C_4^n$, $\neg C_4^n$, $K_0^n$ and $\neg K_0^n$. For Boolean co-operations we proceed with a similar but slightly more complicated partition of the set of all co-operations, described by Denecke and Saengsura in [23]. We recall again that any $n$-ary co-operation $f$ on set $\{0, 1\}$ can be uniquely expressed as a pair $(f_1, f_2)$ of mappings, $f_1 : A \to \{1, \ldots, n\}$ and $f_2 : A \to A$. There are four choices for the action of $f_2$ on the pair $(0, 1)$, and we shall consider two cases for the behaviour of $f_1$, depending on whether $f_1(0)$ equals $f_1(1)$ or not. This results in a partition of $cO^n(A)$ into the following eight pairwise disjoint subsets:

$$F_0 := \{f \in cO^n(A) \mid f_1(0) = f_1(1), f_2(0) = 0 = f_2(1)\},$$
$$F_1 := \{f \in cO^n(A) \mid f_1(0) = f_1(1), f_2(0) = 1 = f_2(1)\},$$
$$F_2 := \{f \in cO^n(A) \mid f_1(0) = f_1(1), f_2(0) = 1, f_2(1) = 0\},$$
$$I_{\{0,1\}}^{(n)} := \{\iota_i^n \mid 1 \le i \le n\},$$
$$F_0^* := \{f \in cO^n(A) \mid f_1(0) \ne f_1(1), f_2(0) = 0 = f_2(1)\},$$
$$F_1^* := \{f \in cO^n(A) \mid f_1^A(0) \ne f_1(1), f_2(0) = 1 = f_2(1)\},$$
$$F_2^* := \{f \in cO^n(A) \mid f_1^A(0) \ne f_1(1), f_2(0) = 1, f_2(1) = 0\}, \quad \text{and}$$
$$F_3^* := \{f \in cO^n(A) \mid f_1^A(0) \ne f_1(1), f_2(0) = 0, f_2(1) = 1\}.$$

From this decomposition of $cO^n(A)$ a wealth of information is now available by careful case-by-case analysis of the action of $+$ on various subsets. Without going into all the details here (see [23] for complete proofs) we summarize in the table in Figure 11.1 how the subsets interrelate.

| $+$ | $F_0$ | $F_1$ | $F_2$ | $F_0^*$ | $F_1^*$ | $F_3^*$ | $F_2^*$ | $I_{\{0,1\}}^{(n)}$ |
|---|---|---|---|---|---|---|---|---|
| $F_0$ | $F_0$ | $F_1$ | $F_1$ | $F_0$ | $F_1$ | $F_0$ | $F_1$ | $F_0$ |
| $F_1$ | $F_0$ | $F_1$ | $F_0$ | $F_0$ | $F_1$ | $F_1$ | $F_0$ | $F_1$ |
| $F_2$ | $F_0$ | $F_1$ | $I_{\{0,1\}}^{(n)}$ | $F_0^*$ | $F_1^*$ | $F_2^*$ | $F_3^*$ | $F_2$ |
| $F_0^*$ | $F_0$ | $F_1$ | $F_1$ | $F_0$ | $F_1$ | $F_0$ | $F_1$ | $F_0$ |
| $F_1^*$ | $F_0$ | $F_1$ | $F_0$ | $F_0$ | $F_1$ | $F_1$ | $F_0$ | $F_1$ |
| $F_3^*$ | $F_0$ | $F_1$ | $F_2$ | $F_0^*$ | $F_1^*$ | $F_3^*$ | $F_2^*$ | $I_{\{0,1\}}^{(n)}$ |
| $F_2^*$ | $F_0$ | $F_1$ | $I_{\{0,1\}}^{(n)}$ | $F_0^*$ | $F_1^*$ | $F_2^*$ | $F_3^*$ | $F_2$ |
| $I_{\{0,1\}}^{(n)}$ | $F_0$ | $F_1$ | $F_2$ | $F_0^*$ | $F_1^*$ | $F_3^*$ | $F_2^*$ | $I_{\{0,1\}}^{(n)}$ |

Fig. 11.1    Addition of $n$-ary Boolean Co-operations

Referring back to the list of all clones of Boolean co-operations from Section 10.6, we can show that $F_0 = M_0^{(n)} \setminus I_{\{0,1\}}^{(n)}$, $F_1 = M_1^{(n)} \setminus I_{\{0,1\}}^{(n)}$, $F_2 = D_c^{(n)} \setminus I_{\{0,1\}}^{(n)}$, $F_0^* = C_3^{(n)} \setminus M_0^{(n)}$, $F_1^* = C_4^{(n)} \setminus M_1^{(n)}$, $F_2^* = cO^n(\{0,1\}) \setminus (L_c^{(n)} \cup C_{1c}^{(n)} \cup C_{0c}^{(n)})$, and $F_3^* = C_2^{(n)} \setminus I_{\{0,1\}}^{(n)}$.

Using this information we can prove the following result regarding the order of elements in $(cO^n(A); +)$. The proof involves a systematic checking of the claim for the eight subsets, and we leave the details to the reader to verify.

**Proposition 11.6.1.** *Let $A = \{0, 1\}$ and let $n \ge 1$. For any $n$-ary Boolean co-operation $f$, either $f + f = f$ or $f + f + f = f$ or $f + f = f + f + f$. Hence any $n$-ary Boolean co-operation has order two.*

The same systematic checking shows that any element of the four sets $F_0$, $F_1$, $F_3^*$ and $I_{\{0,1\}}^{(n)}$ is idempotent, and moreover that no other elements are idempotent. Let us denote by $E(cO^n(\{0,1\}))$, or simply $E$ for short, the set of all idempotents. Any idempotent is regular by definition, but we can also check that elements of the additional sets $F_2$ and $F_2^*$ are regular, while elements of the remaining two sets $F_0^*$ and $F_1^*$ are not regular. This gives the following result.

**Proposition 11.6.2.** *Let* $A = \{0,1\}$ *and let* $n \geq 1$. *An* $n$-*ary Boolean co-operation* $f$ *is idempotent iff* $f$ *is in* $E = F_0 \cup F_1 \cup F_3^* \cup I_{\{0,1\}}^{(n)}$, *and is regular iff* $f$ *is in* $E \cup F_2 \cup F_2^*$.

For convenience we shall use the name $Reg(cO^n)$ for the set of regular Boolean co-operations. A semigroup is called *orthodox* if its idempotent elements form a subsemigroup.

**Theorem 11.6.3.** *Let* $A = \{0,1\}$ *and let* $n \geq 1$. $(Reg(cO^n); +)$ *is an orthodox subsemigroup of* $(cO^n(\{0,1\}); +)$.

**Proof:** Verification that both $E$ and $Reg(cO^n)$ form subsemigroups of the semigroup of all $n$-ary Boolean co-operations proceeds by a lengthy examination of cases, using Proposition 11.6.2. We omit the details here. ∎

We can prove another fact about the set $E$ of idempotent Boolean co-operations. By checking the four partition subsets which make up the set $E$, we can show that elements of this set satisfy the identity $x + y + z + x \approx x + y + x + z + x$ which characterizes the variety of regular bands.

**Proposition 11.6.4.** *Let* $A = \{0,1\}$ *and let* $n \geq 1$. *The semigroup* $(E; +)$ *of idempotent* $n$-*ary Boolean co-operations is a regular band.*

It follows from the table in Figure 11.1 that any element $f$ of $F_0$ or $F_1$ is a right-zero element, satisfying $g + f = f$ for any $n$-ary Boolean co-operation $g$. This means that $F_0 \cup F_1$ forms the universe of a right-zero semigroup. Similarly, any element $f$ from $F_3^* \cup I_{\{0,1\}}^{(n)}$ is a left-identity element for $cO^n(A)$, since $f + g = g$ for any $n$-ary Boolean co-operation $g$ in this case. As a special case, we see that the set $F_3^* \cup I_{\{0,1\}}^{(n)}$ also forms the universe of a right-zero semigroup.

The following additional results about bands were proved in [23].

**Proposition 11.6.5.** *Let* $A = \{0,1\}$ *and let* $n \geq 1$.

(i) $(F_0 \cup F_3^* \cup I_{\{0,1\}}^{(n)}; +)$ and $(F_1 \cup F_3^* \cup I_{\{0,1\}}^{(n)}; +)$ are both normal bands.

(ii) A subsemigroup $(S; +)$ of $(cO^n(\{0,1\}); +)$ is a right-zero semigroup iff it is a rectangular band.

**Proof:** (i) It is straightforward to verify that both the given sets satisfy the required normal equation $f + h + l + g = f + l + h + g$ for all $f, h, l, g$. (ii) Certainly every right-zero semigroup is a rectangular band. Conversely, the rectangular identity $f + g + f = f$ for all $f, g \in S$ forces $S$ to be a subset of one of the right-zero semigroups $F_0 \cup F_1$ or $F_3^* \cup I_{\{0,1\}}^{(n)}$. ∎

Additional results along these lines were proved in [23], including characterizations of which subsemigroups of $(cO^n(A); +)$ are commutative, semilattices, or constant semigroups.

## 11.7 Elements of Semigroups of $n$-ary Co-operations

We now consider a finite set $A$, of cardinality at least two, and the semigroup $(cO^n(A); +)$ of all $n$-ary co-operations defined on $A$, for some fixed $n \geq 1$. In this section we classify all the idempotent and regular elements of this semigroup, and then determine the order of any element.

**Theorem 11.7.1.** *Let $A$ be a finite set of size at least two, and let $n \geq 1$. let $f$ be an $n$-ary co-operation on $A$.*

(i) *$f$ is an idempotent element of $(cO^n(A); +)$ iff $a \in f^{-1}((i, a))$ for all $(i, a) \in Im f$.*

(ii) *$f$ is a regular element of $(cO^n(A); +)$ iff for any two elements $(i, a), (j, b)$ of $Im f$, if $a = b$ then $i = j$.*

**Proof:** (i) First let $f$ be an idempotent co-operation, and let $(i, a) \in Im f$. Then there is an element $x \in A$ such that $f(x) = (i, a)$. Since $(f + f)(x) = f(x)$ and $(f + f)(x) = f(f_2(x)) = f(a)$, we see that $f(x) = f(a) = (i, a)$, making $a \in f^{-1}((i, a))$.

Conversely, suppose that $a \in f^{-1}((i, a))$ for all $(i, a) \in Im f$ holds, and let $x \in A$. Then $f(x) = (f_1(x), f_2(x))$ is in $Im f$, and by assumption $f_2(x) \in f^{-1}(f_1(x), f_2(x))$. Then $(f + f)(x) = f(f_2(x)) = (f_1(x), f_2(x)) = f(x)$, showing that $f$ is idempotent.

(ii) If $f$ is a regular element, then $f + g + f = f$ for some $n$-ary co-operation $g$. Let $(i, a)$ and $(j, b)$ be any two pairs in $Im f$ such that $a = b$. By definition there exist elements $x$ and $y$ in $A$ such that $f(x) = (i, a)$ and $f(y)$

$= (j, b)$; in particular, $f_1(x) = i$, $f_2(x) = a$, $f_1(y) = j$ and $f_2(y) = b$. The assumption $a = b$ then gives $f_2(x) = f_2(y)$. Since $(f + g)(x) = g(f_2(x))$ by definition of the composition, we have

$$
\begin{aligned}
f_1(x) &= (f + g + f)_1(x) &= f_1(g_2(f_2(x))) \\
&= f_1(g_2(f_2(y))) &= (f + g + f)_1(y) \\
&= f_1(y).
\end{aligned}
$$

Conversely, assume that $a = b$ implies $i = j$ for any pairs $(i, a), (j, b) \in Imf$. For each pair $(i, a) \in Imf$ there is an element $d_a \in f^{-1}((i, a))$. We define a co-operation $g : A \to A^{\sqcup n}$ by the rule that

$$
g(x) = \begin{cases} (i, d_x) & if \quad x \in \{a \mid (i, a) \in Imf\} \\ (j, x) & if \quad x \notin \{a \mid (i, a) \in Imf\} \end{cases} \quad \text{for some } j \in \{1, \dots, n\}.
$$

We claim now that $(f + g + f)(x) = f(x)$ for any $x \in A$. We must have $f(x) = (l, b)$ for some $(l, b)$, so that $f_1(x) = l$ and $f_2(x) = b$. Then

$$
\begin{aligned}
(f + g + f)(x) &= ((f + g) + f)(x) &= f((f + g)_2(x)) \\
&= f(g_2(f_2(x))) &= f(g_2(b)) \\
&= f(d_b) &= (l, b) \\
&= f(x).
\end{aligned}
$$

This shows that $f$ is regular. ∎

We have seen in Section 11.6 that in the Boolean case, the sets of idempotent and regular elements both form subsemigroups of the semigroup of all $n$-ary co-operations. When the size of $A$ is at least three, it is still true that the set of regular elements is closed under $+$, and hence forms a subsemigroup. It is no longer true however that the sum of two idempotents is again idempotent, as the following example illustrates.

Let $A = \{a, b, c\}$ and let $f, g \in cO^n(A)$ be defined by the following table:

|   | $f$ | $g$ |
|---|---|---|
| $a$ | $(i, a)$ | $(k, a)$ |
| $b$ | $(i, a)$ | $(l, b)$ |
| $c$ | $(j, c)$ | $(l, b)$ |

for some $i, j, k, l \in \{1, \dots, n\}$. By the characterization from Theorem 11.7.1 both $f$ and $g$ are idempotents. But $(f + g)(c) = g(f_2(c)) = g(c) = (l, b)$

and $(f + g)(b) = g(f_2(b)) = g(a) = (k, a)$, so $b$ is not in $(f + g)^{-1}(l, b)$: therefore $f + g$ is not idempotent.

Next we consider the order of an $n$-ary co-operation $f$. Let us use the notation $kf$, for a natural number $k$, to denote the sum of $k$ copies of $f$. Since $(if)(a) = f((f_2)^{i-1}(a))$ for any natural number $i \geq 2$ and any $a \in A$, the order of $f$ is determined by the order of $f_2$. Let $\lambda(f_2)$ be the index and let $r$ be the period of $f_2$. Then $f_2^{\lambda(f_2)}(a) = f_2^{\lambda(f_2)+r}(a)$ for all $a \in A$. Thus $o(f_2) = \lambda(f_2) + r - 1$, and we have

$$
\begin{aligned}
((\lambda(f_2) + r + 1)f)(a) &= f(f_2^{\lambda(f_2)+r}(a)) \\
&= f(f_2^{\lambda(f_2)}(a)) \\
&= ((\lambda(f_2) + 1)f)(a).
\end{aligned}
$$

This means that $o(f) \leq \lambda(f_2) + r$ and $o(f) \leq o(f_2) + 1$. For the index $m$ of the co-operation $f$ we have $m \leq \lambda(f_2) + 1$.

Now suppose that $f$ is regular. Then for any positive natural numbers $s$ and $t$ we have $f_2^s = f_2^t$ iff $sf = tf$, since for each $a \in A$,

$$
\begin{aligned}
f_2^s(a) = f_2^t(a) &\Leftrightarrow (f_1(f_2^{s-1}(a)), f_2^s(a)) = (f_1(f_2^{t-1}(a)), f_2^t(a)) \\
&\Leftrightarrow (sf)(a) = (tf)(a).
\end{aligned}
$$

This tells us that when $f$ is regular, its order $o(f)$ equals $o(f_2)$ and its index $m$ equals $\lambda(f_2)$. For the order of a non-regular element, we use the following fact.

**Lemma 11.7.2.** *Let $A$ be a finite set of cardinality $m$. Then for every $f \in cO^n(A)$ the co-operation $mf$ is regular.*

**Proof:** Let $k \geq 1$. Since $(k + 1)f = f + kf$ and $Im(f + kf) \subseteq Imkf$, it follows from Theorem 11.7.1(ii) that $(k + 1)f$ is regular if $kf$ is regular. Thus if $mf$ is not regular, then none of $(m - 1)f$, $(m - 2)f$, ..., $f$ can be regular either. By Theorem 11.7.1(ii) again, non-regularity of $(m - 1)f$ means that there exist pairs $(i, c)$, $(j, c) \in Im((m - 1)f)$ with $i \neq j$. Let $a, b \in A$ such that $((m - 1)f)(a) = (i, c)$ and $((m - 1)f)(b) = (j, c)$. Then

$$
\begin{aligned}
(mf)(a) &= f(((m - 1)f)_2(a)) \\
&= f(c) \\
&= f(((m - 1)f)_2(b)) \\
&= (mf)(b).
\end{aligned}
$$

Now let $\varphi : Im((m-1)f) \to Im(mf)$ be defined by $\varphi((m-1)f)(x) = (mf)(x)$. We know that $\varphi$ maps the different elements $(i,c),(j,c) \in Im((m-1)f)$ both to $f(c) \in Im(mf)$. By definition $Im(mf) \subseteq Im((m-1)f)$, but now we see that this is a proper containment. Similarly, one can show that $Im((m-1)f) \subset \ldots \subset Imf$. But this forces $|Im(mf)| < \ldots < |Imf| = m$, implying that $|Im(mf)| = 1$. Therefore $mf$ is regular, a contradiction. ∎

This lemma shows us that for any non-regular co-operation $f$, there is at least one positive natural number $k$ such that $kf$ is regular. We can therefore consider the least such positive integer.

**Definition 11.7.3.** Let $A$ be a finite set, and let $n \geq 1$. Let $f$ be a non-regular element of $cO^n(A)$. Let $\beta(f)$ be the least positive integer such that $\beta(f)f$ is regular.

For every co-operation $f \in cO^n(A)$ we have $\beta(f) \leq \lambda(f_2) + 1$, and we can now describe two possible cases.

**Corollary 11.7.4.** *Let $A$ be a finite set, and let $n \geq 1$. Let $f$ be a non-regular element of $cO^n(A)$. Then*

(i) *If $\beta(f) \leq \lambda(f_2)$, then $o(f) = o(f_2)$.*
(ii) *If $\beta(f) = \lambda(f_2) + 1$, then $o(f) = o(f_2) + 1$.*

**Proof**: (i) Assume that $\beta(f) \leq \lambda(f_2)$. By the argument in the proof of the previous lemma, this implies that $\lambda(f_2)f$ is regular. It follows that for any $s,t \in \mathbb{N}$ we have $f_2^{\lambda(f_2)+s} = f_2^{\lambda(f_2)+t}$ iff $(\lambda(f_2) + s)f = (\lambda(f_2) + t)f$. Therefore $o(f) = o(f_2)$.
(ii) Assume that $\beta(f) = \lambda(f_2) + 1$. Then $(\lambda(f_2) + 1)f$ is regular and by the minimality of $\beta(f)$ we get that $(\lambda(f_2)f)$ is not regular. Let $r$ be the period of $f_2$. This implies that $f_2^{\lambda(f_2)+r} = f_2^{\lambda(f_2)}$, but $(\lambda(f_2) + r)f \neq \lambda(f_2)f$. Therefore $o(f) = \lambda(f_2) + r = o(f_2) + 1$. ∎

For unary co-operations, the semigroup $(cO_A^1; +)$ is isomorphic to the full transformation semigroup $(H_A; \circ)$ on set $A$. Cayley's Theorem then tells us that every abstract semigroup is isomorphic to a semigroup of co-operations.

## 11.8    Bands of $n$-ary Co-operations

As we did for semigroups of operations in Section 11.4, we can describe which bands or idempotent semigroups can occur as subsemigroups of $(cO^n(A); +)$. We assume throughout this section that $A$ is a finite set and $n \geq 1$. Results for left- and right-zero bands, rectangular and normal bands and semilattices were proved by Denecke and Saengsura in [24].

The characterization of left- and right-zero bands involves some basic information about kernels and images of co-operations. The following results are straightforward, and we leave the proofs as exercises.

**Lemma 11.8.1.** *Let $A$ be a finite set and let $n \geq 1$. For any $n$-ary co-operations $f$, $g$, $h$ and $l$ on $A$, the following properties hold:*

(i) *If $f + g = g$ then $Ker f \subset Ker g$.*
(ii) *If $f + g = f$ then $Im f \subseteq Im g$.*
(iii) *$f + g + f = f$ iff $g_2(a) \in f_2^{-1}(a)$ for all $a \in Im f_2$.*
(iv) *$f + h + l + g = f + l + h + g$ iff $((l_2 \circ h_2)(a), (h_2 \circ l_2)(a)) \in Ker g$ for all $a \in Im f_2$.*

**Proposition 11.8.2.** *Let $A$ be a finite set and let $n \geq 1$. Let $S$ be a subsemigroup of idempotents from $cO^n(A)$. Then*

(i) *$(S; +)$ is a right-zero semigroup iff $Ker f = Ker g$ for all $f, g \in S$.*
(ii) *$(S; +)$ is a left-zero semigroup iff $Im f = Im g$ for all $f, g \in S$.*

**Proof:** (i) It follows immediately from Lemma 11.8.1(i) that any two elements in a right-zero semigroup must have the same kernel. Conversely, suppose that $f$ and $g$ are elements of $S$ with $Ker f = Ker g$. Since $f$ is idempotent, we have $f(a) = (f + f)(a) = f(f_2(a))$ for any $a \in A$, and thus $(a, f_2(a)) \in Ker f = Ker g$. Therefore $g(a) = g(f_2(a)) = (f + g)(a)$ for all $a \in A$, and we have $f + g = g$. Similarly $g + f = f$, and so $(S; +)$ is a a right-zero semigroup.

(ii) Again one direction comes immediately from Lemma 11.8.1(ii). Conversely, suppose that any two elements $f, g \in S$ have the same image, and let $a \in A$. Since $f(a) = (f_1(a), f_2(a)) \in Im f$, we have also $(f_1(a), f_2(a)) \in Im g$. Since $g$ is idempotent, by Theorem 11.7.2 we have $f_2(a) \in g^{-1}(f_1(a), f_2(a))$ and therefore $(f + g)(a) = g(f_2(a)) = (f_1(a), f_2(a)) = f(a)$. Thus $f + g = f$, and $(S; +)$ is a left-zero semigroup. ∎

A set of co-operations will be called *rectangular* if it has the property that $f + g + f = f$ for any $f$ and $g$ in the set. But note that a rectangular set of co-operations need not be a semigroup. The following example from [23] illustrates this. Let $A$ be the six-element set $\{a, b, c, d, e, f\}$, and define three co-operations $\alpha$, $\beta$ and $\gamma$ on $A$ by letting $\alpha$ map each of $b$, $c$ and $d$ to $(i, c)$ and each of $a$, $e$ and $f$ to $(j, f)$; $\beta$ map each of $a$, $b$ and $f$ to $(k, a)$ and the remaining three elements to $(l, d)$; and $\gamma$ map each of $a$, $b$ and $c$ to $(m, b)$ and the remaining three elements to $(o, e)$, for some indices $i, j, k, l, m, o \in \{1, \ldots, n\}$. One can check using Lemma 11.8.1 that each of the three co-operations is an idempotent, and that the set $\{\alpha, \beta, \gamma\}$ is a rectangular set. However it is not closed under $+$, since $(\alpha + \beta) + \gamma + (\alpha + \beta) \neq \alpha + \beta$.

Parts (iii) and (iv) of Lemma 11.8.1 also lead directly to characterizations of rectangular and normal bands of idempotents from $cO^n(A)$.

**Theorem 11.8.3.** *Let $A$ be a finite set and let $n \geq 1$. Let $S$ be a subsemigroup of idempotents from $cO^n(A)$.*

(i) *$S$ is a rectangular band iff for every $f, g \in S$ we have $g_2(a) \in f_2^{-1}(a)$ for all $a \in Im f_2$ and $f_2(b) \in g_2^{-1}(b)$ for all $b \in Im g_2$.*

(ii) *$S$ is a normal band iff for all $f, h, l, g \in S$ we have $((l_2 \circ h_2)(a), (h_2 \circ l_2)(a)) \in Ker g$ for all $a \in Im f_2$.*

**Theorem 11.8.4.** *Let $A$ be a finite set and let $n \geq 1$. Let $S$ be a subsemigroup of idempotents from $cO^n(A)$. Then $S$ is a semilattice iff any two co-operations $f, g \in S$ satisfy the following conditions:*

(i) *$Im(f + g) = Im(g + f)$, and*

(ii) *For each $(i, b) \in Im(f + g)$, $f_2(x) \in g^{-1}(i, b)$ iff $g_2(x) \in f^{-1}(i, b)$.*

**Proof:** Commutativity in a semilattice $(S; +)$ guarantees that $Im(f + g) = Im(g + f)$ for any $f$ and $g$ in $S$. Let $(i, b) \in Im(f + g)$ and $f_2(x) \in g^{-1}(i, b)$. Then $f(g_2(x)) = (g + f)(x) = (f + g)(x) = g(f_2(x)) = (i, b)$. Therefore $g_2(x) \in f^{-1}(i, b)$. Dually then $g_2(x) \in f^{-1}(i, b)$ implies that $f_2(x) \in g^{-1}(i, b)$.

Conversely, assume that the co-operations in $S$ satisfy the two given conditions. To show that $f + g = g + f$, let $x \in A$, and write $(f + g)(x) = (i, b)$ for some pair $(i, b)$. Since $(f + g)(x) = g(f_2(x))$ and $(f + g)(x) = (i, b)$, we get $g(f_2(x)) = (i, b)$ and hence $f_2(x) \in g^{-1}(i, b)$. Condition (ii) implies $g_2(x) \in f^{-1}(i, b)$. Therefore $(g + f)(x) = f(g_2(x)) = (i, b) = (f + g)(x)$. Therefore commutativity holds, and $S$ is a semilattice. ∎

## 11.9 Green's Relations $\mathcal{L}$ and $\mathcal{R}$

The final piece of our study of semigroups of co-operations on a finite set $A$ is to look at the two Green's relations $\mathcal{L}$ and $\mathcal{R}$. As we have mentioned before, two operations are $\mathcal{L}$-related iff they have the same images, and are $\mathcal{R}$-related iff they have the same kernel. A similar result holds for $\mathcal{L}$ for co-operations, although the situation for $\mathcal{R}$ is a little more complicated.

**Theorem 11.9.1.** *Let $A$ be a finite set, and $n \geq 1$. Two co-operations $f$ and $g$ in $cO^n(A)$ are $\mathcal{L}$-related iff $Imf = Img$.*

**Proof:** Let $(f, g) \in \mathcal{R}$. If $f = g$, then certainly $Imf = Img$. If $f \neq g$, there are $n$-ary co-operations $h, l \in cO^n(A)$ such that $h + f = g$ and $l + g = f$. From this it follows that $Imf \subseteq Img$ and $Img \subseteq Imf$, so that $Imf = Img$. Conversely, if $Imf = Img$, then for each $(i, a) \in Imf = Img$ we can choose $d_a \in f^{-1}((i, a))$ and $d'_a \in g^{-1}((i, a))$. This allows us to define co-operations $h, l \in cO^n(A)$, by

$$h(x) = (i, d'_a) \quad \text{for all } x \in f^{-1}((i, a)) \quad \text{and}$$
$$l(y) = (i, d_a) \quad \text{for all } y \in g^{-1}((i, a)).$$

The co-operations $h$ and $l$ are well-defined and we have $(h+g)(x) = g(h_2(x))$ $= g(d'_a) = (i, a) = f(x)$ and $(l + f)(y) = f(l_2(y)) = f(d_a) = (i, a) = g(y)$ and thus $h + g = f$ and $l + f = g$, showing that $f$ and $g$ are $\mathcal{L}$-related. ∎

**Theorem 11.9.2.** *Let $A$ be a finite set, and $n \geq 1$. Let $f$ and $g$ be regular co-operations on $A$. Then $(f, g) \in \mathcal{R}$ iff $Kerf = Kerg$.*

**Proof:** If $(f, g) \in \mathcal{L}$ and $f \neq g$, there are $n$-ary co-operations $u, v \in cO^n(A)$ such that $f = g + u$ and $g = f + v$. Let $(a, b) \in Kerf$. Then $f(a) = f(b)$ implies $f_2(a) = f_2(b)$, which in turn implies $v(f_2(a)) = v(f_2(a))$. Therefore $(f + v)(a) = (f + v)(b)$, so $g(a) = g(b)$ and $(a, b) \in Kerg$. This shows that $Kerf \subseteq Kerg$, and dually we get $Kerg \subseteq Kerf$.

Conversely, assume that $Kerf = Kerg$. We form the quotient set $A/Kerf$, which is of some finite size, say $p$, with $1 \leq p \leq |A|$. Since $|A/Kerf| = |Imf| = |Img|$ we can write $Imf = \{(j_1, a_1), \ldots, (j_p, a_p)\}$ and $Img = \{(k_1, b_1), \ldots, (k_p, b_p)\}$ for some $j_1, \ldots, j_p, k_1, \ldots, k_p \in \{1, \ldots, n\}$ and $a_1, \ldots, a_p, b_1, \ldots, b_p \in A$. Now we define two co-operations $\alpha$ and $\beta$ by setting $\alpha(a_i) = (k_i, b_i)$ for all $i = 1, \ldots, p$ and for any other $x \in A$, $\alpha(x) = (r, x)$ for some $r \in \{1, \ldots, n\}$; and setting $\beta(b_i) = (j_i, a_i)$ for all

$i = 1, \ldots, p$ and $\beta(y) = (s, y)$ for any other $y \in A$, for some $s \in \{1, \ldots, n\}$. Because of the regularity of $f$ and $g$ the co-operations $\alpha$ and $\beta$ are well-defined. We claim that $f + \alpha = g$ and $g + \beta = f$. Let $c \in A$. Then there is an element $i \in \{1, \ldots, p\}$ such that $(c, d_i) \in \ker f$ and we have $(f + \alpha)(c) = \alpha(f_2(c)) = \alpha(f_2(d_i)) = \alpha(a_i) = (k_i, b_i) = g(d_i) = g(c)$ and thus $f + \alpha = g$. Dually we get $g + \beta = f$, and so we have $(f, g) \in \mathcal{R}$. ∎

Denecke and Saengsura showed in [24] that the condition of regularity is necessary in Theorem 11.9.2. They used the regularity condition from Section 11.8 to show that if $g$ is any co-operation and $f$ is a non-regular co-operation such that $Kerf = Kerg$ and $f \neq g$, then $f$ and $g$ are not $\mathcal{R}$-related. The non-regularity of $f$ guarantees that there are elements $a, b, x \in A$ and integers $i \neq j \in \{1, \ldots, n\}$ such that $f(a) = (i, x)$ and $f(b) = (j, x)$, with $(a, b)$ not in $Kerf$. Let $\alpha$ be any co-operation. Then $(f + \alpha)(a) = \alpha(f_2(a)) = \alpha(x)$ and $(f + \alpha)(b) = \alpha(f_2(b)) = \alpha(x)$. It follows that $(f + \alpha)(a) = (f + \alpha)(b)$. Since $(a, b) \notin Kerg$ and $Kerf = Kerg$, we have also $g(a) \neq g(b)$ and thus $g \neq f + \alpha$ for any $\alpha \in cO^n(A)$. Thus $f$ and $g$ cannot be $\mathcal{R}$-related.

Since we have shown that in the Boolean case all the $n$-ary co-operations except those in $F_0^*$ or $F_1^*$ are regular, we have the following corollary.

**Corollary 11.9.3.** *Let $A$ be the two-element set $\{0, 1\}$, and let $n \geq 1$. Two elements $f$ and $g$ in $cO^n(\{0, 1\}) \setminus (F_0^* \cup F_1^*)$ are related by $\mathcal{R}$ if and only if $Kerf = Kerg$.*

It is easy to see that the $\mathcal{R}$-class of a Boolean co-operation $f$ is $\{f\}$ for any $f \in F_0 \cup F_1$, while the $\mathcal{L}$-class of $f$ is $\{f, g\}$ if $f, g \in F_0 \cup F_1$ with $f_1(0) = g_1(1)$ and $f_1(1) = g_1(0)$.

## 11.10   Exercises for Chapter 11

1. Prove that the operation $+$ defined in Section 11.1 is associative, on each of $O^n(A)$ and $cO^n(A)$.

2. Prove Lemma 11.2.3: Let $A$ be a finite set.

(i) For any $g \in O^1(A)$, the equivalence class $[g]_\sim$ contains at most one idempotent element.

(ii) If $f$ is idempotent, then for all $g \in [f]_\sim$ we have $Kerg \subseteq Kerf$ and

$g \circ f = f \circ g = f$.

(iii) If $Imf = A$, then $[f]_\sim = \{f\}$.

(iv) If $f$ is idempotent and $|Imf| = |A| - 1$, then $[f]_\sim = \{f\}$.

3. Prove Theorem 11.3.2: Let $A$ be a finite set of size at least two, and let $n \geq 2$. For any $n$-ary operations $f$ and $g$ on $A$, the following properties hold:

(i) $f \in Idem(O^n(A))$ iff for all $a \in Imf$ we have $f(a, \ldots, a) = a$.

(ii) $(f, g) \in \mathcal{R}_{O^n(A)}$ iff $Kerf = Kerg$.

(iii) $(f, g) \in \mathcal{L}_{O^n(A)}$ iff $Imf = Img$.

(iv) $\mathcal{S} \subseteq (O^n(A); +)$ is a right-zero semigroup iff $S \subseteq Idem(O^n(A))$ and for all $f, g \in S$ we have $Imf = Img$.

4. Prove that an $n$-ary operation $f$ on a finite base set $A$ is a regular element of $(O^n(A); +)$ if and only if $Imf = \{f(\underline{a}) \mid a \in A\}$.

5. Let $g$ and $h$ be $n$-ary operations on a finite set $A$, with $n \geq 2$. Prove that $g \sim_n h$ iff $g^{\otimes n} \sim h^{\otimes n}$.

6. Let $f$ be an idempotent $n$-ary operation on a base set $A$, and let $F_f$ be the set of operations defined in Section 11.4. Prove that $g \in F_f$ iff the following two conditions are satisfied:

(i) For all $a \in Imf$ we have $[a]_{g|Imf} = \{a\}$.

(ii) The pair $((x_1, \ldots, x_n), (y_1, \ldots, y_n))$ is in the kernel of $f$ iff the pair $(((g(x_1, \ldots, x_n), \ldots, g(x_1, \ldots, x_n)), (g(y_1, \ldots, y_n), \ldots, g(y_1, \ldots, y_n)))$ is in the kernel of $g$.

7. Prove Proposition 11.5.1: If $\mathcal{S} \subseteq (O^n(A); +)$ is a semilattice, then $Im(f + g) = Imf \cap Img$ for every $f, g \in S$.

8. Let $f, g, h_1, h_2 \in Idem(O^n(A))$. Prove that $f + h_1 + h_2 + g = f + h_2 + h_1 + g$ if and only if for all $b \in Img$ we have $((h_1 + h_2)(\underline{b}), (h_2 + h_1)(\underline{b})) \in kerf$. (See Section 11.4 for notation.)

9. Verify any of the properties of the operation $+$ on sets of Boolean operations from Lemma 11.5.1.

10. Prove Theorem 11.5.3(i), that an $n$-ary Boolean operation $f$ is idempotent if and only if $f \in C_4^n \cup \{c_0^n, c_1^n\}$.

11. Prove Theorem 11.5.3(ii), that an $n$-ary Boolean operation $f$ is a regular element of $O^n(A)$ if and only if $f \in C_4^n \cup \neg C_4^n \cup \{c_0^n, c_1^n\}$.

12. Prove the following Corollary of Theorem 11.5.3. For any $n$-ary Boolean operation $f$ on the set $A = \{0, 1\}$, the following conditions are equivalent:

(i) $f$ is a regular element of $(O^n(A); +)$
(ii) $2f = f$ or $2f = \neg f$
(iii) $2f = f$ or $3f = f$.

13. Let $A = \{0, 1\}$, and let $S$ be a subsemigroup of $(O^n(A); +)$. Prove that an element $f$ of $S$ is regular in $S$ if and only if it is regular as an element of $O^n(A)$. (Use Exercise 12.)

14. Let $A = \{0, 1\}$ and $n \geq 1$, and let $S$ be a subsemigroup of $(O^n(A); +)$. Prove that the following conditions are equivalent:

(i) $S$ is a regular semigroup.
(ii) Either $S \subseteq C_4^n \cup \{c_0^n, c_1^n\}$, or $S \subseteq C_4^n \cup \{c_0^n, c_1^n\} \cup \neg C_4^n$ and for all $f \in S$ the negation $\neg f$ is also in $S$.

15. Let $A = \{0, 1\}$ and $n \geq 1$. A subsemigroup $S \subseteq O^n(A)$ is called an $f$-semigroup if $f \in S$ and $g + g = f$ for all $g \in S$. Prove that $S$ is an $f$-semigroup of $O^n(A)$ iff $S$ is one of the zero semigroups from Proposition 11.5.6 or a semigroup of the form $\{f, \neg f\}$ for some $f \in C_4^n$.

16. Let $A = \{0, 1\}$ and $n \geq 1$. Use the result of Exercise 14 to show that any semigroup of Boolean operations from $O^n(A)$ is a union of $f$-semigroups.

17. Prove that the operation $*$ defined for the four-part semigroup in Section 11.5 is associative.

18. Let $A = \{0, 1\}$ and let $n \geq 1$. Referring to the notation for sets of $n$-ary Boolean co-operations from Section 11.6, prove the following:

a) Any $n$-ary Boolean co-operation $f$ from $F_0$, $F_1$, $F_3^*$ or $I_{\{0,1\}}^{(n)}$ is idempotent.

b) The elements from part a) are the only idempotent $n$-ary Boolean co-operations.

c) Any $n$-ary Boolean co-operation from $E$, $F_2$ or $F_2^*$ is regular in $(cO^n(A); +)$.

d) The elements from part c) are the only regular elements of $(cO^n(A); +)$.

19. Let $A = \{0,1\}$ and let $n \geq 1$. Prove that the semigroup $(E; +)$ of idempotent $n$-ary Boolean co-operations is a regular band.

20. Let $A = \{0,1\}$ and let $n \geq 1$.

a) Prove that any set of the following three types is the universe of a semilattice of $n$-ary Boolean co-operations: $S = \{f\}$ for some idempotent co-operation $f$; or $S = \{f,g\}$ for some $f \in F_0$ and some $g \in F_3^* \cup I_{\{0,1\}}^{(n)}$ such that $f_1(0) = g_1(0)$; or $S = \{f,g\}$ for some $f \in F_1$ and some $g \in F_3^* \cup I_{\{0,1\}}^{(n)}$ such that $f_1(1) = g_1(1)$.

b) Prove that the three types of sets from part a) are the only sets which form subuniverses of semilattices of $n$-ary Boolean co-operations.

# Chapter 12

# Cohyperidentities and M-solid Classes of Coalgebras

In this chapter we look at a special kind of identities, called cohyperidentities, formed from coterms of a coalgebra. In parallel with the theory of hyperidentities for algebras and varieties, we define cohyperidentities for coalgebras. These concepts will then be used in Section 12.3 to solve the completeness problem for clones of Boolean co-operations and to separate pairs of clones of co-operations. These results are based on the work of Denecke and Saengsura in [26].

## 12.1  Clones of Coterms

In Chapter 9 we defined coterms of type $\tau$ and induced coterm co-operations for coalgebras of type $\tau$. We use the notation $cT_\tau^{(n)}$ for the set of all $n$-ary coterms of type $\tau$, for $n \geq 1$, and $(cT_\tau^{(n)})_{n\geq1}$ for the sequence of all coterms of type $\tau$. We have defined a family of superposition operations $(S_m^n)_{m,n\geq1}$ on this sequence, as follows.

**Definition 12.1.1.** The operations $S_m^n : cT_\tau^{(n)} \times (cT_\tau^{(m)})^n \to cT_\tau^{(m)}, m, n \geq 1$, are defined, by induction on the complexity of coterms, as follows:

(i) $S_m^n(e_i^n, t_1, \ldots, t_n) := t_i$  for $1 \leq i \leq n$.

(ii) $S_{n_i}^{n_i}(f_i, e_1^{n_i}, \ldots, e_{n_i}^{n_i}) := f_i$  for any $n_i$-ary co-operation symbol $f_i$.

(iii) $S_m^{n_j}(g_j, t_1, \ldots, t_{n_j}) := g_j[t_1, \ldots, t_{n_j}]$ for any $n_j$-ary co-operation symbol $g_j$.

(iv) $S_m^n(f_i[s_1, \ldots, s_{n_i}], t_1, \ldots, t_n)$
$$:= f_i[S_m^n(s_1, t_1, \ldots, t_n), \ldots, S_m^n(s_{n_i}, t_1, \ldots, t_n)],$$
where $s_1, \ldots, s_{n_i}$ are $n$-ary coterms and $t_1, \ldots, t_n$ are $m$-ary coterms of type $\tau$.

These operations, along with the injection coterm operations $e_j^n$ for $1 \leq j \leq n$ and $n \geq 1$, give us a many-sorted algebra

$$c\mathcal{T}_\tau \;=\; ((cT_\tau^{(n)})_{n \geq 1}, (S_m^n)_{m,n \geq 1}, (e_j^n)_{1 \leq j \leq n}).$$

This algebra also satisfies the three clone axioms (C1), (C2) and (C3) from Section 10.1, and so is a clone. The proof of this is very similar to the proof of Theorem 10.5.2, and we leave the details as an exercise.

**Theorem 12.1.2.** *The many-sorted algebra $c\mathcal{T}_\tau$ is a clone.*

Now let $A$ be any set, and let $\mathcal{A} = (A; (f_i^A)_{i \in I})$ be a coalgebra of type $\tau$. We have seen in Chapter 9 that each coterm of type $\tau$ induces a co-operation on $A$. For $n \geq 1$, let $cT_\tau^{A(n)}$ be the set of all $n$-ary term co-operations on $A$, induced by $n$-ary coterms of type $\tau$, and let $cT_\tau^A = \bigcup_{n \geq 1} cT_\tau^{A(n)}$ be the set of all induced term co-operations on $A$.

We have used the notation $cO^n(A)$ for the set of all $n$-ary co-operations defined on set $A$. This set, along with the injections and the superposition operations $comp_m^n : cO^n(A) \times (cO^m(A))^n \to cO^m(A)$ defined by $comp_m^n(f^A, g_1^A, \ldots, g_n^A) := f^A[g_1^A, \ldots, g_n^A]$, forms a multi-based algebra

$$((cO^n(A))_{n \geq 1}, (comp_m^n)_{m,n \geq 1}, (\imath_i^{n,A})_{1 \leq i \leq n})$$

which is also a clone, called the *clone of all co-operations* on $A$. In [23] it was also proved that the sequence $(cT_\tau^{A(n)})_{n \geq 1}$ is the universe of a subalgebra $c\mathcal{T}_\tau^A$ of the clone of all co-operations on $A$, and that this subalgebra is the algebra generated by the set $\{f_i^A \mid i \in I\}$ of fundamental co-operations. (Here the set $\{f_i^A \mid i \in I\}$ must be regarded as a many-sorted set, sorted according to the arities of the co-operations.)

**Proposition 12.1.3.** *Let $\mathcal{A} = (A; (f_i^A)_{i \in I})$ be any coalgebra of type $\tau$. The clone*

$$c\mathcal{T}_\tau^A = ((cT_\tau^{A(n)})_{n \geq 1}, (comp_m^n)_{m,n \geq 1}, (\imath_i^A)_{1 \leq i \leq n})$$

*of all induced term co-operations of $\mathcal{A}$ is a homomorphic image of the algebra $c\mathcal{T}_\tau$.*

**Proof:** We define a (many-sorted) mapping $\varphi : (cT_\tau^{(n)})_{n \geq 1} \to (cT_\tau^{A(n)})_{n \geq 1}$ in such a way that every co-operation symbol $t \in cT_\tau^{(n)}$ is mapped to the corresponding $n$-ary induced co-operation $t^A$. This mapping is well-defined and we claim that

$$
\begin{aligned}
(*) \quad \varphi(S_m^n(t, s_1, \ldots, s_n)) &= S_m^n(t, s_1, \ldots, s_n)^{\mathcal{A}} \\
&= comp_m^n(t^{\mathcal{A}}, s_1^{\mathcal{A}}, \ldots, s_n^{\mathcal{A}}) \\
&= comp_m^n(\varphi(t), \varphi(s_1), \ldots, \varphi(s_n)).
\end{aligned}
$$

We prove the equation (*) by induction on the complexity of the coterm $t$. If $t = f_i$ is a co-operation symbol or a symbol $e_j^n$, then (*) is simply part of the definition of induced co-operations from Chapter 9. The definition also shows the compatibility of $\varphi$ with the nullary operations. Now suppose inductively that $t = f_i[l_1, \ldots, l_{n_i}]$ for some $f_i$ and some coterms $l_1, \ldots, l_{n_i}$ satisfying (*). Then

$$
\begin{aligned}
&\varphi(S_m^n(t, s_1, \ldots, s_n)) \\
=\ & S_m^n(f_i[l_1, \ldots, l_{n_i}], s_1, \ldots, s_n)^{\mathcal{A}} \\
=\ & (S_m^n(S_n^{n_i}(f_i, l_1, \ldots, l_{n_i}), s_1, \ldots, s_n))^{\mathcal{A}} \\
=\ & [S_m^{n_i}(f_i, S_m^n(l_1, s_1, \ldots, s_n), \ldots, S_m^n(l_{n_i}, s_1, \ldots, s_n))]^{\mathcal{A}} \\
=\ & comp_m^{n_i}(f_i^{\mathcal{A}}, S_m^n(l_1, s_1, \ldots, s_n)^{\mathcal{A}}, \ldots, S_m^n(l_{n_i}, s_1, \ldots, s_n)^{\mathcal{A}}) \\
=\ & comp_m^{n_i}(f_i^{\mathcal{A}}, comp_m^n(l_1^{\mathcal{A}}, s_1^{\mathcal{A}}, \ldots, s_n^{\mathcal{A}}), \ldots, comp_m^n(l_{n_i}^{\mathcal{A}}, s_1^{\mathcal{A}}, \ldots, s_n^{\mathcal{A}})) \\
=\ & comp_m^n(comp_m^{n_i}(f_i^{\mathcal{A}}, l_1^{\mathcal{A}}, \ldots, l_{n_i}^{\mathcal{A}}), s_1^{\mathcal{A}}, \ldots, s_n^{\mathcal{A}})) \\
=\ & comp_m^n(\varphi(t), \varphi(s_1), \ldots, \varphi(s_n)). \qquad \blacksquare
\end{aligned}
$$

We conclude this introductory section by recalling the terminology from Chapter 9 for coidentities. An equation $s \approx t$ of coterms is called a *coidentity* in $\mathcal{A}$ if the two induced co-operations $s^{\mathcal{A}}$ and $t^{\mathcal{A}}$ are equal on $A$.

As an example of a coidentity we consider the two-element coalgebra $\mathcal{C}_{1c}$ from Chapter 10, with base set $A = \{0, 1\}$ and two binary co-operations $d$ and $h_4$. Let $F_1$ and $F_2$ be co-operation symbols for $d$ and $h_4$, respectively. Then it is easy to see that $\mathcal{C}_{1c}$ satisfies the coidentity $F_2 \approx F_1[F_2[e_1^2, e_1^2], F_2]$.

## 12.2    Cohypersubstitutions and Cohyperidentities

Cohyperidentities are special coidentities satisfied by algebras. For a coidentity to be satisfied as a cohyperidentity, it must hold in the algebra and have the additional stronger property that consistent replacement of the co-operation symbols in the identity by co-operations of the same arity results in equations which still hold as coidentities. Our development parallels that of the theory of *hyperidentities* and *solid* varieties, in the setting of algebras and identities; more information on this theory may be found in [30] or in [55]. In this section we introduce the necessary definitions and results. We start with cohypersubstitutions, and recall the notation $cT_\tau$ for the set of all coterms of type $\tau$.

**Definition 12.2.1.** A *cohypersubstitution of type* $\tau$ is a mapping $\sigma : \{f_i \mid i \in I\} \to cT_\tau$, mapping each co-operation symbol $f_i$ of the type to a coterm of the type, of the same arity as $f_i$. Let $Cohyp(\tau)$ be the set of all cohypersubstitutions of type $\tau$. Any cohypersubstitution $\sigma$ extends to a mapping $\hat{\sigma} : cT_\tau \to cT_\tau$, by the following inductive definition:

(i) $\hat{\sigma}[e_j^n] := e_j^n$ for every $n \geq 1$ and $1 \leq j \leq n$,
(ii) $\hat{\sigma}[f_i] := \sigma(f_i)$ for every $i \in I$,
(iii) $\hat{\sigma}[f_i[t_1, \ldots, t_{n_i}]] := S_n^{n_i}(\sigma(f_i), \hat{\sigma}[t_1], \ldots, \hat{\sigma}[t_{n_i}])$ for $t_1, \ldots, t_{n_i} \in cT_\tau^{(n)}$.

It is easy to see that such mappings $\hat{\sigma}$ are compatible with the operations $S_m^n$ of the many-sorted algebra $cT_\tau$.

**Proposition 12.2.2.** *For any cohypersubstitution $\sigma \in Cohyp(\tau)$, the extension $\hat{\sigma}$ is an endomorphism of $cT_\tau$.*

**Proof:** We show that for every coterm $t \in cT_\tau^{(n)}$ and every $s_1, \ldots, s_n$ in $cT_\tau^{(m)}$, the equation $\hat{\sigma}[S_m^n(t, s_1, \ldots, s_n)] = S_m^n(\hat{\sigma}[t], \hat{\sigma}[s_1], \ldots, \hat{\sigma}[s_n])$ is satisfied. The claim follows by definition if $t$ is an injection $e_j^n$ or an $n_i$-ary co-operation symbol $f_i$. Inductively, let $t = f_i[g_1, \ldots, g_{n_i}]$ for some $g_1, \ldots, g_{n_i} \in cT_\tau^{(n)}$ for which the equation holds. Then

$$
\begin{aligned}
&\hat{\sigma}[S_m^n(t, s_1, \ldots, s_n)] \\
=\ &\hat{\sigma}[S_m^{n_i}(f_i(g_1, \ldots, g_{n_i}), s_1, \ldots, s_n)] \\
=\ &S_m^{n_i}(\sigma(f_i), \hat{\sigma}[S_m^n(g_1, s_1, \ldots, s_n)], \ldots, \hat{\sigma}[S_m^n(g_{n_i}, s_1, \ldots, s_n)]) \\
=\ &S^{n_i}(\sigma(f_i), S_m^n(\hat{\sigma}[g_1], \hat{\sigma}[s_1], \ldots, \hat{\sigma}[s_n]), \ldots, \\
&\quad S_m^n(\hat{\sigma}[g_n], \hat{\sigma}[s_1], \ldots, \hat{\sigma}[s_n])) \\
=\ &S_m^n(S_n^{n_i}(\sigma(f_i), \hat{\sigma}[g_1], \ldots, \hat{\sigma}[g_{n_i}]), \hat{\sigma}[s_1], \ldots, \hat{\sigma}[s_n]) \text{ by (C1)} \\
=\ &S_m^n(\hat{\sigma}[t], \hat{\sigma}[s_1], \ldots, \hat{\sigma}[s_n]).
\end{aligned}
$$

The compatibility with the nullary operations also follows by definition. ■

Next we show how a monoid may be constructed on the set $Cohyp(\tau)$. A binary operation denoted by $\circ_{coh}$ is defined by setting $\sigma_1 \circ_{coh} \sigma_2 := \hat{\sigma}_1 \circ \sigma_2$, where $\circ$ denotes the usual composition of mappings. We also let $\sigma_{id}$ be the cohypersubstitution defined by $\sigma_{id}(f_i) := f_i$ for all $i \in I$. It is easy to show that $\sigma_{id}$ acts as an identity element for the operation $\circ_{coh}$, and that $(\sigma_1 \circ_{coh} \sigma_2)^\wedge = \hat{\sigma}_1 \circ \hat{\sigma}_2$ for any two cohypersubstitutions $\sigma_1$ and $\sigma_2$. Since the composition $\circ$ of mappings is associative, it follows that the operation $\circ_{coh}$ is also an associative one. We have thus proved the following theorem.

**Theorem 12.2.3.** *The structure* $(Cohyp(\tau); \circ_{coh}, \sigma_{id})$ *is a monoid.*

Extensions of cohypersubstitutions are significant in that they can be applied to both coidentities and coalgebras. For any equation $s \approx t$ of coterms and a cohypersubstitution $\sigma$, we shall use $\hat{\sigma}[s \approx t]$ as an abbreviation for the equation $\hat{\sigma}[s] \approx \hat{\sigma}[t]$. We can extend this definition additively to a set $\Sigma$ of equations, letting $\sigma(\Sigma) := \{\hat{\sigma}[s \approx t] \mid s \approx t \in \Sigma\}$. We can also extend to sets of cohypersubstitutions: for any submonoid $\mathcal{M}$ of $Cohyp(\tau)$, we set $\chi_M^E[\Sigma] := \bigcup_{\sigma \in M} \sigma(\Sigma)$. The mapping $\chi_M^E$ is then a mapping of the power set of $cT_\tau \times cT_\tau$ to itself.

Cohyperidentities can also be applied to coalgebras, to produce something called a derived coalgebra. We recall that $\sigma(f_i)^{\mathcal{A}}$ is the co-operation induced on a coalgebra $\mathcal{A}$ by the coterm $\sigma(f_i)$.

**Definition 12.2.4.** Let $\mathcal{A} = (A; (f_i^{\mathcal{A}})_{i \in I})$ be a coalgebra of type $\tau$ and let $\sigma \in Cohyp(\tau)$. The *coalgebra derived from* $\mathcal{A}$ *by the cohypersubstitution* $\sigma$ is defined as $\sigma(\mathcal{A}) := (A; (\sigma(f_i)^{\mathcal{A}})_{i \in I})$.

This definition shows that the derived coalgebra $\sigma(\mathcal{A})$ is a coalgebra of the same type as $\mathcal{A}$. The fundamental co-operations of $\sigma(\mathcal{A})$ are given by $f_i^{\sigma(\mathcal{A})} := \sigma(f_i)^{\mathcal{A}}$, for $i \in I$. The same pattern holds for any coterm $t \in cT_\tau$.

**Lemma 12.2.5.** *Let* $\mathcal{A}$ *be a coalgebra of type* $\tau$. *Let* $t$ *be a coterm and* $\sigma$ *be a cohypersubstitution of type* $\tau$. *Then* $t^{\sigma(\mathcal{A})} = \hat{\sigma}[t]^{\mathcal{A}}$.

**Proof:** We proceed by induction on the complexity of the coterm $t$. If $t$ is a symbol $e_j^n$ for $1 \le j \le n$, then $t^{\sigma(\mathcal{A})} = (e_i^n)^{\sigma(\mathcal{A})} = \iota_i^{n,\sigma(\mathcal{A})} = \iota_i^{n,\mathcal{A}} = (e_i^n)^{\mathcal{A}} = (\hat{\sigma}[e_i^n])^{\mathcal{A}} = (\hat{\sigma}[t])^{\mathcal{A}}$. If $t$ is a co-operation symbol $f_i$ for some $i \in I$, then $t^{\sigma(\mathcal{A})} = f_i^{\sigma(\mathcal{A})} = \sigma(f_i)^{\mathcal{A}}$ by definition. Inductively, if $t = f_i[g_1, \ldots, g_{n_i}]$ for some $g_1, \ldots, g_{n_i} \in cT_\tau^{(n)}$ for which $g_j^{\sigma(\mathcal{A})} = \hat{\sigma}[g_j]^{\mathcal{A}}$ for all $j = 1, \ldots, n$, then

$$
\begin{aligned}
t^{\sigma(\mathcal{A})} &= (f_i[g_1, \ldots, g_{n_i}])^{\sigma(\mathcal{A})} \\
&= f_i^{\sigma(\mathcal{A})}[g_1^{\sigma(\mathcal{A})}, \ldots, g_{n_i}^{\sigma(\mathcal{A})}] \\
&= (\sigma(f_i))^{\mathcal{A}}[\hat{\sigma}[g_1]^{\mathcal{A}}, \ldots, \hat{\sigma}[g_{n_i}]^{\mathcal{A}}] \\
&= (\sigma(f_i)(\hat{\sigma}[g_1], \ldots, \hat{\sigma}[g_{n_i}]))^{\mathcal{A}} \\
&= \hat{\sigma}[f_i[g_1, \ldots, g_{n_i}]]^{\mathcal{A}} \\
&= \hat{\sigma}[t]^{\mathcal{A}}.
\end{aligned}
$$

∎

The action of cohypersubstitutions on algebras also extends additively to classes of algebras and sets of cohypersubstitutions. For a class $K$ of coalgebras of type $\tau$ and a cohypersubstitution $\sigma$, we define $\sigma(K) = \{\sigma(\mathcal{A}) \mid \mathcal{A} \in K\}$. For a submonoid $\mathcal{M}$ of $Cohyp(\tau)$ we set $\chi_M^A(K) = \bigcup_{\sigma \in M} \sigma(K)$. The mapping $\chi_M^A$ thus maps sets of coalgebras to sets of coalgebras. The following fact is a consequence of Lemma 12.2.5 and the definition that $s \approx t$ is a coidentity of an algebra $\mathcal{A}$ iff $s^{\mathcal{A}} = t^{\mathcal{A}}$.

**Corollary 12.2.6.** *Let $\mathcal{A}$ be a coalgebra and $\mathcal{A}$ be a monoid of cohypersubstitutions, both of type $\tau$. For any equation $s \approx t$ of coterms of type $\tau$,*

$$(*) \quad \chi_M^A[\mathcal{A}] \text{ satisfies } s \approx t \text{ as a coidentity } \text{iff} \text{ } \mathcal{A} \text{ satisfies } \chi_M^E[s \approx t] \text{ as}$$
*a coidentity.*

The condition (*) is called the *conjugate property* for the two operations $\chi_M^E$ and $\chi_M^A$. These two operators are both closure operators (see Exercise 12.3), and are defined additively. This combination of properties is significant in the development of our theory.

**Theorem 12.2.7.** *Let $\mathcal{M} \leq Cohyp(\tau)$ be a monoid of cohypersubstitutions. Then $(\chi_M^A, \chi_M^E)$ forms a conjugate pair of additive closure operators.*

**Definition 12.2.8.** *Let $s \approx t$ be an equation of coterms $s$ and $t$ of type $\tau$ and let $\mathcal{A}$ be a coalgebra of type $\tau$. Let $\mathcal{A}$ be a submonoid of cohypersubstitutions of type $\tau$. We say that $\mathcal{A}$ satisfies $s \approx t$ as an $M$-cohyperidentity if $\mathcal{A}$ satisfies $\hat{\sigma}[s] \approx \hat{\sigma}[t]$ as a coidentity for every $\sigma \in M$. In the special case that $\mathcal{M}$ is the whole monoid $Cohyp(\tau)$, an $M$-cohyperidentity is called simply a cohyperidentity.*

We now have a relation of "satisfaction" between classes of coalgebras and sets of coterm equations of any type $\tau$. As we saw with the relation of satisfaction of identities by algebras in Section 1.4, we can use this relation to define a Galois connection. Let $K$ be a class of coalgebras of type $\tau$ and $\Sigma$ a set of coequations of type $\tau$. For any fixed submonoid $\mathcal{M}$ of $Cohyp(\tau)$, we define

$$MHcoid \, K \;=\; \{s \approx t \mid \forall \mathcal{A} \in K \, (\mathcal{A} \text{ satisfies } \chi_M^E[s \approx t])\},$$
and
$$MHcomod \, \Sigma \;=\; \{\mathcal{A} \mid \forall s \approx t \in \Sigma \, (\mathcal{A} \text{ satisfies } \chi_M^E[s \approx t])\}.$$

It is clear that the pair $(MHcoid, MHcomod)$ forms a Galois connection. In the algebra-identity situation, the corresponding Galois-closed sets were the varieties of algebras and the equational theories. In the algebra-hyperidentity setting, we get $M$-solid varieties and hyperequational theories (see [30]). Here, the Galois-closed sets are called $M$-*solid classes* of coalgebras and $M$-*cohyperequational theories*, respectively.

The general theory of conjugate pairs of additive closure operators (see [55]) shows that the collection of all $M$-solid classes of coalgebras forms a complete lattice, as does the collection of all $M$-cohyperequational classes. The theory also gives the following characterization of $M$-solid classes of coalgebras.

**Theorem 12.2.9.** ([55]) *let $M$ be a monoid of cohypersubstitutions. Let $K$ be an $M$-solid class of coalgebras of type $\tau$, so that $K = MHcomod\ MHcoid\ K$. Then the following conditions are equivalent:*

(i) $K = MHcomod\ MHcoid\ K$.

(ii) $\chi_M^A(K) = K$.

(iii) $Coid\ K = MHcoid\ K$.

(iv) $\chi_M^E[Coid\ K] = Coid\ K$.

## 12.3 Separation of Clones of Boolean Co-operations by Cohyperidentities

We now consider what it means for a clone of co-operations to satisfy a coidentity. Let $C \subseteq cO(A)$ be a clone of co-operations defined on a base set $A$, and let $F^A$ be a subset of $C$ which is a generating system for $C$. We can think of $F^A$ as an indexed set, writing $F^A = \{f_i^A \mid i \in I\}$. Using this set we define a coalgebra with universe $A$, by $\mathcal{A} = (A; (f_i^A)_{i \in I})$. Then we can ask whether this algebra satisfies a given coidentity $s \approx t$ as a cohyperidentity.

**Definition 12.3.1.** Let $C \subseteq cO(A)$ be a clone of co-operations on a set $A$, with $\{f_i^A \mid i \in I\}$ as a generating system for $C$. Let $\mathcal{A} = (A; (f_i^A)_{i \in I})$ be the coalgebra of type $\tau$ which has this generating system as its set of fundamental co-operations. Then we say that the clone $C$ *satisfies the coidentity* $s \approx t$ if the coalgebra $\mathcal{A} = (A; (f_i^A)_{i \in I})$ satisfies $s \approx t$ as a cohyperidentity.

In the remainder of this section we shall consider clones of Boolean co-operations, on the base set $A = \{0, 1\}$. We recall from Section 10.6 that there are exactly 12 such clones, forming the lattice shown in Figure 10.8. The smallest Boolean clone is the projection clone $I_{\{0,1\}}$, and the largest is the whole clone $cO(\{0,1\})$. There are four atoms in the lattice, the clones $M_0$, $M_1$, $C_2$ and $D_c$; and three co-atoms, the clones $C_{0c}$, $L_c$, and $C_{1c}$. We also showed in Section 10.6 exactly which binary co-operations are in which of the 12 clones.

Our first result shows a coidentity satisfied by the three maximal Boolean clones, but not by the largest clone $cO(\{0,1\})$.

**Theorem 12.3.2.** *Every maximal clone of Boolean co-operations satisfies the coidentity*

$$(*) \quad F[F[F, e_1^2], e_2^2] \approx F,$$

*but the clone $cO(\{0,1\})$ of all Boolean co-operations does not satisfy (\*) as a coidentity.*

**Proof:** For the three maximal clones of Boolean co-operations, $C_{0c}, C_{1c}$ and $L_c$, it can be shown that the first two are isomorphic to each other, and hence satisfy the same coidentities (see Exercise 12.4). Thus we need only show that $C_{0c}$ and $L_c$ satisfy (\*). Since the symbol $F$ in (\*) is a binary co-operation symbol, we consider in the table in Figure 12.1 the result of substituting for $F$ each of the binary Boolean co-operations which occur in $C_{0c}$ and $L_c$.

| $F$ | $F[F, e_1^2]$ | $F[F[F, e_1^2], e_2^2]$ |
|-----|-----|-----|
| $i_1^2$ | $i_1^2$ | $i_1^2$ |
| $i_2^2$ | $i_1^2$ | $i_2^2$ |
| $h_5$ | $h_1$ | $h_5$ |
| $h_8$ | $i_1^2$ | $h_8$ |
| $h$ | $g_1$ | $h$ |
| $h_1$ | $h_1$ | $h_1$ |
| $d$ | $i_1^2$ | $d$ |
| $g_1$ | $h_1$ | $g_1$ |
| $h_6$ | $h_2$ | $h_6$ |
| $h_7$ | $w$ | $h_7$ |
| $h_2$ | $h_2$ | $h_2$ |
| $w$ | $i_1^2$ | $w$ |

Fig. 12.1   Substitution of Binary Boolean Co-operations in (\*)

A comparison of columns one and three of the table then shows that (*) is satisfied by $C_{0c}$, $C_{1c}$ and $L_c$. It can be checked however that the two Boolean co-operations $h_{10}$ and $g_2$ do not satisfy (*), which shows that the clone $cO(\{0,1\})$ does not satisfy (*). ∎

A set of Boolean co-operations is said to be *complete* if it generates the whole clone $cO(\{0,1\})$. A single co-operation which generates the whole clone is also called a *Sheffer* co-operation (see [15]). As noted in Chapter 10, the sets $\{h_{10}\}$ and $\{g_2\}$ are complete one-element sets of co-operations. Since the lattice of all Boolean co-operations is dually atomic (and finite), a set $C$ of Boolean co-operations is complete iff $C$ is not a subset of any of the three maximal clones $C_{0c}$, $C_{1c}$ or $L_c$. This fact gives the following completeness criterion for Boolean co-operations:

**Proposition 12.3.3.** *A set $C$ of Boolean co-operations is complete iff $C$ does not satisfy the coidentity $F[F[F, e_1^2], e_2^2] \approx F$.*

If $C_1$ and $C_2$ are clones of Boolean co-operations with $C_1 \subseteq C_2$, any coidentity satisfied in $C_2$ is also satisfied in $C_1$. But it may be possible to find a coidentity satisfied in $C_1$ but not in $C_2$. We call such a coidentity a *separating coidentity* for the pair $(C_1, C_2)$. Our next aim is to separate by coidentities any two non-isomorphic clones $C_1$ and $C_2$ for which $C_1 \not\subseteq C_2$. Let us note first that the 12 Boolean clones of co-operations, there are three isomorphic pairs: $C_{0c}$ and $C_{1c}$; $C_3$ and $C_4$; and $M_0$ and $M_1$. This leaves us with nine clones to consider (up to isomorphism). It may also be possible to use the same coidentity to separate different pairs of clones. Following the approach used in [20] for separating clones of Boolean operations, we define a partial order relation between pairs of clones.

**Definition 12.3.4.** For any two pairs $(C_1, C_2)$ and $(C_1', C_2')$ of clones of Boolean co-operations, we define
$$(C_1, C_2) \preceq (C_1', C_2') \quad :\Leftrightarrow \quad \exists C_1'' \ (\ C_1'' \cong C_1' \text{ with } C_1 \subseteq C_1'' \ ) \text{ and}$$
$$\exists C_2'' \ (\ C_2'' \cong C_2' \text{ with } C_2'' \subseteq C_2 \ ).$$

We shall say that two pairs $(C_1, C_2)$ and $(C_1', C_2')$ are *isomorphic* if $C_1$ is isomorphic to $C_1'$ and $C_2$ is isomorphic to $C_2'$.

The relation $\preceq$ is clearly reflexive and transitive, and is antisymmetric when restricted to classes of isomorphic pairs of clones. The key fact here is that if $(C_1, C_2) \preceq (C_1', C_2')$ and there exists a coidentity $\varepsilon$ satisfied by $C_1'$

but not by $C_2'$, then the coidentity is also satisfied by $C_1$ but not by $C_2$. This allows us to consider certain special pairs of clones.

**Definition 12.3.5.** Let $(C_1, C_2)$ be a pair of clones of Boolean co-operations, neither one isomorphic to a subclone of the other. Then $(C_1, C_2)$ is called a *D-pair* if for any pair $(K_1, K_2)$ of clones of Boolean operations with $(C_1, C_2) \preceq (K_1, K_2)$, one of $K_1$ or $K_2$ is isomorphic to a subclone of the other.

We proceed to find all *D*-pairs of clones of Boolean co-operations. First, let $m$ be any clone which is minimal in the lattice, that is, any atom in the lattice of all clones of Boolean co-operations. Dually, let $M$ be any maximal clone. Since there are four atoms and three co-atoms, we can form 12 such pairs $(M, m)$. But as we saw above, there are three sets of isomorphic clones, which eliminates half of the pairs, leaving us with six such pairs. If $m$ is not contained in $M$, then $(M, m)$ is in fact a *D*-pair; from the lattice diagram in Figure 10.8 we can see that (up to isomorphism) there are exactly two such pairs, $(C_{0c}, D_c)$ and $(L_c, C_2)$. This leaves us with four pairs $(M, m)$ for which $m \subseteq M$, corresponding to four intervals $m \subseteq M$, the intervals $M_0 \subset C_{0c}$, $M_o \subset L_c$, $C_2 \subset C_{0c}$ and $D_c \subset L_c$. For each of these four intervals we need to find all *D*-pairs in the interval. By a case-by-case analysis of the possible pairs in these intervals, we obtain (up to isomorphism) six more *D*-pairs: $(C_{0c}, M_c)$, $(C_3, C_2)$, $(D_c, M_0)$, $(M_c, D_c)$, $(C_2, M_0)$ and $(L_c, C_3)$.

We have now identified all eight *D*-pairs of clones, and we want to find a separating coidentity, if possible, for each such *D*-pair. Since every clone of Boolean co-operations is generated by two fundamental co-operations, of arities at most two, we shall consider coidentities of type $(2, 2)$ or $(1, 1)$.

**Definition 12.3.6.** Let $s$ and $t$ be coterms of type $(2, 2)$. We say that a *D*-pair $(H_1, H_2)$ of clones of Boolean co-operations is *separated by* $s \approx t$, if $H_1$ satisfies $s \approx t$ as a coidentity, but $H_2$ does not. In this case $s \approx t$ is called a *separating coidentity* for the *D*-pair $(H_1, H_2)$.

We can now exhibit such separating coidentities for six of our eight *D*-pairs of clones of Boolean co-operations.

**Theorem 12.3.7.** *Let $F$ be a binary co-operation symbol and $L$ and $G$ be unary co-operation symbols. The following equations are separating coidentities for the D-pairs indicated:*

(i) $F[F, F[e_2^2, e_2^2]] \approx F[F, F]$  for $(L_c, C_3)$,

(ii) $L[G] \approx G[L]$  for $(C_{0c}, M_c)$,

(iii) $F[F, F[e_1^2, e_1^2]] \approx e_1^2$  for $(D_c, M_0)$,

(iv) $F[F[e_1^2, e_1^2], e_2^2] \approx F$  for $(M_c, D_c)$ and $(C_{0c}, D_c)$, and

(v) $F[e_1^2, e_1^2] \approx e_1^2$  for $(C_2, M_0)$.

**Proof**: (i) We recall first from Chapter 10 that there are eight binary co-operations in $L_c$: $i_1^2$, $i_2^2$, $h_1$, $h_2$, $h_5$, $h_6$, $h_7$ and $w$ (see Figure 10.5 for notation). The table in Figure 12.2 now shows the calculations needed to check our coidentity:

| $F$ | $F[e_2^2, e_2^2]$ | $F[F, F[e_2^2, e_2^2]]$ | $F[F, F]$ |
|---|---|---|---|
| $i_1^2$ | $i_2^2$ | $i_1^2$ | $i_1^2$ |
| $i_2^2$ | $i_2^2$ | $i_2^2$ | $i_2^2$ |
| $h_1$ | $h_5$ | $h_1$ | $h_1$ |
| $h_2$ | $h_6$ | $h_2$ | $h_2$ |
| $h_5$ | $h_5$ | $h_5$ | $h_5$ |
| $h_6$ | $h_6$ | $h_6$ | $h_6$ |
| $h_7$ | $h_7$ | $i_2^2$ | $i_2^2$ |
| $w$ | $h_7$ | $i_1^2$ | $i_1^2$ |

Fig. 12.2   Coidentity Calculations

Equality of the third and fourth columns shows that the clone $L_c$ satisfies the coidentity $F[F, F[e_2^2, e_2^2]] \approx F[F, F]$. But the clone $C_3$ does not satisfy this co-identity, since substitution of the co-operation $g_1$ for $F$ leads to $g_1$ on the left side but $h_1$ on the right. The remaining parts of the theorem are checked similarly, and we omit the details.  ∎

Having separated six of our eight $D$-pairs of clones, we show finally that the last two pairs cannot be separated by coidentities. We do this by means of the next lemma.

**Lemma 12.3.8.** *Let $F$ and $G$ be binary co-operation symbols and let $s_1, s_2, t_1, t_1$ be coterms of type $\tau = (2, 2)$. If the smallest clone $I_{\{0,1\}}$ satisfies the coidentity $F[s_1, s_2] \approx G[t_1, t_2]$, then so does the clone $C_2$.*

**Proof**: Suppose that $C_2$ does not satisfy the coidentity $F[s_1, s_2] \approx G[t_1, t_2]$. We know from Chapter 10 that the binary co-operations of the clone $C_2$ are $i_1^2$, $i_2^2$, $d$ and $h_8$. If we replace one of the binary co-operation symbols $F$ or $G$

in the coidentity by $d$ or $h_8$, then $F[s_1, s_2]^{\mathcal{A}} \neq G[t_1, t_2]^{\mathcal{A}}$ for the coalgebra $\mathcal{A}$ $= (\{0,1\}; h_8, d)$. Since $F[s_1, s_2]^{\mathcal{A}}$ cannot be a binary co-operation, therefore $F[s_1, s_2]^{\mathcal{A}}(0) \neq G[t_1, t_2]^{\mathcal{A}}(0)$ or $F[s_1, s_2]^{\mathcal{A}}(1) \neq G[t_1, t_2]^{\mathcal{A}}(1)$.

Let us consider first the case that $F[s_1, s_2]^{\mathcal{A}}(0) \neq G[t_1, t_2]^{\mathcal{A}}(0)$. Since for any binary co-operation $H^{\mathcal{A}}$ from $I_{\{0,1\}}$ we have $H^{\mathcal{A}}(0) = (i, 0)$, for $i \geq 1$, we must have $F[s_1, s_2]^{\mathcal{A}}(0) = (i, 0)$ and $G[t_1, t_2]^{\mathcal{A}}(0) = (j, 0)$ for some $i \neq j$. Since $d(0) = (0, 0) = \iota_1^2(0)$ and $h_8(0) = (1, 0) = \iota_2^2(0)$, the result will be the same if we replace each occurrence of $d$ and $h_8$ in $F[s_1, s_2]^{\mathcal{A}}$ and $G[t_1, t_2]^{\mathcal{A}}$ by $\iota_1^2$ and $\iota_2^2$, respectively. But this shows that the clone $I_{\{0,1\}}$ does not satisfy the coidentity $F[s_1, s_2] \approx G[t_1, t_2]$ either.

In the second case, that $F[s_1, s_2]^{\mathcal{A}}(1) \neq G[t_1, t_2]^{\mathcal{A}}(1)$, we use the fact that $H^{\mathcal{A}}(1) = (i, 1)$ for $i \geq 1$, for any binary co-operations from $I_{\{0,1\}}$. Then replacing $h_8$ and $d$ by $\iota_1^2$ leads to the same conclusion.     ∎

This lemma deals with coidentities in which both sides consist of a binary coterm $F[s_1, s_2]$ or $G[t_1, t_2]$. A coidentity not of this form must have an injection symbol on one side, since $F[e_1^2, e_2^2] \approx F$ for every binary co-operation symbol $F$. In this case a similar argument can be made as in the previous proof, for one side of the coidentity.

**Corollary 12.3.9.** *There is no separating coidentity for the clone pair* $(I_{\{0,1\}}, C_2)$.

**Theorem 12.3.10.** *There are no separating coidentities for either of the D-pairs* $(C_3, C_2)$ *and* $(L_c, C_2)$.

**Proof:** Any coidentity satisfied in $C_3$ or $L_c$ is also satisfied in the subclone $I_{\{0,1\}}$. But by Corollary 12.3.9, any such coidentity is also satisfied in $C_2$.
     ∎

We have now completely answered the question of which pairs $(C_1, C_2)$ of clones of Boolean co-operations, with $C_2$ not isomorphic to a subclone of $C_1$, can be separated by coidentities. This result shows yet another difference between clones of Boolean operations and clones of Boolean co-operations. Any pair $(C_1, C_2)$ of clones of Boolean operations with $C_2$ not isomorphic to a subclone of $C_1$ can be separated by identities (see [20]), but as we have seen this is not true for clones of Boolean co-operations.

## 12.4    Exercises for Chapter 12

1. Prove Theorem 12.1.1, that the many-sorted algebra $c\mathcal{T}_\tau$ satisfies the three clone axioms.

2. Prove that the operation $\circ_{coh}$ defined on the set $Cohyp(\tau)$ of all cohypersubstitutions of type $\tau$ satisfies $\sigma_1 \circ_{coh} \sigma_2 := \hat{\sigma}_1 \circ \sigma_2$, where $\circ$ denotes the usual composition of mappings.

3. Let $\mathcal{M} \leq Cohyp(\tau)$ be a monoid of cohypersubstitutions. Prove that the operators $\chi_M^A$ and $\chi_M^E$ are closure operators.

4. Let $C_{0c}$ and $C_{01}$ be two of the maximal clones of Boolean co-operations (see Section 10.8 for notation). Define a mapping $\varphi : C_{0c} \to C_{1c}$ which maps each $f \in cO^n(\{0,1\})$ with $f(0) = (i,0)$ to the corresponding 1-preserving co-operation $f' \in cO^n(\{0,1\})$ with $f'(1) = (i,1)$. Prove that $\varphi$ is a bijection and is compatible with the clone operations, making $C_{0c}$ and $C_{1c}$ isomorphic to each other.

5. Verify that the unary hyperidentity $L[G] \approx G[L]$ is satisfied by the clone $C_{0c}$, but not by the clone $M_c$.

# Bibliography

[1] Aczel, P. and Mendler, N. (1989). A final coalgebra theorem, in *Pitt, D. H., Ryeheard, D. E., Dybjer, P., Pitts, A. M. and Popigne, A. (eds), Proceedings of the 1989 Summer Conference on Category Theory and Computer Science, Springer Lecture Notes in Computer Science*, Vol. **389**, pp. 357–365.

[2] Aczel, P. (1997). Lectures on Semantics: The initial algebra and final coalgebra perspectives, Four lectures given at the 1995 "Logic of Computation", Advanced Study Institute, International Summer School at Marktoberdorf, in *Logic of Computation, edited by H. Schwichtenberg*, (Springer).

[3] Adámek, J., Herrlich, H. and Stecker, G. (1990). *Abstract and Concrete Categories*, (John Wiley).

[4] Adámek, J. and Porst, H.-E. (2001). From varieties of algebras to covarieties of coalgebras, in *Corradini et al. (eds), Coalgebraic Methods in Computer Science (CMCS'01), Electronic Notes in Theoretical Computer Science*, Vol. **44**, Elsevier 2001.

[5] Arbib, M. A. and Manes, G., E. (1980). Machines in a category, *J. Pure Appl. Alg.* **19**, pp. 9–20.

[6] Arbib, M. A. and Manes, G., E. (1982). Parametrized data types do not need highly constrained parameters, *Information and Control* **52**, 2, pp. 139–158.

[7] Backhouse, R. C. and Hoogendijk, P. F. (1999). Final Dialgebras: From Categories to Allegories, *ITA* **33**, 4/5, pp. 401–429.

[8] Berman, J. (1980). A proof of Lyndon's finite basis theorem, *Discrete Math.* **29**, pp. 229–233.

[9] Birkhoff, G. (1935). The structure of abstract algebras, *Proc. Cambridge Philosophical Society*, **31**, pp. 433-454.

[10] Birkhoff, G. and Lipson, J. D. (1970). Heterogeneous algebras, *J. Combinat. Theory*, **8**, pp. 115–133.

[11] Brinkmann, H. B. and Puppe, D. (1966). *Kategorien und Funktoren, Lecture Notes in Mathematics*, **18**, (Springer, Berlin, Heidelberg, New York).

[12] Burris, S. and Sankappanavar, H. P., (1981). *A course in universal algebra*, (Springer, New York).

265

[13] Butkote, R., Denecke, K. and Ratanaprasert, Ch. (2008). Semigroup Properties of $n$-ary Operations on Finite Sets, *Asian-European J. of Math.*, Vol. 1, 1, pp. 27–44.

[14] Chajda, I., Eigenthaler, G. and Länger, H. (2003). Congruence classes in universal algebra, *Research and Exposition in Mathematics*, **26**, (Heldermann Verlag, Lemgo).

[15] Csákány, B. (1985). Completeness in coalgebras, *Acta Sci. Math.*, **48**, pp. 75–84.

[16] Corradini, A., Grosse-Rhode, M. and Heckel, R., (1998). Structured transition system as lax coalgebras, in: *B. Jacobs, L. Moss, H. Reichel, J. Rutten, eds, Coalgebraic methods in Computer Science (CMCS'98)*, Electronic Notes in Theoretical Computer Science, Vol. **11**, pp. 23–42.

[17] Denecke, K. (1982). *Preprimal Algebras*, (Akademie-Verlag, Berlin).

[18] Denecke, K., Lau, D., Pöschel, R. and Schweigert, D. (1991). Hyperidentities, hyperequational classes, and clone congruences, in *Contribution to General Algebra* 7, (Verlag Hölder-Pichler-Tempsky Wien), pp. 97–118.

[19] Denecke, K. and Mal'cev, I. A. (1994). Separation of clones by hyperidentities I, (Russian), *Siberian Mathematical Journal*, **35**, 2, pp. 310–316.

[20] Denecke, K., Mal'cev, I. A. and Reschke, M. (1995). On separation of Boolean clones by means of hyperidentities, *Siberian Mathematical Journal*, Vol. **36**, 5, pp. 1094–1066.

[21] Denecke, K. and Pöschel, R. (1988). The characterization of primal algebras by hyperidentities, in *Contributions to General Algebra*, **6**, (Verlag Hölder-Pichler-Tempsky, Wien, Verlag B.G. Teubner Stuttgart), pp. 67–87.

[22] Denecke, K. and Saengsura, K. (2008). Menger Algebras and Clones of Cooperations, *Algebra Colloquium*, **15**:2 pp. 223–234.

[23] Denecke, K. and Saengsura, K. (2007). Semigroup Properties of Boolean Cooperations, *J. Applied Algebra and Discrete Structures*, **5**, pp. 1–20.

[24] Denecke, K. and Saengsura, K. (2007). Semigroup Properties of Cooperations on Finite Sets, preprint 2007, submitted to the Proceedings of the Beijing conference, July 2007.

[25] Denecke, K. and Saengsura, K. (2008). State-based Systems and $(F_1, F_2)$-Coalgebras, preprint.

[26] Denecke, K. and Saengsura, K. (2006). Separation of Clones of Cooperations by Cohyperidentities, preprint, to appear in "Discrete Mathematics".

[27] Denecke, K. and Todorov, K. (1994). *Algebraische Grundlagen der Arithmetik* (Heldermann-Verlag Berlin).

[28] Denecke and K., Todorov, K. (1996). *Allgemeine Algebra und Anwendungen*, (Shaker-Verlag Aachen).

[29] Denecke, K. and Wismath, S. L. (2002). *Universal Algebra and Applications in Theoretical Computer Science*, (Chapman & Hall/CRC, Boca Raton, London, New York, Washington, D.C.).

[30] Denecke, K. and Wismath, S. L. (2000). *Hyperidentities and Clones*, (Gordon and Breach Science Publishers).

[31] Drbohlav, K. (1971). On quasicovarieties, *Acta Fac. Rerum Natur. Univ. Comenian. Math. Mimoriadne Číslo*, pp. 17–20.

[32] Foster, A. L. (1953). Generalized Boolean theory of universal algebras, Part I, *Math. Zeitschr.*, **58**, pp. 306–336, Part II, *Math. Zeitschr.*, **59**, pp. 191–199.

[33] Freyd, P. (1996). Algebra valued functors in general and tensor products in particular, *Colloquium Math.*, **14**, pp. 89-106.

[34] Gécseg, F. and Steinby, M. (1997). *Tree languages, Handbook of formal languages*, Vol. **3**, pp. 1–68, (Springer, Berlin).

[35] Glubudom, P. and Denecke, K. (2006). Power Menger Algebras with Union and Their Endomorphisms, *Nam, Ki-Bong et al., eds. Pragmatic algebra. Delhi: SAS International Publications*, pp. 117–135.

[36] Gluschkow, W. M., Zeitlin, G. J. and Justschenko, J. L. (1980). *Algebra, Sprachen, Programmierung*, (Akademie-Verlag, Berlin).

[37] Grätzer, G. (1979). *Universal Algebra*, 2nd edition, (Springer-Verlag, Berlin, Heidelberg, New York).

[38] Graczyńska, E. and Oziewicz, Z. (1999). Birkhoff's theorems via tree operads, *Bull. Sect. Logic Univ.* **28**, no. 3.

[39] Gumm, H. P. (1999). *Elements of the general theory of coalgebras*, Lecture Notes for LUATCS'99, (Rand Africaans University, Johannesburg).

[40] Gumm, H. P. (2000). *Birkhoff's variety theorem for coalgebras, Contributions to General Algebra*, **13**, (Verlag Johannes Heyn, Klagenfurt), pp. 159–173.

[41] Gumm, H. P. (2001). Functors for Coalgebras, *Algebra Universalis*, **45**, pp. 135–147.

[42] Gumm, H. P. and Schröder, T. (2001). Covarieties and complete covarieties, *Theoretical Computer Science*, **260**, pp. 71–86.

[43] Gumm, H. P. and Schröder, T. (2002). Coalgebras of bounded type, *Mathematical Structures in Computer Science*, **12**, pp. 565–578.

[44] Gumm, H. P. (2003). *Universelle Coalgebra* in: *Th. Ihringer, Allgemeine Algebra*, (Heldermann Verlag, Lemgo).

[45] Heise, W. and Quattrocchi, P. (1995). *Informations-und Codierungstheorie* (Springer Verlag, Berlin, Heidelberg, New York).

[46] Higgins, P. J. (1963). Algebras with a scheme of operators, *Math. Nachr.*, **27**, pp. 115–132.

[47] Hilbert, D. and Bernays, P. (1934, 1939). *Grundlagen der Mathematik*, Vol.1, 194, Vol.2, (Berlin).

[48] Howie, J. M. (2006). *Fundamentals of Semigroup Theory*, (Oxford Science Publications).

[49] Ihringer, Th. (2003). *Allgemeine Algebra mit einem Anhang über Universelle Coalgebra von H. P. Gumm, Berliner Studienreihe zur Mathematik*, Band **10**, (Heldermann Verlag).

[50] Ihringer, Th. (1999) *Diskrete Mathematik*, (Verlag B.G. Teubner, Stuttgart).

[51] Jablonskij, S.V. (1958). Functional constructions in multivalued logics (Russian), *Trudy Inst. Mat. Steklov*, **51** , pp. 5–142.

[52] Knoebel, A. (1985). The equational classes generated by single functionally precomplete algebras, *Memoirs of the Amer. Math. Soc.*, **57**, 332, (Provi-

dence, Rhode Island).

[53] Kurz, A. (2000). *Logics for Coalgebras and Applications to Computer Science*, Dissertation, (Ludwig-Maximilians Universität München).

[54] Kurz, A. (2006). *Logics for Coalgebras and Applications to Computer Science*, (Manuscript).

[55] Koppitz, J. and Denecke, K. (2006). *M-solid Varieties*, (Springer).

[56] Lau, D. (1991). On closed subsets of Boolean functions, A new proof for Post's theorem, *J. Inform. Process. Cybernet.*, *EIK*, **27**, pp. 167–178.

[57] Lau, D. (2006). *Function Algebras on Finite Sets*, (Springer).

[58] Lawvere, F. W. (1963). Functorial semantics of algebraic theories, *Proc. Nat. Acad. Sci.*, **50**, pp. 869–872.

[59] Mal'cev, A.I. (1966). Iterative Algebras and Post's Varieties (Russian), *Algebra and Logic*, **5**, 2, pp. 5–24.

[60] MacLane, S. (1998). *Categories for the Working Mathematician*, Second edition, (Springer).

[61] Nakawaya, A. T. (2000). Algebra-Coalgebra Structures and Bialgebras, *Lecture Notes in Computer Science*, Vol. **1827**, (Springer), pp. 37–61.

[62] Neumann, J. v. (1925). *Eine Axiomatisierung der Mengenlehre.* J. f. reine und angewandte Math., Vol. **154** (1925), 219-240.

[63] Ngom, A., Reischer, C., Simovici, D. A. and Stojmenovic, I. (1996). A Survey of Set Logic Algebra, *Mult. Val. logic*, Vol. **1**, pp. 1–34.

[64] Pattinson, D. (2005). *An Introduction to the Theory of Coalgebras, Lecture Notes*, (LMU München).

[65] Polák, L. (1996). On Hyperassociativity, *Algebra Universalis*, Vol. **36**, 3, pp. 363–378.

[66] Poll, E. and Zwanenburg, J. (2001). From Algebras and Coalgebras to Dialgebras, *Electr. Notes Theor. Comput. Sci. (ENTS)*, **44**, 1.

[67] Pöschel, R. and Kalushnin, L. A. (1979). *Funktionen- und Relationenalgebren*, (VEB Deutscher Verlag der Wissenschaften, Berlin).

[68] Post, E. L. (1921). Introduction to a general theory of elementary propositions, *Amer. J. Math.*, **43**, pp. 163–185.

[69] Post, E. L. (1941). *The two-valued iterative systems of mathematical logic*, Ann. Math. Studies **5**, (Princeton Univ. Press).

[70] Quackenbush, R. W. (1979). A new proof of Rosenberg's primal algebra characterization theorem, in *Finite Algebra and Multiple-valued logic, (Proc. Conf. Szeged, 1979)*, *Colloq. Math. Soc. J. Bolyai*, (North-Holland Amsterdam), vol. **28**, pp. 603–634.

[71] Reschke, M. and Denecke, K. (1989). Ein neuer Beweis für die Ergebnisse von E. L. Post über abgeschlossene Klassen Boolescher Funktionen, *J. Inform. Process. Cybernet., (EIK)*, **25**, pp. 361–380.

[72] Rößiger, M. (2000). *Coalgebras, clone theory, and modal logic*, Dissertation, )Technische Universitt Dresden, Mathematik und Naturwissenschaften).

[73] Rosenberg, I. G. (1970). Über die funktionale Vollständigkeit in den mehrwertigen Logiken, *Rozpr. ČSAV, Řada Mat. Přír. Věd., Praha* **80**, 1, pp. 3–93.

[74] Rosenberg, I. G. (1965). La structure des fonctions de plusieurs variables

sur un ensemble fini, *C.R. Acad. Sci. Paris Ser. A-B*, **260**, pp. 3817–3819.

[75]  Rutten, J. J. M. M. (2000). Universal Coalgebra, a theory of systems, *Theoretical Computer Science* **249**, pp. 3–80.

[76]  Schubert, H. (1970). *Kategorien I, II*, (Akademie-Verlag Berlin).

[77]  Słupecki, J., (1972). Completeness criterion for systems of many-valued propositional calculus, *Studia Logica*, **30**, pp. 153–157.

[78]  Steinby, M. (1998). General varieties of tree languages, *Theoret. Comput. Sci.*, **205**, No. 1-2, pp. 1–23.

[79]  Szendrei, A. (1986). *Clones in universal algebra*, (Les presses de l'Université de Montréal, Montréal).

[80]  Taylor, W. (1973). Characterizing Mal'cev conditions, *Algebra Universalis*, **3**, pp. 351–397.

[81]  Taylor, W. (1980). Mal'cev conditions and spectra, *J. Austral. Math. Soc. (A)*, **29**, pp. 143–152.

[82]  Taylor, W. (1981). Hyperidentities and hypervarieties, *Aequationes Mathematicae*, **23**, pp. 111–127.

[83]  Taylor, W. (1991). Abstract Clone Theory, in *Algebras and Orders*, (Kluwer Academic Publishers, Dordrecht, Boston, London), pp. 507–530.

[84]  Turi, D. and Plotkin, G. (1997). Towards a mathematical operational semantics, in *Proc. of the LICS'97*, pp. 280–305.

[85]  Venema, Y. (2006). Algebras and Coalgebras, in: Handbook of Modal Logic, Part. 2, (Elsevier Science Inc, New York), pp. 331–426.

# Glossary

# Index